Lial/Miller/Greenwell

Student's Solution Manual
to accompany
Finite Mathematics, Fifth Edition

Prepared with the assistance of

Brian Hayes
Triton College

HarperCollins*College*Publishers

Student's Solution Manual to accompany *Lial /Miller/Greenwell: FINITE MATHEMATICS, FIFTH EDITION*

Copyright © 1993 by HarperCollins*CollegePublishers*

ISBN 0-673-46756-2

92 93 94 95 9 8 7 6 5 4 3 2 1

PREFACE

This book provides complete solutions for many of the exercises in <u>Finite Mathematics</u>, fifth edition, by Margaret L. Lial, Charles D. Miller, and Raymond N. Greenwell. Solutions are included for odd-numbered exercises. Solutions are not provided for exercises with open-response answers or for those that require the use of a computer. Sample tests are provided to help you determine if you have mastered the concepts in a given chapter.

This book should be used as an aid as you work to master your course work. Try to solve the exercises that your instructor assigns before you refer to the solutions in this book. Then, if you have difficulty, read these solutions to guide you in solving the exercises. The solutions have been written so that they are consistent with the methods used in the textbook.

You may find that some of the solutions are presented in greater detail than others. Thus, if you cannot find an explanation for a difficulty that you encountered in one exercise, you may find the explanation in the solution for a similar exercise elsewhere in the exercise set.

Solutions that require graphs will refer you to the answer section of the textbook. These graphs are not included in this book.

In addition to solutions, you will find a list of suggestions on how to be successful in mathematics. A careful reading will be helpful for many students.

The following people have made valuable contributions to the production of this <u>Student's Solution Manual</u>: Brian Hayes, editor; Judy Martinez and Sheri Minkner, typists; Therese Brown and Charles Sullivan, artists; and Carmen Eldersveld, proofreader.

We also want to thank Tommy Thompson of Seminole Community College for his suggestions for the essay "To the Student: Success in Mathematics" that follows this preface.

TO THE STUDENT: SUCCESS IN MATHEMATICS

The main reason students have difficulty with mathematics is that they don't know how to study it. Studying mathematics *is* different from studying subjects like English or history. The key to success is regular practice.

This should not be surprising. After all, can you learn to play the piano or to ski well without a lot of regular practice? The same thing is true for learning mathematics. Working problems nearly every day is the key to becoming successful. Here is a list of things you can do to help you succeed in studying mathematics.

1. *Attend class regularly.* Pay attention in class to what your instructor says and does, and make careful notes. In particular, note the problems the instructor works on the board and copy the complete solutions. Keep these notes separate from your homework to avoid confusion when you read them over later.

2. Don't hesitate to ask questions in class. It is not a sign of weakness, but of strength. There are always other students with the same question who are too shy to ask.

3. *Read your text carefully.* Many students read only enough to get by, usually only the examples. Reading the complete section will help you to be successful with the homework problems. Most exercises are keyed to specific examples or objectives that will explain the procedures for working them.

4. Before you start on your homework assignment, rework the problems the instructor worked in class. This will reinforce what you have learned. Many students say, "I understand it perfectly when you do it, but I get stuck when I try to work the problem myself."

5. Do your homework assignment only *after* reading the text and reviewing your notes from class. Check your work with the answers in the back of the book. If you get a problem wrong and are unable to see why, mark that problem and ask your instructor about it. Then practice working additional problems of the same type to reinforce what you have learned.

6. Work as neatly as you can. Write your symbols clearly, and make sure the problems are clearly separated from each other. Working neatly will help you to think clearly and also make it easier to review the homework before a test.

7. After you have completed a homework assignment, look over the text again. Try to decide what the main ideas are in the lesson. Often they are clearly highlighted or boxed in the text.

8. Use the chapter test at the end of each chapter as a practice test. Work through the problems under test conditions, without referring to the text or the answers until you are finished. You may want to time yourself to see how long it takes you. When you have finished, check your answers against those in the back of the book and study those problems that you missed. Answers are referenced to the appropriate sections of the text.

9. Keep any quizzes and tests that are returned to you and use them when you study for future tests and the final exam. These quizzes and tests indicate what your instructor considers most important. Be sure to correct any problems on these tests that you missed, so you will have the corrected work to study.

10. Don't worry if you do not understand a new topic right away. As you read more about it and work through the problems, you will gain understanding. Each time you look back at a topic you will understand it a little better. No one understands each topic completely right from the start.

CONTENTS

1 LINEAR MODELS

1.1 Linear Equations and Inequalities in One Variable   1

1.2 Linear Functions   4

1.3 Slope and Equations of a Line   9

1.4 Linear Mathematical Models   16

1.5 Constructing Mathematical Models   22

Chapter 1 Review Exercises   35

Extended Application: Marginal Cost—Booz, Allen & Hamilton   43

Chapter 1 Test   45

Chapter 1 Test Answers   46

2 SYSTEMS OF LINEAR EQUATIONS AND MATRICES

2.1 Solution of Linear Systems by the Echelon Method   47

2.2 Solution of Linear Systems by the Gauss–Jordan Method   54

2.3 Addition and Subtraction of Matrices   62

2.4 Multiplication of Matrices   65

2.5 Matrix Inverses   68

2.6 Input–Output Models   76

Chapter 2 Review Exercises   79

Extended Application: Leontief's Model of the American Economy   86

Chapter 2 Test   87

Chapter 2 Test Answers   88

3   LINEAR PROGRAMMING: THE GRAPHICAL METHOD

3.1   Graphing Linear Inequalities   89

3.2   Solving Linear Programming Problems Graphically   93

3.3   Applications of Linear Programming   98

Chapter 3 Review Exercises   102

Chapter 3 Test   107

Chapter 3 Test Answers   108

4   LINEAR PROGRAMMING: THE SIMPLEX METHOD

4.1   Slack Variables and the Pivot   109

4.2   Solving Maximization Problems   113

4.3   Nonstandard Problems; Minimization   122

4.4   Duality   136

Chapter 4 Review Exercises   142

Extended Application: Making Ice Cream   147

Extended Application: Merit Pay—The Upjohn Company   147

Chapter 4 Test   148

Chapter 4 Test Answers   150

5   SETS AND PROBABILITY

5.1   Sets   151

5.2   Applications of Venn Diagrams   153

5.3   Introduction to Probability   161

5.4   Basic Concepts of Probability   164

5.5   Conditional Probability; Independent Events   170

5.6   Bayes' Theorem   175

Chapter 5 Review Exercises 181

Extended Application: Medical Diagnosis 186

Chapter 5 Test 187

Chapter 5 Test Answers 189

6 COUNTING PRINCIPLES: FURTHER PROBABILITY TOPICS

6.1 The Multiplication Principle; Permutations 190

6.2 Combinations 193

6.3 Probability Applications of Counting Principles 197

6.4 Bernoulli Trials 202

6.5 Probability Distributions; Expected Value 206

Chapter 6 Review Exercises 212

Extended Application: Optimal Inventory for a Service Truck 216

Chapter 6 Test 217

Chapter 6 Test Answers 219

7 STATISTICS

7.1 Frequency Distributions; Measures of Central Tendency 220

7.2 Measures of Variation 223

7.3 The Normal Distribution 229

7.4 The Binomial Distribution 234

Chapter 7 Review Exercises 238

Chapter 7 Test 245

Chapter 7 Test Answers 247

8 MARKOV CHAINS

8.1 Basic Properties of Markov Chains 248

8.2 Regular Markov Chains 254

8.3 Absorbing Markov Chains 260

Chapter 8 Review Exercises 265

Extended Application: Cavities and Restoration 271

Chapter 8 Test 273

Chapter 8 Test Answers 275

9 GAME THEORY

9.1 Decision Making 276

9.2 Strictly Determined Games 277

9.3 Mixed Strategies 279

9.4 Game Theory and Linear Programming 284

Chapter 9 Review Exercises 299

Extended Application: Decision Making in Life Insurance 309

Chapter 9 Test 311

Chapter 9 Test Answers 313

10 MATHEMATICS OF FINANCE

10.1 Simple Interest and Discount 314

10.2 Compound Interest 316

10.3 Annuities 320

10.4 Present Value of an Annuity; Amortization 325

Chapter 10 Review Exercises 330

Extended Application: Present Value 334

Chapter 10 Test 336

Chapter 10 Test Answers 337

Student's Solution Manual
to accompany
Finite Mathematics, Fifth Edition

CHAPTER 1   LINEAR MODELS

Section 1.1

1. $4x - 1 = 15$

    First use the addition property of equality to add 1 to each side of the equation.

    $4x - 1 = 15$
    $4x = 16$

    Then use the multiplication property of equality to divide both sides by 4.

    $x = 4$

    The solution is 4.

3. $3m + 2 = -m + 7$

    To get the variable terms on one side of the equation and the constants on the other side, add m to both sides of the equation, and then subtract 2 from both sides.

    $4m + 2 = 7$
    $4m = 5$

    Finally, divide both sides by 4.

    $m = \dfrac{5}{4}$

5. $.2m - .5 = .1m + .7$

    Subtract .1m from both sides, and then add .5 to both sides.

    $.1m - .5 = .7$
    $.1m = 1.2$

    Divide both sides by .1 (which is the same as multiplying both sides by 10).

    $m = 12$

7. $\dfrac{5}{6}k - 2k + \dfrac{1}{3} = \dfrac{2}{3}$

    Begin by multiplying both sides of the equation by 6, because then no fractions will appear.

    $5k - 12k + 2 = 4$

    Combine the two like terms on the left side.

    $-7k + 2 = 4$

    Subtract 2 from both sides, and then divide both sides by -7.

    $-7k = 2$
    $k = -\dfrac{2}{7}$

9. $2x - (x + 3) = 7 - x$

    Use the distributive property to clear parentheses.

    $2x - x - 3 = 7 - x$
    $x - 3 = 7 - x$
    $2x - 3 = 7$
    $2x = 10$
    $x = 5$

11. $3r + 2 - 5(r + 1) = 6r + 4$
    $3r + 2 - 5r - 5 = 6r + 4$
    $-2r - 3 = 6r + 4$
    $-8r - 3 = 4$
    $-8r = 7$
    $r = -\dfrac{7}{8}$

13. $5(3x - 2) = 7(x + 2)$

$15x - 10 = 7x + 14$

$8x = 24$

$x = 3$

15. $\dfrac{x}{3} - 7 = x - \dfrac{3x}{4}$

$4x - 84 = 12x - 9x$

$4x - 84 = 3x$

$x = 84$

17. $2x - 5 \le 15$

Add 5 to both sides of the inequality.

$2x \le 20$

Divide both sides by 2.

$x \le 10$

The solution is $x \le 10$.

19. $6 - 4m \ge 12$

Subtract 6 from both sides.

$-4m \ge 6$

Divide both sides by -4, which requires simultaneously reversing the inequality from $\ge$ to $\le$.

$m \le -\dfrac{6}{4}$

Simplify the fraction.

$m \le -\dfrac{3}{2}$

21. $5k + 2 < 2k - 3$

$3k < -5$

$k < -\dfrac{5}{3}$

23. $9 - 3z > 8z + 12$

$-3 > 11z$

$-\dfrac{3}{11} > z$

or $z < -\dfrac{3}{11}$

25. $\dfrac{4}{5}x + 3 \le x - \dfrac{1}{5}$

$4x + 15 \le 5x - 1$

$16 \le x$

or $x \ge 16$

27. $3(t - 2) + 5 \ge t - 4$

$3t - 6 + 5 \ge t - 4$

$2t - 1 \ge -4$

$2t \ge -3$

$t \ge -\dfrac{3}{2}$

29. $5 - 3p + 2(p - 4) \le 4p$

$5 - 3p + 2p - 8 \le 4p$

$-3 - p \le 4p$

$-3 - 5p \le 0$

$-5p \le 3$

$p \ge -\dfrac{3}{5}$

31. $2(k - 5) + 3 < -(k + 1)$

$2k - 10 + 3 < -k - 1$

$3k - 7 < -1$

$3k < 6$

$k < 2$

33. $2(x - a) + b = 3x + a$

$2x - 2a + b = 3x + a$

Get all terms with x on one side of the equation and all terms without x on the other side.

$$2x - 3x = 2a + a - b$$
$$-x = 3a - b$$

Divide both sides by -1.

$$x = b - 3a$$

35. $ax + b = 3(x - a)$
$$ax + b = 3x - 3a$$

Get all terms with x on one side.

$$ax - 3x = -3a - b$$

Use the distributive property to rewrite the left side.

$$(a - 3)x = -3a - b$$

Divide both sides by $a - 3$.

$$x = \frac{-3a - b}{a - 3}$$

Simplify the fraction by multiplying the numerator and the denominator each by -1.

$$x = \frac{3a + b}{3 - a}$$

37.
$$x = a^2x - ax + 3a - 3$$
$$3 - 3a = a^2x - ax - x$$
$$3 - 3a = x(a^2 - a - 1)$$
$$\frac{3 - 3a}{a^2 - a - 1} = x$$
$$x = \frac{3 - 3a}{a^2 - a - 1}$$

39. $a^2x + 3x = 2a^2$
$$x(a^2 + 3) = 2a^2$$
$$x = \frac{2a^2}{a^2 + 3}$$

45. Let x represent the amount Joe invested at 10%. Since Joe received $52,000 profit, $52,000 - x$ is the amount he invested at 8%. For simple interest, $I = Prt$, and the total interest is the sum of the two interest amounts.

$$x(10\%)(1) + (52,000 - x)(8\%)(1) = 4580$$
$$.10x + .08(52,000 - x) = 4580$$
$$.10x + 4160 - .08x = 4580$$
$$.02x + 4160 = 4580$$
$$.02x = 420$$
$$x = 21,000$$

Thus, Joe invested $21,000 at 10%.

47. Since Mary paid 40% income tax on her royalties of $48,000, the amount of tax she paid is $.40(48,000) = 19,200$. Therefore, she has $48,000 - 19,200 = 28,800$ left to invest. Let x represent the amount invested at $10\frac{1}{2}$%, so $28,800 - x$ will be the amount invested at $7\frac{1}{2}$%. Use $I = Prt$ to get the total interest as follows.

$$.105x + .075(28,800 - x) = 2550$$
$$.105x + 2160 - .075x = 2550$$
$$.030x + 2160 = 2550$$
$$.030x = 390$$
$$x = 13,000$$

Mary invested $13,000 at $10\frac{1}{2}$%.

49. Let x represent the additional amount invested at 6%. Then x + 20,000 represents the total amount invested; the interest on this amount is .072(x + 20,000). Use I = Prt as follows.

$$.09(20,000) + .06x = .072(x + 20,000)$$
$$1800 + .06x = .072x + 1440$$
$$.06x = .072x - 360$$
$$-.012x = -360$$
$$x = 30,000$$

$30,000 must be invested at 6%.

51. (a) R = 5x - 100

(b) C = 125 + 4x

(c) $$R > C$$
$$5x - 100 > 125 + 4x$$
$$x - 100 > 125$$
$$x > 225$$

Therefore, the company must produce and sell more than 225 cassettes to make a profit.

53. A liter of 92 octane gasoline is 92% isooctane, and a liter of 98 octane gasoline is 98% isooctane. Let x represent the number of liters of 92 octane gasoline. Then 12 - x will be the number of liters of 98 octane gasoline.

$$.92x + .98(12 - x) = .96(12)$$
$$.92x + 11.76 - .98x = 11.52$$
$$-.06x = -.24$$
$$x = 4$$

It will take 4 liters of 92 octane (mixed with 8 liters of 98 octane) to produce 12 liters of 96 octane.

Section 1.2

1. Since no vertical line cuts the graph at more than one point, the graph is that of a function.

3. This is not a function, since some vertical lines cut the graph at more than one point.
For example, the y-axis (which is a vertical line) cuts the graph at two points.

5. No vertical line cuts the graph at more than one point, so this is a function.

7. f(x) = -2x - 4

(a) f(4) = -2(4) - 4 = -12

(b) f(-3) = -2(-3) - 4 = 2

(c) f(0) = -2(0) - 4 = -4

(d) f(a) = -2a - 4

9. f(x) = 6

This is a constant function.

(a) f(4) = 6

(b) f(-3) = 6

(c) f(0) = 6

(d) f(a) = 6

11. $f(x) = 2x^2 + 4x$

(a) $f(4) = 2(4)^2 + 4(4)$
$= 32 + 16 = 48$

(b) $f(-3) = 2(-3)^2 + 4(-3)$
$= 18 - 12 = 6$

(c) $f(0) = 0$

(d) $f(a) = 2a^2 + 4a$

13. $f(x) = (x + 1)(x + 2)$

(a) $f(4) = (4 + 1)(4 + 2)$
$= (5)(6) = 30$

(b) $f(-3) = (-3 + 1)(-3 + 2)$
$= (-2)(-1) = 2$

(c) $f(0) = (0 + 1)(0 + 2)$
$= (1)(2) = 2$

(d) $f(a) = (a + 1)(a + 2)$
$= a^2 + 3a + 2$

For Exercises 15-29, see the answer graph in the back of the textbook.

15. $y = 2x + 1$

To find the y-intercept, let $x = 0$.

$y = 2(0) + 1$
$y = 1$

To find the x-intercept, let $y = 0$.

$0 = 2x + 1$
$-1 = 2x$
$-\frac{1}{2} = x$

These intercepts give us the ordered pairs $(0, 1)$ and $(-1/2, 0)$. Graph these two points and connect them with a straight line. A third point may be used as a check.

(For example, let $x = 1$.

$y = 2(1) + 1$
$y = 3$

The ordered pair $(1, 3)$ is a third point on the same line.)

17. $y = 4x$

To find the y-intercept, let $x = 0$.

$y = 4(0) = 0$

The ordered pair is $(0, 0)$.
Since this point (the origin) is both the x-intercept and y-intercept, another point is needed.
Let $x = 1$.

$y = 4(1) = 4$

The ordered pair is $(1, 4)$.
Use the points $(0, 0)$ and $(1, 4)$ to draw the line. A third point may be used as a check.

19. $3y + 4x = 12$

Find the y-intercept.

$3y + 4(0) = 12$
$3y = 12$
$y = 4$

The ordered pair is $(0, 4)$.
Find the x-intercept.

$3(0) + 4x = 12$
$4x = 12$
$x = 3$

The ordered pair is $(3, 0)$.
Draw the line through these two points.

21. y = -2

The graph is a horizontal line having y-intercept -2. The graph has no x-intercept. Draw a line parallel to the x-axis and passing through the point (0, -2).

23. x + 5 = 0

     x = -5

Some ordered pairs that satisfy this equation are (-5, -1), (-5, 0), (-5, 1) and (-5, 3).
The x-intercept is -5; there is no y-intercept. The graph is a vertical line passing through (-5, 0).

25. 8x + 3y = 10

Find the y-intercept.

$$8(0) + 3y = 10$$
$$3y = 10$$
$$y = \frac{10}{3}$$

The ordered pair is (0, 10/3).
Find the x-intercept.

$$8x + 3(0) = 10$$
$$8x = 10$$
$$x = \frac{5}{4}$$

The ordered pair is (5/4, 0).
Draw the line through these two points.

27. y = 2x

Find the y-intercept.

$$y = 2(0) = 0$$

The ordered pair is (0, 0).

Both x- and y-intercepts are at the origin, so another point is needed. Choose any other value of x, for example, x = 1.

$$y = 2(1) = 2$$

The ordered pair is (1, 2).
Draw the line through these two points.

29. x + 4y = 0

Find the y-intercept.

$$0 + 4y = 0$$
$$y = 0$$

The ordered pair is (0, 0).
Since the x-intercept is also at the origin, choose another value of x (or y) to find a second point. For example, let x = 4.

$$4 + 4y = 0$$
$$4y = -4$$
$$y = -1$$

The ordered pair is (4, -1).
Draw the line through these two points.

33. S(t) = 1050 + 50t

(a) For 1991, t = 0.

$$S(0) = 1050 + 50(0) = 1050$$

The formula approximates the catalog sales in thousands of dollars, so the estimated sales for 1991 are $1,050,000.

(b) For 1994, t = 1994 − 1991 = 3.

$$S(3) = 1050 + 50(3) = 1200$$

The estimated sales for 1994 are $1,200,000.

(c) For 1997, t = 1997 − 1991 = 6.

$$S(6) = 1050 + 50(6) = 1350$$

The estimated sales for 1997 are $1,350,000.

(d) Let S(t) = 1600 and solve for t.

$$1050 + 50t = 1600$$
$$50t = 550$$
$$t = 11$$

The estimated sales will reach $1,600,000 eleven years after 1991, in 2002.

35. $p = 16 - \frac{5}{4}q$

(a) 0 can openers means q = 0.

$$p = 16 - \frac{5}{4}(0) = 16$$

The price for one can opener is $16 when 0 can openers are demanded.

(b) 400 can openers means q = 4.

$$p = 16 - \frac{5}{4}(4) = 11$$

The price is $11 when 400 can openers are demanded.

(c) 800 can openers means q = 8.

$$p = 16 - \frac{5}{4}(8) = 6$$

The price is $6 when 800 can openers are demanded.

(d) Price $6 means p = 6.

$$6 = 16 - \frac{5}{4}q$$
$$-10 = -\frac{5}{4}q$$
$$8 = q$$

800 can openers are demanded when the price is $6.

(e) Price $11 means p = 11.

$$11 = 16 - \frac{5}{4}q$$
$$-5 = -\frac{5}{4}q$$
$$4 = q$$

400 can openers are demanded when the price is $11.

(f) Price $16 means p = 16.

$$16 = 16 - \frac{5}{4}q$$
$$0 = -\frac{5}{4}q$$
$$0 = q$$

0 can openers are demanded when the price is $16.

(g) $p = 16 - \frac{5}{4}q$

This is a linear equation, so the graph will be a portion of a line. When q = 0, p = 16, so (0, 16) is one ordered pair, and when p = 0, $q = \frac{64}{5}$, so $\left(\frac{64}{5}, 0\right)$ is another ordered pair; draw the segment joining these two points. (See the answer graph in the back of the textbook.)

(h) $p = \frac{3}{4}q$

Price $0 means p = 0.

$$0 = \frac{3}{4}q$$

$$0 = q$$

The supply is 0 can openers when the price is $0.

(i) $p = \frac{3}{4}q$

Price $10 means p = 10.

$$10 = \frac{3}{4}q$$

$$\frac{40}{3} = q$$

When the price is $10, the supply is $\frac{40}{3}$ hundred can openers, or about 1333 can openers.

(j) $p = \frac{3}{4}q$

Price $20 means p = 20.

$$20 = \frac{3}{4}q$$

$$\frac{80}{3} = q$$

When the price is $20, the supply is $\frac{80}{3}$ hundred can openers, or about 2667 can openers.

(k) $p = \frac{3}{4}q$

This is a linear equation with domain q ≥ 0, so the graph will be a portion of a line. When q = 0, p = 0, so (0, 0) is one ordered pair, and when q = 4,

p = 3, so (4, 3) is another ordered pair; draw the ray through these two points, starting at (0, 0).

(See the answer graph in the back of the textbook.)

(1) To find the equilibrium quantity, set the two expressions that relate price to supply and price to demand equal to each other and solve.

$$16 - \frac{5}{4}q = \frac{3}{4}q$$

$$64 - 5q = 3q$$

$$64 = 8q$$

$$8 = q$$

The equilibrium quantity is 800 can openers.

(m) The equilibrium quantity is 800 can openers. Let q = 8 in each of the price equations to find the equilibrium price.

$$p = 16 - \frac{5}{4}(8) = 16 - 10 = 6$$

$$p = \frac{3}{4}(8) = 6$$

Observe that the same price, $6, results from both equations. $6 is the equilibrium price.

37. $p = S(q) = \frac{2}{5}q$ and

$p = D(q) = 100 - \frac{2}{5}q$

(a) The graph of the supply function is a ray with endpoint (0, 0) and passing through (125, 50).

The graph of the demand function is a segment joining the end-points (0, 100) and (250, 0). (See the answer graphs in the back of the textbook.)

(b) The equilibrium quantity (also called the equilibrium demand and the equilibrium supply) is found when the prices from both supply and demand are equal. Set the two expressions for p equal to each other and solve.

$$\frac{2}{5}q = 100 - \frac{2}{5}q$$

$$\frac{4}{5}q = 100$$

$$q = 125$$

The equilibrium demand is 125 units.

(c) To find the equilibrium price, let q = 125 in either expression for p.

$$p = S(125) = \frac{2}{5}(125) = 50$$

$$p = D(125) = 100 - \frac{2}{5}(125)$$

$$= 100 - 50 = 50$$

The equilibrium price is $50.

39. y = .07x + 135

(a) Let x = 0.

$$y = .07(0) + 135 = 135$$

When the weekly revenue is $0, the franchise fee for that week is $135.

(b) Let x = 1000.

$$y = .07(1000) + 135$$
$$= 70 + 135 = 205$$

When the revenue is $1000, the fee is $205.

(c) Let x = 2000.

$$y = .07(2000) + 135$$
$$= 140 + 135 = 275$$

When the revenue is $2000, the fee is $275.

(d) Let x = 3000.

$$y = .07(3000) + 135$$
$$= 210 + 135 = 345$$

When the revenue is $3000, the fee is $345.

(e) The equation is linear and the domain is x ≥ 0, so the graph will be a portion of a line. Draw the ray with endpoint (0, 135) and passing through (1000, 205), (2000, 275), and (3000, 345). (See the answer graph in the back of the textbook.)

Section 1.3

1. (−8, 6), (2, 4)

$$m = \frac{\Delta y}{\Delta x} = \frac{4 - 6}{2 - (-8)} = \frac{-2}{10} = -\frac{1}{5}$$

or

$$m = \frac{6 - 4}{-8 - 2} = \frac{2}{-10} = -\frac{1}{5}$$

3. $(-1, 4)$, $(2, 6)$

$$m = \frac{\Delta y}{\Delta x} = \frac{6 - 4}{2 - (-1)} = \frac{2}{3}$$

5. The origin and $(-4, 6)$

The coordinates of the origin are $(0, 0)$.

$$m = \frac{6 - 0}{-4 - 0} = \frac{6}{-4} = -\frac{3}{2}$$

7. $(-2, 9)$, $(-2, 11)$

$$m = \frac{11 - 9}{-2 - (-2)} = \frac{2}{0},$$

which is undefined.

9. $(3, -6)$, $(-5, -6)$

$$m = \frac{-6 - (-6)}{-5 - 3} = \frac{0}{-8} = 0$$

11. $y = 3x + 4$

$y = mx + b$ is the equation of a line in slope-intercept form, where m is the slope and b is the y-intercept. Thus $y = 3x + 4$ has slope $m = 3$, and y-intercept $b = 4$.

13. $y + 4x = 8$

To put this equation in slope-intercept form, solve for y.

$$y + 4x = 8$$
$$y = -4x + 8$$

The slope is $m = -4$ and the y-intercept is $b = 8$.

15. $3x + 4y = 5$

$$4y = -3x + 5$$
$$y = -\frac{3}{4}x + \frac{5}{4}$$

The slope is $m = -\frac{3}{4}$ and the y-intercept is $b = \frac{5}{4}$.

17. $3x + y = 0$

$$y = -3x + 0$$

The slope is $m = -3$ and the y-intercept is $b = 0$.

19. $2x + 5y = 0$

$$5y = -2x$$
$$y = -\frac{2}{5}x$$

The slope is $m = -\frac{2}{5}$ and the y-intercept is $b = 0$.

21. $y = 8$

Write this equation in slope-intercept form by inserting the term 0x.

$$y = 0x + 8$$

The slope is $m = 0$ and the y-intercept is $b = 8$.

23. $y + 2 = 0$

$$y = -2$$
$$y = 0x - 2$$

The slope is $m = 0$ and the y-intercept is $b = -2$.

25. x = -8

This is the equation of a vertical line, so it has undefined slope. The line has -8 as an x-intercept, but it does not have a y-intercept.

For Exercises 27–37, see the answer graph in the back of the textbook.

27. $(-4, 2)$, $m = \frac{2}{3}$

Graph the point $(-4, 2)$. To find a second point, use the slope, $m = \frac{\Delta y}{\Delta x} = \frac{2}{3}$. If x changes by 3 units $(\Delta x = 3)$, then y will change by 2 units $(\Delta y = 2)$. Start at $(-4, 2)$ and move 2 units up and 3 units to the right. Once the second point, $(-1, 4)$, is located, draw the line through the two points.

29. $(-5, -3)$, $m = -2$

Graph the point $(-5, -3)$. To find a second point, move 2 units down and 1 unit to the right since

$$m = -2 = \frac{-2}{1} = \frac{\Delta y}{\Delta x}.$$

The second point is $(-4, -5)$; draw the line through these two points.

31. $(8, 2)$, $m = 0$

The graph is the horizontal line passing through $(8, 2)$. Draw this line, which is parallel to the x-axis and 2 units above it.

33. $(6, -5)$, undefined slope

The graph is the vertical line passing through $(6, -5)$. Draw this line, which is parallel to the y-axis and 6 units to the right of it.

35. $(0, -2)$, $m = \frac{3}{4}$

Graph the point $(0, -2)$. To find a second point, move 3 units up and 4 units to the right. The second point is $(4, 1)$; draw the line through these two points.

37. $(5, 0)$, $m = \frac{1}{4}$

Graph the point $(5, 0)$. To find a second point, move 1 unit up and 4 units to the right. The second point is $(9, 1)$; draw the line through these two points.

39. $b = 4$, $m = -\frac{3}{4}$

The equation $y = mx + b$ becomes $y = -\frac{3}{4}x + 4$.

41. $b = \frac{5}{4}$, $m = \frac{3}{2}$

The equation $y = mx + b$ becomes $y = \frac{3}{2}x + \frac{5}{4}$.

43. Through $(-4, 1)$, $m = 2$

Substitute the values $x_1 = -4$, $y_1 = 1$, and $m = 2$ into the point-slope form.

$$y - y_1 = m(x - x_1)$$
$$y - 1 = 2[x - (-4)]$$
$$y - 1 = 2(x + 4)$$
$$y - 1 = 2x + 8$$

Put this equation in standard form.

$$-2x + y = 9$$
$$2x - y = -9$$

45. Through $(0, 3)$, $m = -3$

Since this line passes through the point $(0, 3)$, the y-intercept is 3. The slope-intercept form of the equation of this line is therefore

$$y = -3x + 3.$$

Put this equation in standard form.

$$3x + y = 3$$

47. Through $(3, 2)$, $m = \frac{1}{4}$

Substitute $x_1 = 3$, $y_1 = 2$, and $m = 1/4$ into the point-slope form.

$$y - y_1 = m(x - x_1)$$
$$y - 2 = \frac{1}{4}(x - 3)$$

Multiply both sides of this equation by 4 to clear fractions.

$$4(y - 2) = x - 3$$
$$4y - 8 = x - 3$$

Put this equation in standard form.

$$-x + 4y = 5$$
$$x - 4y = -5$$

49. Through $(-1, 1)$ and $(2, 5)$

First, find the slope.

$$m = \frac{5 - 1}{2 - (-1)} = \frac{4}{3}$$

Now use $m = 4/3$ and either of the two points in the point-slope form. Let $(x_1, y_1) = (2, 5)$.

$$y - y_1 = m(x - x_1)$$
$$y - 5 = \frac{4}{3}(x - 2)$$

Multiply both sides of this equation by 3.

$$3(y - 5) = 4(x - 2)$$
$$3y - 15 = 4x - 8$$
$$-4x + 3y = 7$$
$$4x - 3y = -7$$

51. Through $\left(\frac{1}{2}, \frac{5}{3}\right)$ and $\left(3, \frac{1}{6}\right)$

Find the slope.

$$m = \frac{\frac{1}{6} - \frac{5}{3}}{3 - \frac{1}{2}} = \frac{-\frac{3}{2}}{\frac{5}{2}} = -\frac{3}{5}$$

Use $m = -\frac{3}{5}$ and $(x_1, y_1) = \left(3, \frac{1}{6}\right)$ in the point-slope form.

$$y - y_1 = m(x - x_1)$$
$$y - \frac{1}{6} = -\frac{3}{5}(x - 3)$$
$$y - \frac{1}{6} = -\frac{3}{5}x + \frac{9}{5}$$

Multiply both sides of this equation by 30.

$$30y - 5 = -18x + 54$$
$$18x + 30y = 59$$

53. Through $(-8, 4)$ and $(-8, 6)$
Find the slope.

$$m = \frac{6 - 4}{-8 - (-8)} = \frac{2}{0},$$

which is undefined. A line with
undefined slope is vertical, and
vertical lines have equations of the
form $x = k$. Since the x–coordinate
of the two given ordered pairs is
-8, the desired equation is

$$x = -8.$$

55. Through $(-1, 3)$ and $(0, 3)$
Find the slope.

$$m = \frac{3 - 3}{-1 - 0} = \frac{0}{-1} = 0$$

Use $m = 0$ and $(x_1, y_1) = (0, 3)$ in
the point–slope form.

$$y - y_1 = m(x - x_1)$$
$$y - 3 = 0(x - 0)$$
$$y - 3 = 0$$
$$y = 3$$

57. x–intercept 3, y–intercept -2
The line passes through the points
$(3, 0)$ and $(0, -2)$; find its slope.

$$m = \frac{-2 - 0}{0 - 3} = \frac{-2}{-3} = \frac{2}{3}$$

Use $m = \frac{2}{3}$ and $b = -2$ in the slope–
intercept form.

$$y = mx + b$$
$$y = \frac{2}{3}x - 2$$
$$3y = 2x - 6$$
$$-2x + 3y = -6$$
$$2x - 3y = 6$$

59. Through $(-1, 4)$, parallel to
$x + 3y = 5$.
Find the slope of the line
$x + 3y = 5$.

$$x + 3y = 5$$
$$3y = -x + 5$$
$$y = -\frac{1}{3}x + \frac{5}{3}$$

This line has slope $m = -\frac{1}{3}$, and any
line parallel to it will also have
that slope.

Use $m = -\frac{1}{3}$, $x_1 = -1$, and $y_1 = 4$.
in the point–slope form.

$$y - y_1 = m(x - x_1)$$
$$y - 4 = -\frac{1}{3}[x - (-1)]$$
$$y - 4 = -\frac{1}{3}(x + 1)$$
$$y - 4 = -\frac{1}{3}x - \frac{1}{3}$$
$$3y - 12 = -x - 1$$
$$x + 3y = 11$$

61.

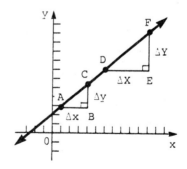

Choose 2 pairs of points, A and C, D
and F, and complete right triangles
as shown. Angle CAB = angle FDE

because when two parallel lines are cut by a transversal, corresponding angles are equal. Angle B = angle E because they are both right angles. Therefore, triangle CAB is similar to triangle FED because these triangles have two pairs of equal corresponding angles. When two triangles are similar, their corresponding sides are proportional. Therefore,

$$\frac{\Delta y}{\Delta x} = \frac{\Delta Y}{\Delta X}.$$

Since the slope is defined by

$$m = \frac{\Delta y}{\Delta x},$$

the slope will be the same no matter which two distinct points on the line are chosen to compute it.

63. Suppose that $(0, b)$ and (x_1, y_1) are distinct points on the line $y = mx + b$. By the definition of slope, the slope of this line is

$$\frac{\Delta y}{\Delta x} = \frac{y_1 - b}{x_1 - 0} = \frac{y_1 - b}{x_1}.$$

Since m represents the slope in slope-intercept form,

$$m = \frac{y_1 - b}{x_1}.$$

65. Two points on the line are $(-2, 0)$ and $(0, -4)$. Use the slope formula.

$$m = \frac{\Delta y}{\Delta x} = \frac{-4 - 0}{0 - (-2)}$$

$$= \frac{-4}{2} = -2$$

The slope is -2.

67. Two points on the line are $(0, -6)$ and $(1, 0)$.

$$m = \frac{-6 - 0}{0 - 1} = \frac{-6}{-1} = 6$$

The slope is 6.

69. The given information gives us 2 data points, $(20, 13{,}900)$ and $(10, 7500)$.

Let x = number of heaters;
    y = total cost in dollars.

(a) $m = \frac{13{,}900 - 7500}{20 - 10} = \frac{6400}{10} = 640$

$y - 7500 = 640(x - 10)$

$y - 7500 = 640x - 6400$

$y = 640x + 1100$

(b) Let $y = 23{,}500$ and solve for x.

$23{,}500 = 640x + 1100$

$22{,}400 = 640x$

$35 = x$

35 solar heaters can be made for $23,500.

(c) Let $y = 1000$ and solve for x.

$1000 = 640x + 1100$

$-100 = 640x$

$-\frac{100}{640} = x$

Negative numbers are not in the domain of the equation, so this is an impossible situation. No solar heaters can be made for $1000.

71. Let h = height (in centimeters);
        r = length of radius bone
              (in centimeters).

The data points are (24, 167) and (26, 174).

(a) $m = \dfrac{\Delta h}{\Delta r} = \dfrac{174 - 167}{26 - 24}$

$$= \dfrac{7}{2} = 3.5$$

Using m = 3.5 and
$(h_1, r_1) = (24, 167)$ in the point-slope form gives the required equation.

$$h - h_1 = m(r - r_1)$$
$$h - 167 = 3.5(r - 24)$$
$$h - 167 = 3.5r - 84$$
$$h = 3.5r + 83$$

(b) If r = 23,
$$h = 3.5(23) + 83$$
$$= 163.5$$

If r = 27,
$$h = 3.5(27) + 83$$
$$= 177.5.$$

If the radius bones had lengths between 23 and 27 cm, then the females' heights were between 163.5 and 177.5 cm.

(c) Find r when h = 170.

$$170 = 3.5r + 83$$
$$87 = 3.5r$$
$$24.857 = r$$

The length of a radius bone for a woman 170 cm tall is about 25 cm.

73. Let x = percent of two-party vote
        y = percent of sets won.

The data points are (45, 42.5) and (55, 67.5).

(a) $m = \dfrac{\Delta y}{\Delta x} = \dfrac{67.5 - 42.5}{55 - 45}$

$$= \dfrac{25}{10} = 2.5$$

$$y - y_1 = m(x - x_1)$$
$$y - 42.5 = 2.5(x - 45)$$
$$y - 42.5 = 2.5x - 112.5$$
$$y = 2.5x - 70$$

(b) Find y when x = 50.

$$y = 2.5(50) - 70$$
$$= 55$$

The Democrats will win 55% of the seats if they get 50% of the vote.

(c) Find x when y = 60.

$$60 = 2.5x - 70$$
$$130 = 2.5x$$
$$52 = x$$

The Democrats will capture 60% of the seats if they get 52% of the vote.

75. Let x = the number of years after
    1974;

  y = the number of immigrants.

The data points are (0, 86,821) and
(10, 140,289).

(a) $m = \dfrac{\Delta y}{\Delta x} = \dfrac{140,289 - 86,821}{10 - 0}$

$= \dfrac{53,468}{10} = 5346.8$

$y - y_1 = m(x - x_1)$

$y - 86,821 = 5346.8(x - 0)$

$y = 5346.8x + 86,821$

(b) For the year 2001,
    $x = 2001 - 1974 = 27.$

$y = 5346.8(27) + 86,821$

$= 144,363.6 + 86,821$

$= 231,184.6$

The immigration to California in
2001 will be about 231,185.

77. Let t = the number of years after
    1970;

  T = the average global temper-
    ature (in degrees Celsius).

In 1970, when t = 0, the average
global temperature was T = 15°C, so
one data point is (0, 15).
A decade later, when t = 10, the
average temperature rose .3°C to
T = 15.3°C, so another data point is
(10, 15.3).

(a) $m = \dfrac{\Delta T}{\Delta t} = \dfrac{15.3 - 15}{10 - 0} = \dfrac{.3}{10} = .03$

$T - T_1 = m(t - t_1)$

$T - 15 = .03(t - 0)$

$T - 15 = .03t$

$T = .03t + 15$

(b) Find t when T = 19°C.

$19 = .03t + 15$

$4 = .03t$

$133 \approx t$

The sea level will rise by 65 cm
about 133 years after 1970, in
the year 2103.

Section 1.4

1. Let C(x) be the cost (in dollars) of
renting a chain saw for x hours.
Then

$$C(x) = 12 + 1 \cdot x = 12 + x.$$

3. Let P(x) be the cost of (in cents)
of parking for x half-hours.
Then

$$P(x) = 35x + 50.$$

5. Fixed cost: $100; 50 items cost
$1600 to produce.
We are assuming that the cost
function is linear, that is, it
can be expressed as

$$C(x) = mx + b,$$

where b is the fixed cost and m
is the variable cost per item.
Then b = 100 and C(50) = 1600.
Substitute x = 50, b = 100, and
C(50) = 1600 into C(x) = mx + b.

$$C(x) = mx + 100$$
$$C(50) = m \cdot 50 + 100$$
$$1600 = 50m + 100$$
$$1500 = 50m$$
$$30 = m$$

Thus,

$$C(x) = 30x + 100.$$

7. Fixed cost: $1000; 40 items cost $2000 to produce.

$$b = 1000, \quad C(40) = 2000$$

Substitute these values into $C(x) = mx + b$

$$C(x) = mx + 1000$$
$$C(40) = m \cdot 40 + 1000$$
$$2000 = 40m + 1000$$
$$1000 = 40m$$
$$25 = m$$

Hence,

$$C(x) = 25x + 1000.$$

9. Variable cost: $50; 80 items cost $4500 to produce.

$$m = 50, \quad C(80) = 4500$$

$$C(x) = mx + b$$
$$C(80) = 50 \cdot 80 + b$$
$$4500 = 4000 + b$$
$$500 = b$$
$$C(x) = 50x + 500$$

11. $m = 90$, $C(150) = 16,000$

$$C(x) = mx + b$$
$$C(150) = 90 \cdot 150 + b$$
$$16,000 = 13,500 + b$$
$$2500 = b$$
$$C(x) = 90x + 2500$$

13. (a) $m = 3.50$, $C(60) = 300$

$$C(x) = mx + b$$
$$C(60) = 3.50(60) + b$$
$$300 = 210 + b$$
$$90 = b$$

The cost function is

$$C(x) = 3.50x + 90.$$

(b) Each T-shirt is sold for $9, so the revenue function is

$$R(x) = 9x.$$

To find the break-even point, solve the equation $R(x) = C(x)$.

$$R(x) = C(x)$$
$$9x = 3.50x + 90$$
$$5.50x = 90$$
$$x = 16.36$$

Since the number of T-shirts must be a whole number, she must produce and sell 17 shirts to break even.

(c) $P(x) = R(x) - C(x)$
$$= 9x - (3.50x + 90)$$
$$= 5.5x - 90$$

Let P(x) = 500 and solve for x.

500 = 5.5x − 90

590 = 5.5x

107.27 = x

She must produce and sell 108 T-shirts to make a profit of $500.

15. Let x = the number of years after 1981;

y = the total sales in year x.

The given data points are (0, 200,000) and (7, 1,000,000).

(a) $m = \dfrac{1,000,000 - 200,000}{7 - 0}$

$= \dfrac{800,000}{7}$

$y - y_1 = m(x - x_1)$

$y - 200,000 = \dfrac{800,000}{7}(x - 0)$

$y - 200,000 = \dfrac{800,000}{7}x$

$y = \dfrac{800,000}{7}x + 200,000$

(b) In 1992, x = 1992 − 1981 = 11.

Let x = 11 and solve for y.

$y = \dfrac{800,000}{7}(11) + 200,000$

$\approx 1,257,000 + 200,000$

$= 1,457,000$

The sales in 1992 will be about $1,457,000.

(c) Let y = 2,000,000 and solve for x.

$2,000,000 = \dfrac{800,000}{7}x + 200,000$

$1,800,000 = \dfrac{800,000}{7}x$

$12,600,000 = 800,000x$

$15.75 = x$

Sales are expected to reach $2,000,000 15.75 years after 1981, which will be in 1997.

17.   C(100) = 11.02, C(400) = 40.12

As ordered pairs, the data points are (100, 11.02) and (400, 40.12).

(a) $m = \dfrac{40.12 - 11.02}{400 - 100} = \dfrac{29.10}{300} = .097$

$y - y_1 = m(x - x_1)$

$y - 11.02 = .097(x - 100)$

$y - 11.02 = .097x - 9.70$

$y = .097x + 1.32$

$C(x) = .097x + 1.32$

(b) C(1000) = .097(1000) + 1.32

= 97.00 + 1.32

= 98.32

The total cost of producing 1000 cups is $98.32.

(c) C(1001) = .097(1001) + 1.32

= 97.097 + 1.32

= 98.417

The total cost of producing 1001 cups is $98.417.

(d) The marginal cost
of the 1001st cup is

$C(1001) - C(1000)$
$= 98.417 - 98.32$
$= .097.$

The marginal cost
is \$.097, or 9.7¢.

(e) The marginal cost
of any cup is

$C(n + 1) - C(n)$
$= [.097(n + 1) + 1.32]$
$\quad - [.097n + 1.32]$
$= .097n + .097 + 1.32 - .097n$
$\quad - 1.32$
$= .097.$

The marginal cost here is
always \$.097, or 9.7¢.

19. $C(x) = 800 + 20x$

(a) $x = 10$

$\dfrac{C(x)}{x} = \dfrac{800 + 20(10)}{10}$

$\qquad = \dfrac{1000}{10} = 100$

The average cost per item is
\$100.

(b) $x = 50$

$\dfrac{C(x)}{x} = \dfrac{800 + 20(50)}{50}$

$\qquad = \dfrac{1800}{50} = 36$

The average cost per item is
\$36.

(c) $x = 200$

$\dfrac{C(x)}{x} = \dfrac{800 + 20(200)}{200}$

$\qquad = \dfrac{4800}{200} = 24$

The average cost per item is
\$24.

21. $C(x) = 5x + 20,\ R(x) = 15x$

(a) The break-even point occurs when
$C(x) = R(x)$.

$5x + 20 = 15x$
$20 = 10x$
$2 = x$

The break-even point is 2 units.

(b) $\quad P(x) = R(x) - C(x)$
$\qquad = 15x - (5x + 20)$
$\qquad = 10x - 20$
$P(100) = 10(100) - 20 = 980$

The profit from 100 units is
\$980.

(c) Let $P(x) = 500$ and solve for x.

$500 = 10x - 20$
$520 = 10x$
$52 = x$

52 units will produce a profit
of \$500.

23. $C(x) = 50x + 5000,\ R(x) = 60x$

(a) The break-even point occurs when
$C(x) = R(x)$.

$50x + 5000 = 60x$
$5000 = 10x$
$500 = x$

$R(500) = 60(500) = 30,000$

The break-even point is 500 units. The revenue at the break-even point is $30,000.

(b) $P(x) = R(x) - C(x)$
$$= 60x - (50x + 5000)$$
$$= 10x - 5000$$
$$P(800) = 10(800) - 5000 = 3000$$

The profit from 800 units is $3000.

(c) Let $P(x) = 5000$ and solve for x.

$$5000 = 10x - 5000$$
$$10,000 = 10x$$
$$1000 = x$$

1000 units will produce a profit of $5000.

25. $C(x) = 85x + 900$, $R(x) = 105x$,
$0 \le x \le 38$

The break-even point occurs when $C(x) = R(x)$.

$$85x + 900 = 105x$$
$$900 = 20x$$
$$45 = x$$

The break-even point is 45 units. Since this is beyond the number which can be sold, the manufacturer should not produce.

27. $C(x) = 70x + 500$, $R(x) = 60x$

The break-even point occurs when $C(x) = R(x)$.

$$70x + 500 = 60x$$
$$500 = -10x$$
$$-50 = x$$

The break-even point is -50 units. Note that each unit sold brings in $60, while each unit made adds $70 to the cost. Hence the proposition is unprofitable for any positive number of units made. The manufacturer should not produce.

29.   (a) and (b): See the graphs in the back of the textbook.

(c) (2, 66) and (5, 90) are the data points to be used.

$$m = \frac{\Delta y}{\Delta x} = \frac{90 - 66}{5 - 2} = 8$$

Using the point-slope form of the equation and the point (2, 66), we obtain

$$y - 66 = 8(x - 2)$$
$$y - 66 = 8x - 16$$
$$y = 8x + 50.$$

(d)

Year	Actual Sales	Predicted Sales	Difference
0	48	50	-2
1	59	58	1
2	66	66	0
3	75	74	1
4	80	82	-2
5	90	90	0

(e) For year 7, $y = 8 \cdot 7 + 50 = 106$ thousand dollars, or $106,000.

(f) For year 9, $y = 8 \cdot 9 + 50 = 122$ thousand dollars, or $122,000.

31. The given data may be written as the ordered pairs (2, .75) and (3, .93). Since the relationship is linear, write the equation of the line through those two points.

$$m = \frac{.93 - .75}{3 - 2} = .18$$

Use the point–slope form with this slope and the point (2, .75).

$$y - .75 = .18(x - 2)$$
$$y - .75 = .18x - .36$$
$$y = .18x + .39$$

The pony switched from a trot to a gallop when the critical force was 1.16 times its body weight. Let y = 1.16 and solve for x.

$$1.16 = .18x + .39$$
$$.77 = .18x$$
$$4.3 \approx x$$

The pony switched from a trot to a gallop when it reached a speed of approximately 4.3 m/sec.

33. $y = .03x$

(a) $(.03)(10) = .3$ cm

(b) $(.03)(20) = .6$ cm

(c) $(.03)(50) = 1.5$ cm

(d) $(.03)(100) = 3$ cm

(e) The average rate of change of the JND with respect to the original length of the line is

$$\frac{.6 - .3}{20 - 10} = \frac{.3}{10} = .03.$$

35. The data points are (32, 0) and (212, 100).

$$m = \frac{\Delta C}{\Delta F} = \frac{100 - 0}{212 - 32}$$
$$= \frac{100}{180} = \frac{5}{9}$$
$$C - 0 = \frac{5}{9}(F - 32)$$
$$C = \frac{5}{9}(F - 32)$$

37. The data points are (0, 32) and (100, 212).

$$m = \frac{\Delta F}{\Delta C} = \frac{212 - 32}{100 - 0}$$
$$= \frac{180}{100} = \frac{9}{5}$$
$$F - 32 = \frac{9}{5}(C - 0)$$
$$F - 32 = \frac{9}{5}C$$
$$F = \frac{9}{5}C + 32$$

39. $y = \frac{38}{3}x + \frac{5}{6}$

(a) Let $y = 4\frac{1}{4}$ and solve for x.

$$4\frac{1}{4} = \frac{38}{3}x + \frac{5}{6}$$
$$\frac{17}{4} = \frac{38}{3}x + \frac{5}{6}$$

Multiply both sides of the equation by 12.

$$51 = 152x + 10$$
$$41 = 152x$$
$$\frac{41}{152} = x$$

Unfortunately, $\frac{41}{152}$ is not a multiple of $\frac{1}{64}$. To find out how many 64ths is $\frac{41}{152}$, solve the equation

$$\frac{n}{64} = \frac{41}{152}.$$

$$n = 64 \cdot \frac{41}{152}$$

$$= \frac{328}{19} \approx 17.3$$

Since n must be an integer, and a bit <u>at least</u> $4\frac{1}{4}$ in is needed, let n = 18. The drill bit to be ordered should have a diameter of $\frac{18}{64} = \frac{9}{32}$ in.

(b) Let $y = 5\frac{1}{2}$ and solve for x.

$$5\frac{1}{2} = \frac{38}{3}x + \frac{5}{6}$$

$$33 = 76x + 5$$

$$28 = 76x$$

$$\frac{28}{76} = x$$

$$\frac{7}{19} = x$$

Unfortunately, $\frac{7}{19}$ is not a multiple of $\frac{1}{64}$. Solve the following equation.

$$\frac{n}{64} = \frac{7}{19}$$

$$n = 64 \cdot \frac{7}{19}$$

$$= \frac{448}{19} \approx 23.6 \approx 24$$

The drill bit to be ordered should have a diameter of $\frac{24}{64} = \frac{3}{8}$ in.

Section 1.5

3. n = 10, x = \$28,000

(a) Straight–line depreciation uses the formula $D(x) = \frac{1}{n}x$.

For year 1,

$$D = \frac{1}{10}(\$28,000) = \$2800.$$

For year 4,

$$D = \frac{1}{10}(\$28,000) = \$2800.$$

(b) The sum–of–the–years'–digits method uses the formula

$$D(x) = \frac{2(n + 1 - j)}{n(n + 1)}x.$$

For year 1, j = 1, and

$$D = \frac{2(10 + 1 - 1)}{10(10 + 1)}(\$28,000)$$

$$= \frac{20}{110}(\$28,000)$$

$$\approx \$5090.91.$$

For year 4, j = 4, and

$$D = \frac{2(10 + 1 - 4)}{10(10 + 1)}(\$28,000)$$

$$= \frac{14}{110}(\$28,000)$$

$$\approx \$3563.64.$$

5. n = 6, x = $14,600

(a) Straight-line depreciation uses the formula $D(x) = \frac{1}{n}x$.

For year 1,

$D = \frac{1}{6}(\$14,600) \approx \$2433.33$.

For year 4,

$D = \frac{1}{6}(\$14,600) \approx \$2433.33$.

(b) The sum-of-the-years'-digits method uses the formula

$D(x) = \frac{2(n + 1 - j)}{n(n + 1)}x$.

For year 1,

$D = \frac{2(6 + 1 - 1)}{6(6 + 1)}(\$14,600)$

$= \frac{12}{42}(\$14,600)$

$\approx \$4171.43$.

For year 4,

$D = \frac{2(6 + 1 - 4)}{6(6 + 1)}(\$14,600)$

$= \frac{6}{42}(\$14,600)$

$\approx \$2085.71$.

7. n = 4, x = $10,000

(a) Sum-of-the-years'-digits method

$D(x) = \frac{2(n + 1 - j)}{n(n + 1)}x$

For year 1, j = 1, and

$D = \frac{2(4 + 1 - 1)}{4(4 + 1)}(\$10,000)$

$= \frac{8}{20}(\$10,000)$

$= \$4000$.

For year 2, j = 2, and

$D = \frac{2(4 + 1 - 2)}{4(4 + 1)}(\$10,000)$

$= \frac{6}{20}(\$10,000)$

$= \$3000$.

For year 3, j = 3, and

$D = \frac{2(4 + 1 - 3)}{4(4 + 1)}(\$10,000)$

$= \frac{4}{20}(\$10,000)$

$= \$2000$.

For year 4, j = 4, and

$D = \frac{2(4 + 1 - 4)}{4(4 + 1)}(\$10,000)$

$= \frac{2}{20}(\$10,000)$

$= \$1000$.

(b) Annual straight-line depreciation

$D(x) = \frac{1}{n}x$

$= \frac{1}{4}(\$10,000)$

$= \$2500$

9. n = 6, x = $210,000

(a) For year 1,

$D = \frac{2(6 + 1 - 1)}{6(6 + 1)}(\$210,000)$

$= \frac{12}{42}(\$210,000)$

$= \$60,000$.

(b) For year 2,

$D = \frac{2(5)}{42}(\$210,000) = \$50,000$.

(c) For year 3,

$$D = \frac{2(4)}{42}(\$210,000) = \$40,000.$$

(d) For year 4,

$$D = \frac{2(3)}{42}(\$210,000) = \$30,000.$$

(e) By the straight-line method, the annual depreciation for each year would be

$$D = \frac{1}{6}(\$210,000) = \$35,000.$$

11. n = 3, x = $1125

Annual straight-line depreciation

For each year in the life,

$$D(x) = \frac{1}{n}x = \frac{1}{3}(\$1125) = \$375.$$

Sum-of-the-years'-digits depreciation

For year 1,

$$D = \frac{2(3 + 1 - 1)}{3(4)}(\$1125) = \$562.50.$$

For year 2,

$$D = \frac{2(3 + 1 - 2)}{3(4)}(\$1125) = \$375.00.$$

For year 3,

$$D = \frac{2(3 + 1 - 3)}{3(4)}(\$1125) = \$187.50.$$

The table that summarizes this information is as follows. (The accumulated depreciation column is obtained by adding all the depreciations from year 1 to the present year.

Year	Straight-Line Depreciation	Accumulated Depreciation at End of Year	Sum-of-the-Years'-Digits Depreciation	Accumulated Depreciation at End of Year
1	$375.00	$375.00	$562.50	$562.50
2	$375.00	$750.00	$375.00	$937.50
3	$375.00	$1125.00	$187.50	$1125.00
Totals	$1125.00		$1125.00	

13. (a) To find the equation of the least squares line for the given data, first find the required sums. Let x represent the annual store sales (in thousands of dollars) and let y represent the pretax profit (in percent of sales), and prepare a table as follows.

x	y	xy	x^2	y^2
250	9.3	2325	62,500	86.49
375	14.8	5550	140,625	219.09
450	14.2	6390	202,500	201.64
500	15.9	7950	250,000	252.81
600	19.2	11,520	360,000	368.64
650	21.0	13,650	422,500	441.00
2825	94.4	47,385	1,438,125	1569.62

There are six pairs of data, so n = 6. The least squares line is described by the equation

$$y' = mx + b,$$

where $m = \dfrac{n(\Sigma xy) - (\Sigma x)(\Sigma y)}{n(\Sigma x^2) - (\Sigma x)^2}$ and $b = \dfrac{\Sigma y - m(\Sigma x)}{n}$.

In this problem, using the information from the table that has been constructed,

$$m = \frac{n(\Sigma xy) - (\Sigma x)(\Sigma y)}{n(\Sigma x^2) - (\Sigma x)^2}$$

$$= \frac{6(47,385) - (2825)(94.4)}{6(1,438,125) - (2825)^2}$$

$$= \frac{284,310 - 266,680}{8,628,750 - 7,980,625}$$

$$= \frac{17,630}{648,125} \approx .027$$

and

$$b = \frac{\Sigma y - m(\Sigma x)}{n}$$

$$= \frac{94.4 - (.0272)(2825)}{6}$$

$$= \frac{94.4 - 76.84}{6}$$

$$= \frac{17.56}{6} \approx 2.93.$$

Therefore,

$$y' = .027x + 2.93.$$

(b) Let x = 400 and find y'.

$$y' = .027(400) + 2.93$$
$$= 10.8 + 2.93$$
$$= 13.7$$

Sales of $400,000 would produce a profit of 13.7%.

(c) Let y' = 15 and solve for x.

$$15 = .027x + 2.93$$
$$15 - 2.93 = .027x$$
$$x \approx 447$$

Sales of $447,000 would produce a profit of 15%.

(d) Use the table of sums constructed in part (a) to find the coefficient of correlation.

$$r = \frac{n(\Sigma xy) - (\Sigma x)(\Sigma y)}{\sqrt{n(\Sigma x^2) - (\Sigma x)^2} \cdot \sqrt{n(\Sigma y^2) - (\Sigma y)^2}}$$

$$= \frac{6(47,385) - (2825)(94.4)}{\sqrt{6(1,438,125) - (2825)^2} \cdot \sqrt{6(1569.62) - 94.4)^2}}$$

$$= \frac{284,310 - 266,680}{\sqrt{8,628,750 - 7,980,625} \cdot \sqrt{9417.72 - 8911.36}}$$

$$= \frac{17,630}{\sqrt{658,125} \cdot \sqrt{506.36}} \approx .97$$

(e) Since the coefficient of correlation has a value close to 1, there does seem to be a linear relationship between the two sets of data.

15. (a) To find the equation of the least squares line, first find the sums. Let x represent the number of years since 1982, as suggested, and let y represent the number of customer lines (in millions), and prepare a table as follows.

x	y	xy	x^2	y^2
1	5.0	5.0	1	25.00
2	7.8	15.6	4	60.84
3	8.9	26.7	9	79.21
4	9.8	39.2	16	96.04
5	9.6	48.0	25	92.16
6	10.4	62.4	36	108.16
21	51.5	196.9	91	461.41

There are six pairs of data, so n = 6. Find m and b, using the information from the table.

$$m = \frac{n(\Sigma xy) - (\Sigma x)(\Sigma y)}{n(\Sigma x^2) - (\Sigma x)^2}$$

$$= \frac{6(196.9) - (21)(51.5)}{6(91) - (21)^2}$$

$$= \frac{1181.4 - 1081.5}{546 - 441}$$

$$= \frac{99.9}{105} \approx .95$$

$$b = \frac{\Sigma y - m(\Sigma x)}{n}$$

$$= \frac{51.5 - (.95)(21)}{6}$$

$$= \frac{51.5 - 19.95}{6} \approx 5.3$$

So in this problem, y′ = mx + b becomes y′ = .95x + 5.3.

(b) Let x = 1994 − 1982 = 12 and find y′.

$$\begin{aligned} y' &= .95(12) + 5.3 \\ &= 11.4 + 5.3 \\ &= 16.7 \end{aligned}$$

There will be 16.7 million lines in 1994.

(c) Let y′ = 19 and solve for x.

$$\begin{aligned} 19 &= .95x + 5.3 \\ 13.7 &= .95x \\ 14.4 &= x \end{aligned}$$

There will be at least 19 million lines about 14.4 years after 1982, which will be in 1997.

(d) The coefficient of correlation is

$$r = \frac{n(\Sigma xy) - (\Sigma x)(\Sigma y)}{\sqrt{n(\Sigma x^2) - (\Sigma x)^2} \cdot \sqrt{n(\Sigma y^2) - (\Sigma y)^2}}$$

$$= \frac{6(196.9) - (21)(51.5)}{\sqrt{6(91) - (21)^2} \cdot \sqrt{6(461.41) - (51.5)^2}}$$

$$= \frac{1181.4 - 1081.2}{\sqrt{546 - 441} \cdot \sqrt{2768.46 - 2652.25}}$$

$$= \frac{99.9}{\sqrt{105} \cdot \sqrt{116.21}} \approx .90.$$

(e) Since the coefficient of correlation has a value close to 1, the increase in the number of customer lines does appear to be linear with time.

17. n = 10, Σx = 30, Σy = 24, Σxy = 75, Σx² = 100, Σy² = 80

(a) The least squares line is described by the equation y′ = mx + b, where

$$m = \frac{n(\Sigma xy) - (\Sigma x)(\Sigma y)}{n(\Sigma x^2) - (\Sigma x)^2} \text{ and}$$

$$b = \frac{\Sigma y - m(\Sigma x)}{n}.$$

In this problem,

$$m = \frac{10(75) - (30)(24)}{10(100) - (30)^2}$$

$$= \frac{750 - 720}{1000 - 900} = \frac{30}{300} = .3 \text{ and}$$

$$b = \frac{24 - .3(30)}{10} = \frac{24 - 9}{10}$$

$$= \frac{15}{10} = 1.5.$$

Thus,

$$y' = .3x + 1.5.$$

(b)   The coefficient of correlation is

$$r = \frac{n(\Sigma xy) - (\Sigma x)(\Sigma y)}{\sqrt{n(\Sigma x^2) - (\Sigma x)^2} \cdot \sqrt{n(\Sigma y^2) - (\Sigma y)^2}}$$

$$= \frac{10(75) - (30)(24)}{\sqrt{10(100) - (30)^2} \cdot \sqrt{10(80) - (24)^2}}$$

$$= \frac{750 - 720}{\sqrt{1000 - 900} \cdot \sqrt{800 - 576}}$$

$$= \frac{30}{\sqrt{100} \cdot \sqrt{224}} = \frac{30}{10\sqrt{224}} = \frac{3}{\sqrt{224}} \approx .20.$$

(c) $y' = .3x + 1.5$

Find y' when $x = 3$.

$$y' = .3(3) + 1.5$$
$$= .9 + 1.5$$
$$= 2.4$$

The equation predicts than an ear of corn will be 2.4 dm long if 3 tons of fertilizer are used per acre.

(d)   Since the value of r is not close to 1 and not close to −1, there does not seem to be a linear relationship between the two sets of data.

19.   (a)   To find the equation of the least squares line, first find the sums. Let x represent the temperature (in degrees Fahrenheit) and let y represent the number of chirps per second, and prepare the following table.

x	y	xy	x^2	y^2
88.6	20.0	1772.00	7849.96	400.00
71.6	16.0	1145.60	5126.56	256.00
93.3	19.8	1847.34	8704.89	392.04
84.3	18.4	1551.12	7106.49	338.56
80.6	17.1	1378.26	6496.36	292.41
75.2	15.5	1165.60	5655.04	240.25
69.7	14.7	1024.59	4858.09	216.09
82.0	17.1	1402.20	6724.00	292.41
69.4	15.4	1068.76	4816.36	237.16
83.3	16.2	1349.46	6938.89	262.44
79.6	15.0	1194.00	6336.16	225.00
82.6	17.2	1420.72	6822.76	295.00
80.6	16.0	1289.60	6496.36	256.00
83.5	17.0	1419.50	6972.25	289.00
76.3	14.4	1098.72	5821.69	207.36
1200.6	249.8	20,127.47	96,725.86	4200.56

There are fifteen pairs of data, so n = 15. Find m and b, using the information from the table.

$$m = \frac{n(\Sigma xy) - (\Sigma x)(\Sigma y)}{n(\Sigma x^2) - (\Sigma x)^2}$$

$$= \frac{15(20,127.47) - (1200.6)(249.8)}{15(96,725.86) - (1200.6)^2}$$

$$= \frac{301,912.05 - 299,909.88}{1,450,887.9 - 1,441,440.3}$$

$$= \frac{2002.17}{9447.6} \approx .21192 \approx .212$$

$$b = \frac{\Sigma y - m(\Sigma x)}{n}$$

$$= \frac{249.8 - .21192(1200.6)}{15}$$

$$= \frac{249.8 - 254.43115}{15}$$

$$= \frac{-4.63115}{15} \approx -.309$$

So in this problem,

$$y' = mx + b$$

becomes

$$y' = .212x - .309.$$

(b) Let x = 73 and find y'.

$$y' = .212(73) - .309$$
$$= 15.476 - .309$$
$$= 15.167$$
$$\approx 15.2$$

If the temperature is 73°F, the cricket would be expected to make about 15.2 chirps per sec.

(c) Let y' = 18 and solve for x.

$$18 = .212x - .309$$
$$18.309 = .212x$$
$$86.4 \approx x$$

If the cricket chirps 18 times per sec, the temperature would be expected to be about 86.4°F.

(d) The coefficient of correlation is

$$r = \frac{n(\Sigma xy) - (\Sigma x)(\Sigma y)}{\sqrt{n(\Sigma x^2) - (\Sigma x)^2} \cdot \sqrt{n(\Sigma y^2) - (\Sigma y)^2}}$$

$$= \frac{15(20,127.47) - (1200.6)(249.8)}{\sqrt{15(96,725.86) - (1200.6)^2} \cdot \sqrt{15(4200.56 - (249.8)^2}}$$

$$= \frac{2002.17}{\sqrt{9447.6} \cdot \sqrt{608.36}} \approx .835$$

21. (a) To find the least squares equation prepare a table of required sums. Let x represent the number of square feet and let y represent the number of BTUs.

x	y	xy	x^2
150	5000	750,000	22,500
175	5500	962,500	30,625
215	6000	1,290,000	46,225
250	6500	1,625,000	62,500
280	7000	1,960,000	78,400
310	7500	2,325,000	96,100
350	8000	2,800,000	122,500
370	8500	3,145,000	136,900
420	9000	3,780,000	176,400
450	9500	4,275,000	202,500
2970	72,500	22,912,500	974,650

There are ten pairs of data, so n = 10. Find m and b using the information from the table.

$$m = \frac{n(\Sigma xy) - (\Sigma x)(\Sigma y)}{n(\Sigma x^2) - (\Sigma x)^2}$$

$$= \frac{10(22,912,500) - (2970)(72,500)}{10(974,650) - (2970)^2}$$

$$= \frac{229,125,000 - 215,325,000}{9,746,500 - 8,820,900}$$

$$= \frac{13,800,000}{925,600} \approx 14.91$$

$$b = \frac{\Sigma y - m(\Sigma x)}{n}$$

$$= \frac{72,500 - 14.91(2970)}{10}$$

$$= \frac{72,500 - 44,282.7}{10}$$

$$= \frac{28,217.3}{10} \approx 2820$$

So in this problem,

$$y' = mx + b$$

becomes

$$y' = 14.9x + 2820.$$

(b) When x = 150,

$$y' = 14.91(150) + 2820$$
$$= 2236.5 + 2820$$
$$\approx 5060,$$

which is close to the actual value of y = 5000.

When x = 280,

$$y' = 14.91(280) + 2820$$
$$= 4174.8 + 2820$$
$$\approx 6990,$$

which is very close to the actual value of y = 7000.

When x = 420,

$$y' = 14.91(420) + 2820$$
$$= 6262.2 + 2820$$
$$\approx 9080,$$

which is close to the actual value of y = 9000.

(c) When x = 230,

$$y' = 14.91(230) + 2820$$
$$= 3429.3 + 2820$$
$$\approx 6250.$$

About 6250 BTUs would be needed to cool the room. If air conditioners are available only with BTU choices in the table, then at least 6500 BTUs would be needed; 6000 BTUs or fewer would not be sufficient.

23. (a) To find the least squares equation, prepare a table of the required sums. Let x represent the number of years since 1900 and let y represent the death rate.

x	y	xy	x^2
10	14.7	147	100
20	13.0	260	400
30	11.3	339	900
40	10.8	432	1600
50	9.6	480	2500
60	9.5	570	3600
70	9.5	665	4900
80	8.7	696	6400
90	8.6	774	8100
450	95.7	4363	28,500

There are nine pairs of data, so n = 9. Find m and b using the information from the table.

$$m = \frac{n(\Sigma xy) - (\Sigma x)(\Sigma y)}{n(\Sigma x^2) - (\Sigma x)^2}$$

$$= \frac{9(4363) - (450)(95.7)}{9(28,500) - (450)^2}$$

$$= \frac{39,267 - 43,065}{256,500 - 202,500}$$

$$= \frac{-3798}{54,000} \approx -.070$$

$$b = \frac{\Sigma y - m(\Sigma x)}{n}$$

$$= \frac{95.7 - (-.070)(450)}{9}$$

$$= \frac{95.7 + 31.5}{9}$$

$$= \frac{127.2}{9} \approx 14.1$$

So in this problem,

$$y' = mx + b$$

becomes

$$y' = -.070x + 14.1.$$

(b) When $x = 1920 - 1900 = 20$,

$$y' = -.070(20) + 14.1$$
$$= -1.4 + 14.1$$
$$= 12.7,$$

which is very close to the actual value of $y = 13.0$.

When $x = 1950 - 1900 = 50$,

$$y' = -.070(50) + 14.1$$
$$= -3.5 + 14.1$$
$$= 10.6,$$

which is close to the actual value of 9.6.

When $x = 1980 - 1900 = 80$,

$$y' = -.070(80) + 14.1$$
$$= -5.6 + 14.1$$
$$= 8.5,$$

which is very close to the actual value of $y = 8.8$.
The least squares equation is giving good estimates of the death rates, so it seems to be a reasonable fit for the given data.

(c) When $x = 1994 - 1900 = 94$,

$$y' = -.070(94) + 14.17$$
$$= -6.58 + 14.1$$
$$\approx 7.6$$

The equation predicts that the death rate in 1994 will be 7.6.

25. (a) $n = 30$, $x = \$145,000$

Straight-line depreciation uses the formula $D(x) = \frac{1}{n}x$. For each of the thirty years,

$$D = \frac{1}{30}(\$145,000) \approx \$4833.33.$$

The sum-of-year'-digits method uses the formula

$$D(x) = \frac{2(n + 1 - j)}{n(n + 1)}x \text{ for year } j.$$

For year 1, $j = 1$, and

$$D = \frac{2(30 + 1 - 1)}{30(30 + 1)}(\$145,000)$$
$$= \frac{60}{930}(\$145,000) \approx \$9354.84.$$

For year 2, $j = 2$, and

$$D = \frac{2(30 + 1 - 2)}{30(30 + 1)}(\$145,000)$$
$$= \frac{58}{930}(\$145,000) \approx \$9043.01.$$

For year 30, $j = 30$, and

$$D = \frac{2(30 + 1 - 30)}{30(30 + 1)}(\$145,000)$$
$$= \frac{2}{930}(\$145,000) \approx \$311.82.$$

Repeat this procedure for the other twenty-seven years to obtain the second column of the following table.

Year	Straight-Line Depreciation	Sum-of-the-Years'-Digits Depreciation
1	$4833.33	$9354.84
2	$4833.33	$9043.01
3	$4833.33	$8731.18
4	$4833.33	$8419.35
5	$4833.33	$8107.53
6	$4833.33	$7795.70
7	$4833.33	$7483.87
8	$4833.33	$7172.04
9	$4833.33	$6860.22
10	$4833.33	$6548.39
11	$4833.33	$6236.56
12	$4833.33	$5924.73
13	$4833.33	$5612.90
14	$4833.33	$5301.08
15	$4833.33	$4989.25
16	$4833.33	$4677.42
17	$4833.33	$4365.59
18	$4833.33	$4053.76
19	$4833.33	$3741.94
20	$4833.33	$3430.11
21	$4833.33	$3118.28
22	$4833.33	$2806.45
23	$4833.33	$2494.62
24	$4833.33	$2182.80
25	$4833.33	$1870.97
26	$4833.33	$1559.14
27	$4833.33	$1247.31
28	$4833.33	$935.48
29	$4833.33	$623.66
30	$4833.33	$311.82

(b) n = 15, x = $258,000

For each of the fifteen years, the straight-line depreciation is

$$D(x) = \frac{1}{n}x$$

$$= \frac{1}{15}(\$258,000) = \$17,200.$$

The sum-of-the-years'-digits method uses the formula

$$D(x) = \frac{2(n + 1 - j)}{n(n + 1)}x \text{ for year j.}$$

For year 1, j = 1, and

$$D = \frac{2(15 + 1 - 1)}{15(15 + 1)}(\$258,000)$$

$$= \frac{30}{240}(\$258,000) = \$32,250.$$

For year 2, j = 2, and

$$D = \frac{2(15 + 1 - 2)}{15(15 + 1)}(\$258,000)$$

$$= \frac{28}{240}(\$258,000) = \$30,100.$$

For year 15, j = 15, and

$$D = \frac{2(15 + 1 - 15)}{15(15 + 1)}(\$258,000)$$

$$= \frac{2}{240}(\$258,000) = \$2150.$$

Repeat this procedure for the other twelve years to obtain the second column of the following table.

Year	Straight-Line Depreciation	Sum-of-the-Years'-Digits Depreciation
1	$17,200	$32,250
2	$17,200	$30,100
3	$17,200	$27,950
4	$17,200	$25,800
5	$17,200	$23,650
6	$17,200	$21,500
7	$17,200	$19,350
8	$17,200	$17,200
9	$17,200	$15,050
10	$17,200	$12,900
11	$17,200	$10,750
12	$17,200	$8600
13	$17,200	$6450
14	$17,200	$4300
15	$17,200	$2150

27. (a) To find an equation for the least squares line, prepare a table. Let x represent the Math SAT score and let y represent the placement test score.

x	y	xy	x^2	y^2
540	20	10,800	291,600	400
510	16	8160	260,100	256
490	10	4900	240,100	100
560	8	4480	313,600	64
470	12	5640	220,900	144
600	11	6600	360,000	121
540	10	5400	291,600	100
580	8	4640	336,400	64
680	15	10,200	462,400	225
560	8	4480	313,600	64
560	13	7280	313,600	169
500	14	7000	250,000	196
470	10	4700	220,900	100
440	10	4400	193,600	100
520	11	5720	270,400	121
620	11	6820	384,400	121
680	8	5440	462,400	64
550	8	4400	302,500	64
620	7	4340	384,400	49
10,490	210	115,400	5,872,500	2522

There are 19 pairs of data, so n = 19. Find m and b.

$$m = \frac{n(\Sigma xy) - (\Sigma x)(\Sigma y)}{n(\Sigma x^2) - (\Sigma x)^2}$$

$$= \frac{19(115,400) - (10,490)(210)}{19(5,872,500) - (10,490)^2}$$

$$= \frac{2,192,600 - 2,202,900}{111,577,500 - 110,040,100}$$

$$= \frac{-10,300}{1,537,400} \approx -.0067$$

$$b = \frac{\Sigma y - m(\Sigma x)}{n}$$

$$= \frac{210 - (-.0067)(10,490)}{19}$$

$$= \frac{210 + 70.283}{19}$$

$$= \frac{280.283}{19} \approx 14.75$$

An equation for the least squares line is

$$y' = -.0067x + 14.75.$$

(b) When x = 420,

$$y' = -.0067(420) + 14.75$$

$$= -2.814 + 14.75$$

$$= 11.936 \approx 12$$

The equation predicts that a student with a Math SAT score of 420 will get a mathematics placement test score of 12.

(c) When x = 620,

$$y' = -.0067(620) + 14.75$$
$$= -4.154 + 14.75$$
$$= 10.596 \approx 11$$

The equation predicts that a student with a Math SAT score of 620 will get a mathematics placement test score of 11.

(d) The coefficient of correlation is

$$r = \frac{n(\Sigma xy) - (\Sigma x)(\Sigma y)}{\sqrt{n(\Sigma x^2) - (\Sigma x)^2} \cdot \sqrt{n(\Sigma y^2) - (\Sigma y)^2}}$$

$$= \frac{19(115,400) - (10,490)(210)}{\sqrt{19(5,872,500) - (10,490)^2} \cdot \sqrt{19(2522) - (210)^2}}$$

$$= \frac{2,192,600 - 2,202,900}{\sqrt{1,537,400} \cdot \sqrt{3818}}$$

$$= \frac{-10,300}{76,614.58} \approx -.13.$$

(e) Since the value of r is not close to 1 and not close to -1, there does not seem to be a linear relationship between the two sets of test scores.

Chapter 1 Review Exercises

1. For example, 5x + 3 = 7x - 9 and 6(4x - 7) = 3x - 14 are linear equations since they can be written in the form ax = b; $x^2 - 3x + 2 = 0$ and $x^3 = 9x^2$ are non-linear equations. Of course, there are many other examples of each type of equation.

5. 3x + 2 = 8
       3x = 6
        x = 2

7. 2k - 3 = 6 - k
       3k = 9
        k = 3

9. $5n - (n + 1) = 3(n + 2)$

    $5n - n - 1 = 3n + 6$

    $4n - 1 = 3n + 6$

    $n = 7$

11. $\frac{3}{2} + 3z = 5 + \frac{2z}{4}$

    $6 + 12z = 20 + 2z$

    $10z = 14$

    $z = \frac{7}{5}$

13. $4m + 2 \leq 12$

    $4m \leq 10$

    $m \leq \frac{5}{2}$

15. $5 - 2m > 7$

    $-2m > 2$

    $m < -1$

17. $2k + 3 \leq 5k - 4$

    $-3k \leq -7$

    $k \geq \frac{7}{3}$

19. $\frac{2}{3}x + 4 \geq x - \frac{1}{3}$

    $2x + 12 \geq 3x - 1$

    $-x \geq -13$

    $x \leq 13$

21. $2(k + 5) - 3k < k + 4$

    $2k + 10 - 3k < k + 4$

    $-k + 10 < k + 4$

    $-2k < -6$

    $k > 3$

23. $f(x) = 4x - 1$

    (a) $f(6) = 4(6) - 1$

    $= 24 - 1 = 23$

    (b) $f(-2) = 4(-2) - 1$

    $= -8 - 1 = -9$

    (c) $f(-4) = 4(-4) - 1$

    $= -16 - 1 = -17$

    (d) $f(r + 1) = 4(r + 1) - 1$

    $= 4r + 4 - 1$

    $= 4r + 3$

25. $f(x) = -x^2 + 2x - 4$

    (a) $f(6) = -(6)^2 + 2(6) - 4$

    $= -36 + 12 - 4 = -28$

    (b) $f(-2) = -(-2)^2 + 2(-2) - 4$

    $= -4 - 4 - 4 = -12$

    (c) $f(-4) = -(-4)^2 + 2(-4) - 4$

    $= -16 - 8 - 4 = -28$

    (d) $f(r + 1)$

    $= -(r + 1)^2 + 2(r + 1) - 4$

    $= -(r^2 + 2r + 1) + 2r + 2 - 4$

    $= -r^2 - 3$

For Exercises 27-33, see the answer graph in the back of the textbook.

27. $y = 4x + 3$

    Use the slope, $m = \frac{\Delta y}{\Delta x} = \frac{4}{1}$ and the y-intercept, $b = 3$, to draw the graph. Graph the point $(0, 3)$. To find a second point, move 4 units up and one unit to the right. Ordered pairs may also be found by substituting values for x in the equation.

For example,

if $x = 1$, $y = 4(1) + 3 = 7$; and

if $x = -1$, $y = 4(-1) + 3 = -1$.

Draw the line through the points
$(1, 7)$ and $(-1, -1)$.

29. $3x - 5y = 15$

Find the x– and y–intercepts.
When $x = 0$, $-5y = 15$, so $y = -3$.
When $y = 0$, $3x = 15$, so $x = 5$.

Draw the line through the points
$(0, -3)$ and $(5, 0)$.

31. $x + 2 = 0$
or $x = -2$

This is a vertical line 2 units to
the left of the y–axis.

33. $y = 2x$

When $x = 0$, $y = 2(0) = 0$.
When $x = 2$, $y = 2(2) = 4$.

Draw the line through the points
$(0, 0)$ and $(2, 4)$.

35. Through $(-2, 5)$ and $(4, 7)$

$$m = \frac{\Delta y}{\Delta x} = \frac{7 - 5}{4 - (-2)}$$

$$= \frac{2}{6} = \frac{1}{3}$$

37. Through the origin and $(11, -2)$
The origin has coordinates $(0, 0)$.

$$m = \frac{-2 - 0}{11 - 0} = -\frac{2}{11}$$

39. $2x + 3y = 15$

$$3y = -2x + 15$$

$$y = -\frac{2}{3}x + 5$$

$$m = -\frac{2}{3}$$

41. $x + 4 = 9$
$x = 5$

This is a vertical line, so the slope
is undefined.

43. Through $(5, -1)$, slope $= 2/3$

$$y - y_1 = m(x - x_1)$$

$$y - (-1) = \frac{2}{3}(x - 5)$$

$$y + 1 = \frac{2}{3}(x - 5)$$

$$3(y + 1) = 2(x - 5)$$

$$3y + 3 = 2x - 10$$

$$-2x + 3y = -13$$

$$2x - 3y = 13$$

45. Through $(5, -2)$ and $(1, 3)$

$$m = \frac{3 - (-2)}{1 - 5} = \frac{5}{-4} = -\frac{5}{4}$$

$$y - y_1 = m(x - x_1)$$

$$y - 3 = -\frac{5}{4}(x - 1)$$

$$4(y - 3) = -5(x - 1)$$

$$4y - 12 = -5x + 5$$

$$5x + 4y = 17$$

47. Through $(-1, 4)$, undefined slope
If the slope is undefined, the
line must be vertical and its
equation has the form $x = k$.
Since the line passes through
$(-1, 4)$, the equation is $x = -1$.

49. Let x represent the amount invested at 10%. Then 30,000 - x is the amount invested at $8\frac{1}{2}$%.

$$.10x + .085(30,000 - x) = 2820$$
$$.10x + 2550 - .085x = 2820$$
$$.015x = 270$$
$$x = 18,000$$

The amount invested at 10% is $18,000.

51. Supply: $p = S(q) = 6q + 3$
Demand: $p = D(q) = -2q + 19$

(a) p = 10

Supply:	Demand:
$10 = 6q + 3$	$10 = -2q + 19$
$7 = 6q$	$2q = 9$
$\frac{7}{6} = q$	$q = \frac{9}{2}$

(b) p = 15

Supply:	Demand:
$15 = 6q + 3$	$15 = -2q + 19$
$12 = 6q$	$2q = 4$
$2 = q$	$q = 2$

(c) p = 18

Supply:	Demand:
$18 = 6q + 3$	$18 = -2q + 19$
$15 = 6q$	$2q = 1$
$\frac{5}{2} = q$	$q = \frac{1}{2}$

(d) See the answer graph in the back of the textbook.

(e) At the equilibrium price, the supply and the demand are equal.

Supply = Demand
$$6q + 3 = -2q = 19$$
$$8q = 16$$
$$q = 2$$
$$p = S(2) = 6(2) + 3 = 15$$

The equilibrium price is $15.

(f) From (b) or (e), it can be seen that equilibrium supply = equilibrium demand = 2 units.

53. Eight units cost $300; fixed cost is $60.

Since the cost function is assumed to be linear, use $C(x) = mx + b$ with $C(8) = 300$ and $b = 60$.

$$C(x) = mx + b$$
$$C(8) = m(8) + 60$$
$$300 = 8m + 60$$
$$240 = 8m$$
$$30 = m$$

$C(x) = 30x + 60$ is the cost function.

55. Twelve units cost $445; 50 units cost $1585.

$C(12) = 445$, $C(50) = 1585$

(12, 445) and (50, 1585) are the data points.

First find the slope.

$$m = \frac{1585 - 445}{50 - 12} = \frac{1140}{38} = 30$$

Use point-slope form with m = 30 and $(x_1, y_1) = (12, 445)$.

$y - 445 = 30(x - 12)$

$y - 445 = 30x - 360$

$\quad\quad y = 30x + 85$

$C(x) = 30x + 85$ is the cost function.

57. $C(x) = 20x + 100$

$R(x) = 40x$

(a) At the break-even point,

$\quad C(x) = R(x)$.

$\quad\quad 20x + 100 = 40x$

$\quad\quad\quad\quad 100 = 20x$

$\quad\quad\quad\quad\quad 5 = x$

The break-even point occurs when 5 units are produced.

(b) $R(5) = 40(5) = 200$

The revenue at the break-even point is $200.

59. $n = 7$, $x = \$19,000$

(a) The straight-line depreciation for the third year (or any other year of the life) is

$$D(x) = \frac{1}{n}x$$

$$= \frac{1}{7}(\$19,000)$$

$$\approx \$2714.29.$$

(b) The sum-of-the-years'-digits depreciation for the third year ($j = 3$) is

$$D(x) = \frac{2(n + 1 - j)}{n(n + 1)}x$$

$$= \frac{2(7 + 1 - 3)}{7(7 + 1)}(\$19,000)$$

$$= \frac{10}{56}(\$19,000)$$

$$\approx \$3392.86.$$

61. $n = 4$, $x = \$68,000$

The straight-line depreciation for each year is

$$D(x) = \frac{1}{n}x$$

$$= \frac{1}{4}(\$68,000)$$

$$= \$17,000.$$

The sum-of-the-years'-digits depreciation for year j is

$$D(x) = \frac{2(n + 1 - j)}{n(n + 1)}x.$$

For year 1, $j = 1$, and

$$D = \frac{2(4 + 1 - 1)}{4(5)}(\$68,000)$$

$$= \$27,200.$$

For year 2, $j = 2$, and

$$D = \frac{2(4 + 1 - 2)}{4(5)}(\$68,000)$$

$$= \$20,400.$$

For year 3, $j = 3$, and

$$D = \frac{2(4 + 1 - 3)}{4(5)}(\$68,000)$$

$$= \$13,600.$$

For year 4, $j = 4$, and

$$D = \frac{2(4 + 1 - 4)}{4(5)}(\$68,000)$$

$$= \$6800.$$

The table that collects all this information is as follows.

Year	Straight-Line Depreciation	Sum-of-the-Years'-Digits Depreciation
1	$17,000	$27,200
2	$17,000	$20,400
3	$17,000	$13,600
4	$17,000	$6800

63. n = 3, x = $18,000

The straight-line depreciation for each year is

$$D(x) = \frac{1}{n}x$$

$$= \frac{1}{3}(\$18,000)$$

$$= \$6000.$$

The sum-of-the-years'-digits depreciation for year j is

$$D(x) = \frac{2(n + 1 - j)}{n(n + 1)}x.$$

For year 1, j = 1, and

$$D = \frac{2(3 + 1 - 1)}{3(3 + 1)}(\$18,000)$$

$$= \frac{6}{12}(\$18,000)$$

$$= \$9000.$$

For year 2, j = 2, and

$$D = \frac{2(3 + 1 - 2)}{3(3 + 1)}(\$18,000)$$

$$= \frac{4}{12}(\$18,000)$$

$$= \$6000.$$

For year 3, j = 3, and

$$D = \frac{2(3 + 1 - 3)}{3(3 + 1)}(\$18,000)$$

$$= \frac{2}{12}(\$18,000)$$

$$= \$3000.$$

Year	Straight-Line Depreciation	Sum-of-the-Years'-Digits Depreciation
1	$6000	$9000
2	$6000	$6000
3	$6000	$3000

65.

x	y	xy	x^2	y^2
3	4	12	9	16
5	11	55	25	121
7	20	140	49	400
8	23	184	64	529
23	58	391	147	1066

There are four pairs of data, so n = 4.

$$m = \frac{n(\Sigma xy) - (\Sigma x)(\Sigma y)}{n(\Sigma x^2) - (\Sigma x)^2}$$

$$= \frac{4(391) - (23)(58)}{4(147) - (23)^2}$$

$$= \frac{1564 - 1334}{588 - 529} = \frac{230}{59} \approx 3.9$$

$$b = \frac{\Sigma y - m(\Sigma x)}{n} = \frac{58 - (3.9)(23)}{4}$$

$$= \frac{58 - 89.7}{4} = \frac{-31.7}{4} \approx -7.9$$

The equation of the least squares line is

$$y' = mx + b,$$

which in this problem becomes

$$y' = 3.9x - 7.9.$$

67. The coefficient of correlation is

$$r = \frac{n(\Sigma xy) - (\Sigma x)(\Sigma y)}{\sqrt{n(\Sigma x^2) - (\Sigma x)^2} \cdot \sqrt{n(\Sigma y^2) - (\Sigma y)^2}}$$

$$= \frac{4(391) - 23(58)}{\sqrt{4(147) - (23)^2} \sqrt{4(1066) - (58)^2}}$$

$$= \frac{230}{\sqrt{588 - 529} \sqrt{4264 - 3364}}$$

$$= \frac{230}{\sqrt{59} \sqrt{900}} \approx .998.$$

69. (a) See the answer graph in the back of the textbook.

(b) Let x represent the population of each city and let y represent the mean velocity of the pedestrians of each city. There are n = 15 pairs of data.

x	y	xy	x^2	y^2
341,948	4.81	1,644,769.88	1.1693×10^{11}	23.1361
1,092,759	5.88	6,425,422.92	1.1941×10^{12}	34.5744
5491	3.31	18,175.21	3.0151×10^{7}	10.9561
49,375	4.90	241,937.50	2.4379×10^{9}	24.0100
1,340,000	5.62	7,530,800.00	1.7956×10^{12}	31.5844
365	2.67	974.55	1.3323×10^{5}	7.1289
2500	2.27	5675.00	6.2500×10^{6}	5.1529
78,200	3.85	301,070.00	6.1152×10^{9}	14.8225
867,023	5.21	4,517,189.83	7.5173×10^{11}	27.1441
14,000	3.70	51,800.00	1.9600×10^{8}	13.6900
23,700	3.27	77,499.00	5.6169×10^{8}	10.6929
70,700	4.31	304,717.00	4.9985×10^{9}	18.5761
304,500	4.42	1,345,890.00	9.2720×10^{10}	19.5364
138,000	4.39	605,820.00	1.9044×10^{10}	19.2721
2,602,000	5.05	13,140,100.00	6.7704×10^{12}	25.5025
6,930,561	63.66	36,211,840.89	1.0755×10^{13}	285.7794

The coefficient of correlation is

$$r = \frac{n(\Sigma xy) - (\Sigma x)(\Sigma y)}{\sqrt{n(\Sigma x^2) - (\Sigma x)^2} \cdot \sqrt{n(\Sigma y^2) - (\Sigma y)^2}}$$

$$= \frac{15(36,211,840.89) - (6,930,561)(63.66)}{\sqrt{15(1.0755 \times 10^{13}) - (6,930,561)^2} \cdot \sqrt{15(285.7794) - (63.66)^2}}$$

$$= \frac{1.0198 \times 10^{8}}{\sqrt{1.1329 \times 10^{14}} \cdot \sqrt{234.095}} \approx .63$$

(c) Begin by making a table of log x values for the original fifteen values of x (these log x values are listed in part (d) of this exercise), and then plot the fifteen points with the log x values along the horizontal axis. (See the answer graph in the back of the textbook.) The data does appear more linear than the data in part (a) did.

(d)   Let X = log x and let y remain as before.   Again, n = 15.

X = log x	y	Xy	X²	y²
5.53	4.81	26.62	30.62	23.1361
6.04	5.88	35.51	36.46	34.5744
3.74	3.31	12.39	13.98	10.9561
4.69	4.90	23.00	22.03	24.0100
6.13	5.62	34.43	37.54	31.5844
2.56	2.67	6.84	6.57	7.1289
3.40	2.27	7.71	11.55	5.1529
4.89	3.85	18.84	23.94	14.8225
5.94	5.21	30.94	35.26	27.1441
4.15	3.70	15.34	17.19	13.6900
4.37	3.27	14.31	19.14	10.6929
4.85	4.31	20.90	23.52	18.5761
5.48	4.42	24.24	30.07	19.5364
5.14	4.39	22.56	26.42	19.2721
6.42	5.05	32.40	41.16	25.5025
73.33	63.66	326.03	375.45	285.7794

The coefficient of correlation for this data is

$$r = \frac{n(\Sigma Xy) - (\Sigma X)(\Sigma y)}{\sqrt{n(\Sigma X^2) - (\Sigma X)^2} \cdot \sqrt{n(\Sigma y^2) - (\Sigma y)^2}}$$

$$= \frac{15(326.03) - (73.33)(63.66)}{\sqrt{15(375.45) - (73.33)^2} \cdot \sqrt{15(285.7794) - (63.66)^2}}$$

$$= \frac{222.26}{\sqrt{254.46} \cdot \sqrt{234.10}} \approx .91.$$

This value of r is closer to 1 than the value calculated in part (b), due to the fact that this data is more linear than the other was.

(e) $m = \dfrac{n(\Sigma Xy) - (\Sigma X)(\Sigma y)}{n(\Sigma X^2) - (\Sigma X)^2}$

$$= \frac{15(326.03) - (73.33)(63.66)}{15(375.45) - (73.33)^2}$$

$$= \frac{226.26}{254.46} \approx .873$$

$$b = \frac{\Sigma y - m(\Sigma X)}{n}$$

$$= \frac{63.66 - .873(73.33)}{15}$$

$$= \frac{-.3906}{15} \approx -.0255$$

The equation of the least squares line is

$$y' = mX + b,$$

which in this problem becomes

$$y' = .873X - .0255$$

or $\quad y' = .873 \log x - .0255.$

Extended Application

1. The marginal cost of production of product A is approximately $y = .133x + 10.09$, where x is the number of unit produced (in millions). Selling price is $10.73. Let $y = 10.73$ and solve for x.

$$10.73 = .133x + 10.09$$
$$.64 = .133x$$
$$4.8 \approx x.$$

Therefore, the company has selling price equal to marginal cost when they make 4.8 million units.

2. See the answer graph in the back of the textbook.

3. For product B, $y = .0667x + 10.29$, which is always at least 10.29 for $x \geq 0$. The selling price is only $9.65. In the interval under discussion (3.1 to 5.7 million units), the marginal cost always exceeds the selling price. Therefore, there is no production level possible where the marginal cost is equal to the selling price.

4. $y = .133x + 9.46$

    (a) $y = .133(3.1) + 9.46$
    $$= 9.87$$

    At a level of production of 3.1 million units, the marginal cost is $9.87.

    $$y = .133(5.7) + 9.46$$
    $$= 10.22$$

    At a level of production of 5.7 million units, the marginal cost is $10.22.

    (b) See the answer graph in the back of the textbook.

    (c) $.133x + 9.46 = 9.57$
    $$.133x = .11$$
    $$x = .827 \text{ million}$$
    or about .83 million units

This is outside of the domain where
the function applies. Therefore,
there is no known production level at
which the cost will equal the selling
price.

CHAPTER 1   TEST

1. Solve the linear equation $5(y - 3) + 2y = 3(2y + 1)$.

2. Solve the inequality $2r - 10 \leq 6r + 30$.

3. Given $f(x) = -x^2 + 3x - 1$, find each of the following.

   (a) $f(2)$ (b) $f(-3)$ (c) $f(a + 1)$

4. Graph the linear function defined by $3x - 2y = 12$.

5. Find the slope of the line containing the indicated pair of points.

   (a) $(3, -3)$ and $(2, -1)$ (b) $(2, 4)$ and $(2, -1)$ (c) $(2, -1)$ and $(3, -1)$

6. Find the equation in standard form of the line through the points $(4, -3)$ and $(-2, -1)$.

7. Find the linear cost function if the fixed cost is $40 and 20 units cost $240.

8. An asset costs $10,000. If the asset has a life of 5 years, find

   (a) the straight-line depreciation in year 2, and

   (b) the total amount depreciated by the end of year 3.

9. Complete a depreciation table using the sum-of-the-years'-digits method for an asset costing $10,000 which is to be depreciated over 4 years.

10. Find the coefficient of correlation for the following data.

x	7	9	3	6
y	1	3	4	5

CHAPTER 1   TEST ANSWERS

1. y = 18

2. r ≥ -10

3. (a) 1   (b) -19   (c) $-(a + 1)^2 + 3(a + 1) - 1$ or $-a^2 + a + 1$

4.

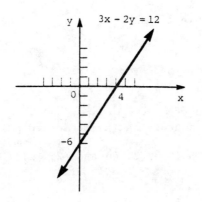

5.(a) -2

(b) Undefined

(c) 0

6. x + 3y = -5

7. C(x) = 10x + 40

8. (a) $2000

(b) $6000

9.

Year	Depreciation
1	$4000
2	$3000
3	$2000
4	$1000

10. r = -.41

CHAPTER 2   SYSTEMS OF LINEAR EQUATIONS AND MATRICES

Section 2.1

1. $x + y = 9$ (1)

   $2x - y = 0$ (2)

Multiply equation (1) by -2 and add the result to equation (2). The new system is

$$x + y = 9 \quad (1)$$

$(-2)R_1 + 1R_2 \rightarrow R_2 \qquad -3y = -18. \quad (3)$

Multiply equation (3) by $-\frac{1}{3}$ to get

$$x + y = 9 \quad (1)$$

$\left(-\frac{1}{3}\right)R_2 \rightarrow R_2 \qquad y = 6. \quad (4)$

Substitute 6 for y in equation (1) to get $x = 3$. The solution of the system is (3, 6).

3. $5x + 3y = 7$ (1)

   $7x - 3y = -19$ (2)

Multiply equation (1) by 7 and equation (2) by -5 and add the results together. The new system is

$$5x + 3y = 7 \quad (1)$$

$7R_1 + (-5)R_2 \rightarrow R_2 \qquad 36y = 144. \quad (3)$

Now make the coefficient of the first term in each row equal to 1. To accomplish this, multiply equation (1) by $\frac{1}{5}$ and equation (3) by $\frac{1}{36}$. We get the system

$\frac{1}{5}R_1 \rightarrow R_1 \qquad x + \frac{3}{5}y = \frac{7}{5} \quad (4)$

$\frac{1}{36}R_2 \rightarrow R_2 \qquad y = 4. \quad (5)$

Back-substitution gives

$$x + \frac{3}{5}(4) = \frac{7}{5}$$

$$x + \frac{12}{5} = \frac{7}{5}$$

$$x = -1.$$

The solution of the system is $(-1, 4)$.

5. $3x + 2y = -6$ (1)

   $5x - 2y = -10$ (2)

Eliminate x in equation (2) to get the system

$$3x + 2y = -6 \quad (1)$$

$5R_1 + (-3)R_2 \rightarrow R_2 \qquad 16y = 0. \quad (3)$

Make the coefficient of the first term in each equation equal 1.

$\frac{1}{3}R_1 \rightarrow R_1 \quad x + \frac{2}{3}y = -2 \quad (4)$

$\frac{1}{16}R_2 \rightarrow R_2 \qquad y = 0 \quad (5)$

Substitute 0 for y in equation (4) to get $x = -2$. The solution of the system is $(-2, 0)$.

7. $2x - 3y = -7$ (1)

   $5x + 4y = 17$ (2)

Eliminate x in equation (2).

$$2x - 3y = -7 \quad (1)$$

$5R_1 + (-2)R_2 \rightarrow R_2 \qquad -23y = -69 \quad (3)$

Make the coefficient of the first term in each equation equal 1.

$\frac{1}{2}R_1 \rightarrow R_1 \quad x - \frac{3}{2}y = -\frac{7}{2} \quad (4)$

$\left(-\frac{1}{23}\right)R_2 \rightarrow R_2 \qquad y = 3 \quad (5)$

Substitute 3 for y in equation (4) to get $x = 1$. The solution of the system is (1, 3).

9. $5p + 7q = 6$ *(1)*

$10p - 3q = 46$ *(2)*

Eliminate p in equation (2).

$$5p + 7q = 6 \quad (1)$$

$(-2)R_1 + 1R_2 \to R_2 \qquad -17q = 34 \quad (3)$

Make the coefficient of the first term in each equation equal 1.

$\frac{1}{5}R_1 \to R_1 \quad p + \frac{7}{5}q = \frac{6}{5}$ *(4)*

$\left(-\frac{1}{17}\right)R_2 \to R_2 \qquad q = -2$ *(5)*

Substitute -2 for q in equation (4) to get p = 4. The solution of the system is (4, -2).

11. $6x + 7y = -2$ *(1)*

$7x - 6y = 26$ *(2)*

Eliminate x in equation (2).

$$6x + 7y = -2 \quad (1)$$

$7R_1 + (-6)R_2 \to R_2 \qquad 85y = -170 \quad (3)$

Make the coefficient of the first term in each equation equal 1.

$\frac{1}{6}R_1 \to R_1 \quad x + \frac{7}{6}y = -\frac{1}{3}$ *(4)*

$\frac{1}{85}R_2 \to R_2 \qquad y = -2$ *(5)*

Substitute -2 for y in equation (4) to get x = 2. The solution for the system is (2, -2).

13. $3x + 2y = 5$ *(1)*

$6x + 4y = 8$ *(2)*

Eliminate x in equation (2).

$$3x + 2y = 5 \quad (1)$$

$(-2)R_1 + 1R_2 \to R_2 \qquad 0 = -2 \quad (3)$

Equation (3) is a false statement. The system is inconsistent and has no solution.

15. $4x - y = 9$ *(1)*

$-8x + 2y = -18$ *(2)*

Eliminate x in equation (2).

$$4x - y = 9 \quad (1)$$

$2R_1 + 1R_2 \to R_2 \qquad 0 = 0 \quad (3)$

The true statement in equation (3) indicates that there are an infinite number of solutions for the system. Solve equation (1) for x.

$$4x - y = 9 \quad (1)$$

$$4x = y + 9$$

$$x = \frac{y + 9}{4} \quad (4)$$

For each value of y, equation (4) indicates that $x = \frac{y + 9}{4}$, and all ordered pairs of the form $\left(\frac{y + 9}{4}, y\right)$ are solutions.

17. The solution of an inconsistent system is no solution.

19. $\frac{x}{2} + \frac{y}{3} = 8$ *(1)*

$\frac{2x}{3} + \frac{3y}{2} = 17$ *(2)*

Rewrite the equations without fractions.

$6R_1 \to R_1 \quad 3x + 2y = 48$ *(3)*

$6R_2 \to R_2 \quad 4x + 9y = 102$ *(4)*

Eliminate x in equation (4).

$$3x + 2y = 48 \quad (5)$$

$(-4)R_1 + 3R_2 \to R_2 \qquad 19y = 114 \quad (6)$

Make each leading coefficient equal 1.

$$\frac{1}{3}R_1 \rightarrow R_1 \qquad x + \frac{2}{3}y = 16 \qquad (7)$$

$$\frac{1}{19}R_2 \rightarrow R_2 \qquad\qquad y = 6 \qquad (8)$$

Substitute 6 for y in equation (7) to get x = 12. The solution of the system is (12, 6).

21. $\frac{x}{2} + y = \frac{3}{2}$ (1)

$\frac{x}{3} + y = \frac{1}{3}$ (2)

Rewrite the equations without fractions.

$$2R_1 \rightarrow R_1 \qquad x + 2y = 3 \qquad (3)$$
$$3R_2 \rightarrow R_2 \qquad x + 3y = 1 \qquad (4)$$

Eliminate x in equation (4).

$$x + 2y = 3 \qquad (3)$$
$$(-1)R_1 + 1R_2 \rightarrow R_2 \qquad y = -2 \qquad (4)$$

Substitute -2 for y in equation (3) to get x = 7. The solution of the system is (7, -2).

23. $x + y + z = 2$ (1)

$2x + y - z = 5$ (2)

$x - y + z = -2$ (3)

Eliminate x in equations (2) and (3).

$$x + y + z = 2 \qquad (1)$$
$$(-2)R_1 + 1R_2 \rightarrow R_2 \qquad -y - 3z = 1 \qquad (4)$$
$$(-1)R_1 + 1R_3 \rightarrow R_3 \qquad -2y \quad\;\; = -4 \qquad (5)$$

Eliminate y in equation (5).

$$x + y + z = 2 \qquad (1)$$
$$-y - 3z = 1 \qquad (4)$$
$$(-2)R_2 + 1R_3 \rightarrow R_3 \qquad 6z = -6 \qquad (6)$$

Make each leading coefficient equal 1.

$$x + y + z = 2 \qquad (1)$$
$$(-1)R_2 \rightarrow R_2 \qquad y + 3z = -1 \qquad (7)$$
$$\frac{1}{6}R_3 \rightarrow R_3 \qquad\qquad z = -1 \qquad (8)$$

Substitute -1 for z in equation (7) to get y = 2. Finally, substitute -1 for z and 2 for y in equation (1) to get x = 1. The solution of the system is (1, 2, -1).

25. $x + 3y + 4z = 14$ (1)

$2x - 3y + 2z = 10$ (2)

$3x - y + z = 9$ (3)

Eliminate x in equations (2) and (3).

$$x + 3y + 4z = 14 \qquad (1)$$
$$(-2)R_1 + 1R_2 \rightarrow R_2 \qquad -9y - 6z = -18 \qquad (4)$$
$$(-3)R_1 + 1R_3 \rightarrow R_3 \qquad -10y - 11z = -33 \qquad (5)$$

Eliminate y in equation (5).

$$x + 3y + 4z = 14 \qquad (1)$$
$$-9y - 6z = -18 \qquad (4)$$
$$10R_2 + (-9)R_3 \rightarrow R_3 \qquad 39z = 117 \qquad (6)$$

Make each leading coefficient equal 1.

$$x + 3y + 4z = 14 \qquad (1)$$
$$\left(-\frac{1}{9}\right)R_2 \rightarrow R_2 \qquad y + \frac{2}{3}z = 2 \qquad (7)$$
$$\frac{1}{39}R_3 \rightarrow R_3 \qquad\qquad z = 3 \qquad (8)$$

Substitute 3 for z in equation (2) to get y = 0. Finally, substitute 3 for z and 0 for y in equation (1) to get x = 2. The solution of the system is (2, 0, 3).

27. $x + 2y + 3z = 8$ (1)

$3x - y + 2z = 5$ (2)

$-2x - 4y - 6z = 5$ (3)

Eliminate x in equations (2) and (3).

$$x + 2y + 3z = 8 \quad (1)$$
$$(-3)R_1 + 1R_2 \rightarrow R_2 \quad -7y - 7z = -19 \quad (4)$$
$$2R_1 + 1R_3 \rightarrow R_3 \quad 0 = 21 \quad (5)$$

Equation (5) is a false statement. The system is inconsistent and has no solution.

29. $2x - 4y + z = -4 \quad (1)$
$x + 2y - z = 0 \quad (2)$
$-x + y + z = 6 \quad (3)$

Eliminate x in equations (2) and (3).

$$2x - 4y + z = -4 \quad (1)$$
$$1R_1 + (-2)R_2 \rightarrow R_2 \quad -8y + 3z = -4 \quad (4)$$
$$1R_1 + 2R_3 \rightarrow R_3 \quad -2y + 3z = 8 \quad (5)$$

Eliminate y in equation (5).

$$2x - 4y + z = -4 \quad (1)$$
$$-8y + 3z = -4 \quad (4)$$
$$1R_2 + (-4)R_3 \rightarrow R_3 \quad -9z = -36 \quad (6)$$

Make each leading coefficient equal 1.

$$\tfrac{1}{2}R_1 \rightarrow R_1 \quad x - 2y + \tfrac{1}{2}z = -2 \quad (7)$$
$$\left(-\tfrac{1}{8}\right)R_2 \rightarrow R_2 \quad y - \tfrac{3}{8}z = \tfrac{1}{2} \quad (8)$$
$$\left(-\tfrac{1}{9}\right)R_3 \rightarrow R_3 \quad z = 4 \quad (9)$$

Substitute 4 for z in equation (8) to get y = 2. Finally, substitute 4 for z and 2 for y in equation (7) to get x = 0. The solution of the system is (0, 2, 4).

31. $x + 4y - z = 6 \quad (1)$
$2x - y + z = 3 \quad (2)$
$3x + 2y + 3z = 16 \quad (3)$

Eliminate x in equations (2) and (3).

$$x + 4y - z = 6 \quad (1)$$
$$(-2)R_1 + 1R_2 \rightarrow R_2 \quad -9y + 3z = -9 \quad (4)$$
$$(-3)R_1 + 1R_3 \rightarrow R_3 \quad -10y + 6z = -2 \quad (5)$$

Eliminate y in equation (5).

$$x + 4y - z = 6 \quad (1)$$
$$-9y + 3z = -9 \quad (4)$$
$$(-10)R_2 + 9R_3 \rightarrow R_3 \quad 24z = 72 \quad (6)$$

Make each leading coefficient equal 1.

$$x + 4y - z = 6 \quad (1)$$
$$\left(-\tfrac{1}{9}\right)R_2 \rightarrow R_2 \quad y - \tfrac{1}{3}z = 1 \quad (7)$$
$$\tfrac{1}{24}R_3 \rightarrow R_3 \quad z = 3 \quad (8)$$

Substitute 3 for z in equation (7) to get y = 2. Finally, substitute 3 for z and 2 for y in equation (1) to get x = 1. The solution of the system is (1, 2, 3).

33. $5m + n - 3p = -6 \quad (1)$
$2m + 3n + p = 5 \quad (2)$
$-3m - 2n + 4p = 3 \quad (3)$

Eliminate m in equations (2) and (3).

$$5m + n - 3p = -6 \quad (1)$$
$$2R_1 + (-5)R_2 \rightarrow R_2 \quad -13n - 11p = -37 \quad (4)$$
$$3R_1 + 5R_3 \rightarrow R_3 \quad -7n + 11p = -3 \quad (5)$$

Eliminate n in equation (5).

$$5m + n - 3p = -6 \quad (1)$$
$$-13n - 11p = -37 \quad (4)$$
$$7R_2 + (-13)R_3 \rightarrow R_3 \quad -220p = -220 \quad (6)$$

Make each leading coefficient equal 1.

$$\tfrac{1}{5}R_1 \rightarrow R_1 \quad m + \tfrac{1}{5}n - \tfrac{3}{5}p = -\tfrac{6}{5} \quad (7)$$
$$\left(-\tfrac{1}{13}\right)R_2 \rightarrow R_2 \quad n + \tfrac{11}{13}p = \tfrac{37}{13} \quad (8)$$
$$\left(-\tfrac{1}{220}\right)R_3 \rightarrow R_3 \quad p = 1 \quad (9)$$

Substitute 1 for p in equation (8) to get n = 2. Finally, substitute 1 for p and 2 for n in equation (7) to get m = -1. The solution of the system is (-1, 2, 1).

35. $a - 3b - 2c = -3$ *(1)*

$3a + 2b - c = 12$ *(2)*

$-a - b + 4c = 3$ *(3)*

Eliminate a in equations (2) and (3).

$$a - 3b - 2c = -3 \quad (1)$$
$$(-3)R_1 + 1R_2 \rightarrow R_2 \quad 11b + 5c = 21 \quad (4)$$
$$1R_1 + 1R_3 \rightarrow R_3 \quad -4b + 2c = 0 \quad (5)$$

Eliminate b in equation (5).

$$a - 3b - 2c = -3 \quad (1)$$
$$11b + 5c = 21 \quad (4)$$
$$4R_2 + 11R_3 \rightarrow R_3 \quad 42c = 84 \quad (6)$$

Make each leading coefficient equal 1.

$$a - 3b - 2c = -3 \quad (1)$$
$$\tfrac{1}{11}R_2 \rightarrow R_2 \quad b + \tfrac{5}{11}c = \tfrac{21}{11} \quad (7)$$
$$\tfrac{1}{42}R_3 \rightarrow R_3 \quad c = 2 \quad (8)$$

Substitute 2 for c in equation (7) to get b = 1. Finally, substitute 2 for c and 1 for b in equation (1) to get a = 4. The solution of the system is (4, 1, 2).

39. $3x + y - z = 0$ *(1)*

$2x - y + 3z = -7$ *(2)*

Eliminate x in equation (2).

$$3x + y - z = 0 \quad (1)$$
$$2R_1 + (-3)R_2 \rightarrow R_2 \quad 5y - 11z = 21 \quad (3)$$

Make each leading coefficient equal 1.

$$\tfrac{1}{3}R_1 \rightarrow R_1 \quad x + \tfrac{1}{3}y - \tfrac{1}{3}z = 0 \quad (4)$$
$$\tfrac{1}{5}R_2 \rightarrow R_2 \quad y - \tfrac{11}{5}z = \tfrac{21}{5} \quad (5)$$

Solve equation (5) for y in terms of z.

$$y = \frac{11}{5}z + \frac{21}{5}$$

Substitute this expression for y in equation (4) and solve the equation for x.

$$x + \frac{1}{3}\left(\frac{11}{5}z + \frac{21}{5}\right) - \frac{1}{3}z = 0$$
$$x + \frac{11}{15}z + \frac{7}{5} - \frac{1}{3}z = 0$$
$$x + \frac{2}{5}z = -\frac{7}{5}$$
$$x = -\frac{2}{5}z - \frac{7}{5}$$

The solution of the system is

$$\left(-\frac{2}{5}z - \frac{7}{5}, \; \frac{11}{5}z + \frac{21}{5}, \; z\right) \text{ or}$$
$$\left(\frac{-2z - 7}{5}, \; \frac{11z + 21}{5}, \; z\right).$$

41. $-x + y - z = -7$ *(1)*

$2x + 3y + z = 7$ *(2)*

Eliminate x in equation (2).

$$-x + y - z = -7 \quad (1)$$
$$2R_1 + 1R_2 \rightarrow R_2 \quad 5y - z = -7 \quad (3)$$

Make each leading coefficient equal 1.

$$(-1)R_1 \rightarrow R_1 \quad x - y + z = 7 \quad (4)$$
$$\tfrac{1}{5}R_2 \rightarrow R_2 \quad y - \tfrac{1}{5}z = -\tfrac{7}{5} \quad (5)$$

Solve equation (5) for y in terms of z.

$$y = \frac{1}{5}z - \frac{7}{5}$$

Substitute this expression for y in equation (4) and solve the equation for x.

$$x - \left(\frac{1}{5}z - \frac{7}{5}\right) + z = 7$$

$$x - \frac{1}{5}z + \frac{7}{5} + z = 7$$

$$x + \frac{4}{5}z = \frac{28}{5}$$

$$x = -\frac{4}{5}z + \frac{28}{5}$$

The solution of the system is

$\left(-\frac{4}{5}z + \frac{28}{5}, \frac{1}{5}z - \frac{7}{5}, z\right)$ or

$\left(\frac{-4z + 28}{5}, \frac{z - 7}{5}, z\right)$.

43. $4x - 3y + z + w = 21$ *(1)*

$-2x - y + 2z + 7w = 2$ *(2)*

$x + 4y + 2z - w = -12$ *(3)*

Eliminate x in equations (2) and (3).

$$4x - 3y + z + w = 21 \quad (1)$$

$1R_1 + 2R_2 \to R_2 \quad -5y + 5z + 15w = 25$ *(4)*

$1R_1 + (-4)R_3 \to R_3 \quad -19y - 7z + 5w = 69$ *(5)*

Eliminate y in equation (5).

$$4x - 3y + z + w = 21 \quad (1)$$

$$-5y + 5z + 15w = 25 \quad (4)$$

$19R_2 + (-5)R_3 \to R_3 \quad 130z + 260w = 130$ *(6)*

Make each leading coefficient equal 1.

$\frac{1}{4}R_1 \to R_1 \quad x - \frac{3}{4}y + \frac{1}{4}z + \frac{1}{4}w = \frac{21}{4}$ *(7)*

$\left(-\frac{1}{5}\right)R_2 \to R_2 \quad y - z - 3w = -5$ *(8)*

$\frac{1}{130}R_3 \to R_3 \quad z + 2w = 1$ *(9)*

Solve equation (9) for z in terms
of w.

$$z = -2w + 1 \quad (10)$$

Substitute this expression for z in
equation (8) and solve for y.

$$y - (-2w + 1) - 3w = -5$$

$$y + 2w - 1 - 3w = -5$$

$$y - w = -4$$

$$y = w - 4 \quad (11)$$

Substitute the results of equations
(10) and (11) in equation (7) and
solve the equation for x.

$$x - \frac{3}{4}(w - 4) + \frac{1}{4}(-2w + 1) + \frac{1}{4}w = \frac{21}{4}$$

$$x - \frac{3}{4}w + 3 - \frac{1}{2}w + \frac{1}{4} + \frac{1}{4}w = \frac{21}{4}$$

$$x - w = 2$$

$$x = w + 2$$

The solution of the system is

$(w + 2, w - 4, -2w + 1, w)$.

45.   Let x = the number of units of ROM
chips; and y = the number of units of
RAM chips.

Use a chart to organize the information.

	ROM	RAM	Totals
Hours on Line A	1	3	15
Hours on Line B	2	1	15

The system to be solved is

$$x + 3y = 15 \quad (1)$$

$$2x + y = 15. \quad (2)$$

Eliminate x in equation (2).

$$x + 3y = 15 \quad (1)$$

$(-2)R_1 + 1R_2 \to R_2 \quad -5y = -15$ *(3)*

Make each leading coefficient equal 1.

$$x + 3y = 15 \quad (1)$$

$\left(-\frac{1}{5}\right)R_2 \to R_2 \quad y = 3$ *(4)*

Substitute 3 for y in equation (1) to
get x = 6. 6 units of ROM chips and
3 units of RAM chips can be produced
in a day.

47.   Let x = the number of fives;

y = the number of tens; and

z = the number of twenties.

Since the number of fives is three times the number of tens, $x = 3y$.

The system to be solved is

$$\begin{aligned} x + y + z &= 70 \quad (1) \\ x - 3y &= 0 \quad (2) \\ 5x + 10y + 20z &= 960. \quad (3) \end{aligned}$$

Eliminate x in equations (2) and (3).

$$\begin{aligned} x + y + z &= 70 \quad (1) \\ R_1 + (-1)R_2 \to R_2 \qquad 4y + z &= 70 \quad (4) \\ (-5)R_1 + 1R_3 \to R_3 \qquad 5y + 15z &= 610 \quad (5) \end{aligned}$$

Eliminate y in equation (5).

$$\begin{aligned} x + y + z &= 70 \quad (1) \\ 4y + z &= 70 \quad (4) \\ (-5)R_2 + 4R_3 \to R_3 \qquad 55z &= 2090 \quad (6) \end{aligned}$$

Make each leading coefficient equal 1.

$$\begin{aligned} x + y + z &= 70 \quad (1) \\ \tfrac{1}{4}R_2 \to R_2 \qquad y + \tfrac{1}{4}z &= \tfrac{35}{2} \quad (7) \\ \tfrac{1}{55}R_3 \to R_3 \qquad z &= 38 \quad (8) \end{aligned}$$

Substitute 38 for z in equation (7) to get $y = 8$. Finally, substitute 38 for z and 8 for y in equation (1) to get $x = 24$.

There are 24 fives, 8 tens, and 38 twenties.

49.   **(a)** Let x = the amount borrowed at 13%; y = the amount borrowed at 14%; and z = the amount borrowed at 12%.

Note that $y = \tfrac{1}{2}x + 2000$, and use the formula for simple interest, $I = Prt$, to obtain $.13x + .14y + .12z = 3240$.

The system to be solved is

$$\begin{aligned} x + y + z &= 25{,}000 \quad (1) \\ -\tfrac{1}{2}x + y &= 2000 \quad (2) \\ .13x + .14y + .12z &= 3240. \quad (3) \end{aligned}$$

Rewrite the system without decimals or fractions.

$$\begin{aligned} x + y + z &= 25{,}000 \quad (1) \\ 2R_2 \to R_2 \qquad -x + 2y &= 4000 \quad (4) \\ 100R_3 \to R_3 \qquad 13x + 14y + 12z &= 324{,}000 \quad (5) \end{aligned}$$

Eliminate x in equations (4) and (5).

$$\begin{aligned} x + y + z &= 25{,}000 \quad (1) \\ 1R_1 + 1R_2 \to R_2 \qquad 3y + z &= 29{,}000 \quad (6) \\ (-13)R_1 + 1R_3 \to R_3 \qquad y - z &= -1000 \quad (7) \end{aligned}$$

Eliminate y in equation (7).

$$\begin{aligned} x + y + z &= 25{,}000 \quad (1) \\ 3y + z &= 29{,}000 \quad (6) \\ 1R_2 + (-3)R_3 \to R_3 \qquad 4z &= 32{,}000 \quad (8) \end{aligned}$$

Make each leading coefficient equal 1.

$$\begin{aligned} x + y + z &= 25{,}000 \quad (1) \\ \tfrac{1}{3}R_2 \to R_2 \qquad y + \tfrac{1}{3}z &= \tfrac{29{,}000}{3} \quad (9) \\ \tfrac{1}{4}R_3 \to R_3 \qquad z &= 8000 \quad (10) \end{aligned}$$

Substitute 8000 for z in equation (9) to get $y = 7000$. Finally, substitute 8000 for z and 7000 for y in equation (1) to get $x = 10{,}000$.

The company borrowed $10,000 at 13%, $7000 at 14%, and $8000 at 12%.

(b) Let x, y, z remain as they were defined in part (a). This time, the system to be solved is

$$\begin{aligned} x + y + z &= 25{,}000 \quad (1) \\ .13x + .14y + .12z &= 3240. \quad (2) \end{aligned}$$

Rewrite the system without decimals.

$$x + y + z = 25{,}000 \quad (1)$$
$$100R_2 \rightarrow R_2 \quad 13x + 14y + 12z = 324{,}000 \quad (3)$$

Eliminate x in equation (3).

$$x + y + z = 25{,}000 \quad (1)$$
$$(-13)R_1 + 1R_2 \rightarrow R_2 \quad y - z = -1000 \quad (4)$$

Solve equation (4) for y in terms of z.

$$y = z - 1000$$

Substitute this expression for y in equation (1) and solve the equation for x.

$$x + (z - 1000) + z = 25{,}000$$
$$x + 2z = 26{,}000$$
$$x = 26{,}000 - 2z$$

We have solved a dependent system, and the parameter is z, the amount borrowed at 12%.

For any (positive) amount borrowed at 12%, the amount borrowed at 13% must be twice that amount less than $26,000, and the amount borrowed at 14% must be $1000 less than the amount at 12%.

(c) No, the conditions given in part (a) cannot be satisfied if the company can borrow only $6000 at 12%, since the solution to part (a) requires the company to borrow $8000 at 12%. (However, the conditions of part (b) would still be satisfied if z = 6000.)

Section 2.2

1. $2x + 3y = 11$
    $x + 2y = 8$

The equations are already in proper form. The augmented matrix obtained from the coefficients and the constants is

$$\begin{bmatrix} 2 & 3 & | & 11 \\ 1 & 2 & | & 8 \end{bmatrix}.$$

3. $2x + y + z = 3$
    $3x - 4y + 2z = -7$
    $x + y + z = 2$

leads to an augmented matrix

$$\begin{bmatrix} 2 & 1 & 1 & | & 3 \\ 3 & -4 & 2 & | & -7 \\ 1 & 1 & 1 & | & 2 \end{bmatrix}.$$

5. We are given the augmented matrix

$$\begin{bmatrix} 1 & 0 & | & 2 \\ 0 & 1 & | & 3 \end{bmatrix}.$$

This is equivalent to the system of equations

$$x \quad = 2$$
$$y = 3, \text{ or}$$

$x = 2, y = 3$.

7. $$\begin{bmatrix} 1 & 0 & 0 & | & 2 \\ 0 & 1 & 0 & | & 3 \\ 0 & 0 & 1 & | & -2 \end{bmatrix}$$

The system associated with this matrix is

$$x \quad = 2$$
$$y \quad = 3$$
$$z = -2, \text{ or}$$

$x = 2, y = 3, z = -2$.

9. Row operations on a matrix correspond to transformations of a system of equations.

11. $\begin{bmatrix} 2 & 3 & 8 & | & 20 \\ 1 & 4 & 6 & | & 12 \\ 0 & 3 & 5 & | & 10 \end{bmatrix}$

$R_1 + (-2)R_2 \rightarrow R_2$ $\begin{bmatrix} 2 & 3 & 8 & | & 20 \\ 2 + (-2)(1) & 3 + (-2)(4) & 8 + (-2)(6) & | & 20 + (-2)(12) \\ 0 & 3 & 5 & | & 10 \end{bmatrix} = \begin{bmatrix} 2 & 3 & 8 & | & 20 \\ 0 & -5 & -4 & | & -4 \\ 0 & 3 & 5 & | & 10 \end{bmatrix}$

13. $\begin{bmatrix} 1 & 4 & 2 & | & 9 \\ 0 & 1 & 5 & | & 14 \\ 0 & 3 & 8 & | & 16 \end{bmatrix}$

$-4R_2 + R_1 \rightarrow R_1$ $\begin{bmatrix} -4(0) + 1 & -4(1) + 4 & -4(5) + 2 & | & -4(14) + 9 \\ 0 & 1 & 5 & | & 14 \\ 0 & 3 & 8 & | & 16 \end{bmatrix} = \begin{bmatrix} 1 & 0 & -18 & | & -47 \\ 0 & 1 & 5 & | & 14 \\ 0 & 3 & 8 & | & 16 \end{bmatrix}$

15. $\begin{bmatrix} 3 & 0 & 0 & | & 18 \\ 0 & 5 & 0 & | & 9 \\ 0 & 0 & 4 & | & 8 \end{bmatrix}$

$\frac{1}{3}R_1 \rightarrow R_1$ $\begin{bmatrix} \frac{1}{3}(3) & \frac{1}{3}(0) & \frac{1}{3}(0) & | & \frac{1}{3}(18) \\ 0 & 5 & 0 & | & 9 \\ 0 & 0 & 4 & | & 8 \end{bmatrix} = \begin{bmatrix} 1 & 0 & 0 & | & 6 \\ 0 & 5 & 0 & | & 9 \\ 0 & 0 & 4 & | & 8 \end{bmatrix}$

17. $x + y = 5$
$x - y = -1$

has augmented matrix

$$\begin{bmatrix} 1 & 1 & | & 5 \\ 1 & -1 & | & -1 \end{bmatrix}.$$

Use row operations as follows.

$-1R + R_2 \rightarrow R_2$ $\begin{bmatrix} 1 & 1 & | & 5 \\ 0 & -2 & | & -6 \end{bmatrix}$

$-\frac{1}{2}R_2 \rightarrow R_2$ $\begin{bmatrix} 1 & 1 & | & 5 \\ 0 & 1 & | & 3 \end{bmatrix}$

$-1R_2 + R_1 \rightarrow R_1$ $\begin{bmatrix} 1 & 0 & | & 2 \\ 0 & 1 & | & 3 \end{bmatrix}$

Read the solution from the last
column of the matrix. The solution
of the system is (2, 3).

19. $3x + 3y = -9$
$2x - 5y = -6$

The augmented matrix is

$$\begin{bmatrix} 3 & 3 & | & -9 \\ 2 & -5 & | & -6 \end{bmatrix}.$$

$2R_1 + (-3)R_2 \rightarrow R_2$ $\begin{bmatrix} 3 & 3 & | & -9 \\ 0 & 21 & | & 0 \end{bmatrix}$

$-7R_1 + R_2 \rightarrow R_1$ $\begin{bmatrix} -21 & 0 & | & 63 \\ 0 & 21 & | & 0 \end{bmatrix}$

$-\frac{1}{21}R_1 \rightarrow R_1$ $\begin{bmatrix} 1 & 0 & | & -3 \\ 0 & 1 & | & 0 \end{bmatrix}$
$\frac{1}{21}R_2 \rightarrow R_2$

Read the solution from the last
column of the matrix. The solution
of the system is (-3, 0).

21. $2x = 10 + 3y$

$2y = 5 - 2x$

First rewrite the system in proper form as follows.

$$2x - 3y = 10$$
$$2x + 2y = 5$$

Now, write the augmented matrix of the system and use row operations.

$$\begin{bmatrix} 2 & -3 & | & 10 \\ 2 & 2 & | & 5 \end{bmatrix}$$

$-R_1 + R_2 \to R_2 \begin{bmatrix} 2 & -3 & | & 10 \\ 0 & 5 & | & -5 \end{bmatrix}$

$5R_1 + 3R_2 \to R_1 \begin{bmatrix} 10 & 0 & | & 35 \\ 0 & 5 & | & -5 \end{bmatrix}$

$\frac{1}{10}R_1 \to R_1 \begin{bmatrix} 1 & 0 & | & \frac{7}{2} \\ 0 & 1 & | & -1 \end{bmatrix}$
$\frac{1}{5}R_2 \to R_2$

The solution of the system is $\left(\frac{7}{2}, -1\right)$.

23. $2x - 5y = 10$

$4x - 5y = 15$

Write the augmented matrix and use row operations.

$$\begin{bmatrix} 2 & -5 & | & 10 \\ 4 & -5 & | & 15 \end{bmatrix}$$

$-2R_1 + R_2 \to R_2 \begin{bmatrix} 2 & -5 & | & 10 \\ 0 & 5 & | & -5 \end{bmatrix}$

$R_1 + R_2 \to R_1 \begin{bmatrix} 2 & 0 & | & 5 \\ 0 & 5 & | & -5 \end{bmatrix}$

$\frac{1}{2}R_1 \to R_1 \begin{bmatrix} 1 & 0 & | & \frac{5}{2} \\ 0 & 1 & | & -1 \end{bmatrix}$
$\frac{1}{5}R_2 \to R_2$

The solution of the system is $\left(\frac{5}{2}, -1\right)$.

25. $2x - 3y = 2$

$4x - 6y = 1$

Write the augmented matrix and use row operations.

$$\begin{bmatrix} 2 & -3 & | & 2 \\ 4 & -6 & | & 1 \end{bmatrix}$$

$-2R_1 + R_2 \to R_2 \begin{bmatrix} 2 & -3 & | & 2 \\ 0 & 0 & | & -3 \end{bmatrix}$

The system associated with the last matrix is

$$2x - 3y = 2$$
$$0x + 0y = -3.$$

Since the second equation is $0 = -3$, the system is inconsistent and therefore has no solution.

27. $6x - 3y = 1$

$-12x + 6y = -2$

Write the augmented matrix of the system and use row operations.

$$\begin{bmatrix} 6 & -3 & | & 1 \\ -12 & 6 & | & -2 \end{bmatrix}$$

$2R_1 + R_2 \to R_2 \begin{bmatrix} 6 & -3 & | & 1 \\ 0 & 0 & | & 0 \end{bmatrix}$

$\frac{1}{6}R_1 \to R_1 \begin{bmatrix} 1 & -\frac{1}{2} & | & \frac{1}{6} \\ 0 & 0 & | & 0 \end{bmatrix}$

This is as far as we can go with the Gauss–Jordan method. To complete the solution, write the equation that corresponds to the first row of the matrix.

$$x - \frac{1}{2}y = \frac{1}{6}$$

Solve this equation for x in terms of y.

$$x = \frac{1}{2}y + \frac{1}{6}$$

The solution of the system is

$$\left(\frac{1}{2}y + \frac{1}{6}, y\right).$$

29.
$$2x - 2y \quad\;\;\; = -2$$
$$y + z = 4$$
$$x \quad\;\; + z = 1$$

Write the augmented matrix and use row operations.

$$\begin{bmatrix} 2 & -2 & 0 & | & -2 \\ 0 & 1 & 1 & | & 4 \\ 1 & 0 & 1 & | & 1 \end{bmatrix}$$

$$R_1 + (-2)R_3 \to R_3 \begin{bmatrix} 2 & -2 & 0 & | & -2 \\ 0 & 1 & 1 & | & 4 \\ 0 & -2 & -2 & | & -4 \end{bmatrix}$$

$$\begin{matrix} 2R_2 + R_1 \to R_1 \\ \\ 2R_2 + R_3 \to R_3 \end{matrix} \begin{bmatrix} 2 & 0 & 2 & | & 6 \\ 0 & 1 & 1 & | & 4 \\ 0 & 0 & 0 & | & 4 \end{bmatrix}$$

This matrix corresponds to the system

$$2x + 2z = 6$$
$$y + z = 4$$
$$0 = 4.$$

The false statement $0 = 4$ indicates that the system is inconsistent and therefore has no solution.

31.
$$4x + 4y - 4z = 24$$
$$2x - y + z = -9$$
$$x - 2y + 3z = 1$$

Write the augmented matrix and use row operations.

$$\begin{bmatrix} 4 & 4 & -4 & | & 24 \\ 2 & -1 & 1 & | & -9 \\ 1 & -2 & 3 & | & 1 \end{bmatrix}$$

$$\begin{matrix} R_1 + (-2)R_2 \to R_2 \\ R_1 + (-4)R_3 \to R_3 \end{matrix} \begin{bmatrix} 4 & 4 & -4 & | & 24 \\ 0 & 6 & -6 & | & 42 \\ 0 & 12 & -16 & | & 20 \end{bmatrix}$$

$$\begin{matrix} -3R_1 + 2R_2 \to R_1 \\ \\ -2R_2 + R_3 \to R_3 \end{matrix} \begin{bmatrix} -12 & 0 & 0 & | & 12 \\ 0 & 6 & -6 & | & 42 \\ 0 & 0 & -4 & | & -64 \end{bmatrix}$$

$$2R_2 - 3R_3 \to R_2 \begin{bmatrix} -12 & 0 & 0 & | & 12 \\ 0 & 12 & 0 & | & 276 \\ 0 & 0 & -4 & | & -64 \end{bmatrix}$$

$$\begin{matrix} -\frac{1}{12}R_1 \to R_1 \\ \frac{1}{12}R_2 \to R_2 \\ -\frac{1}{4}R_3 \to R_3 \end{matrix} \begin{bmatrix} 1 & 0 & 0 & | & -1 \\ 0 & 1 & 0 & | & 23 \\ 0 & 0 & 1 & | & 16 \end{bmatrix}$$

The solution of the system is $(-1, 23, 16)$.

33.
$$y = x - 1$$
$$y = 6 + z$$
$$z = -1 - x$$

First write the system in proper form.

$$-x + y \quad\;\; = -1$$
$$y - z = 6$$
$$x \quad\;\; + z = -1$$

Write the augmented matrix and use row operations.

$$\begin{bmatrix} -1 & 1 & 0 & | & -1 \\ 0 & 1 & -1 & | & 6 \\ 1 & 0 & 1 & | & -1 \end{bmatrix}$$

$$R_1 + R_3 \to R_3 \begin{bmatrix} -1 & 1 & 0 & | & -1 \\ 0 & 1 & -1 & | & 6 \\ 0 & 1 & 1 & | & -2 \end{bmatrix}$$

$$\begin{matrix} -1R_2 + R_1 \to R_1 \\ \\ -1R_2 + R_3 \to R_3 \end{matrix} \begin{bmatrix} -1 & 0 & 1 & | & -7 \\ 0 & 1 & -1 & | & 6 \\ 0 & 0 & 2 & | & -8 \end{bmatrix}$$

$$\begin{matrix} -2R_1 + R_3 \to R_1 \\ 2R_2 + R_3 \to R_2 \end{matrix} \begin{bmatrix} 2 & 0 & 0 & | & 6 \\ 0 & 2 & 0 & | & 4 \\ 0 & 0 & 2 & | & -8 \end{bmatrix}$$

$$\begin{matrix} \frac{1}{2}R_1 \to R_1 \\ \frac{1}{2}R_2 \to R_2 \\ \frac{1}{2}R_3 \to R_3 \end{matrix} \begin{bmatrix} 1 & 0 & 0 & | & 3 \\ 0 & 1 & 0 & | & 2 \\ 0 & 0 & 1 & | & -4 \end{bmatrix}$$

The solution of the system is $(3, 2, -4)$.

35. $3x + 5y - z = 0$

$4x - y + 2z = 1$

$-6x - 10y + 2z = 0$

Write the augmented matrix and use row operations.

$$\begin{bmatrix} 3 & 5 & -1 & 0 \\ 4 & -1 & 2 & 1 \\ -6 & -10 & 2 & 0 \end{bmatrix}$$

$$\begin{matrix} 4R_1 + (-3)R_2 \to R_2 \\ 2R_1 + R_3 \to R_3 \end{matrix} \begin{bmatrix} 3 & 5 & -1 & 0 \\ 0 & 23 & -10 & -3 \\ 0 & 0 & 0 & 0 \end{bmatrix}$$

$$23R_1 + (-5)R_2 \to R_1 \begin{bmatrix} 69 & 0 & 27 & 15 \\ 0 & 23 & -10 & -3 \\ 0 & 0 & 0 & 0 \end{bmatrix}$$

$$\begin{matrix} \frac{1}{69}R_1 \to R_1 \\ \frac{1}{23}R_2 \to R_2 \end{matrix} \begin{bmatrix} 1 & 0 & \frac{9}{23} & \frac{5}{23} \\ 0 & 1 & -\frac{10}{23} & -\frac{3}{23} \\ 0 & 0 & 0 & 0 \end{bmatrix}$$

The row of zeros indicates dependent equations. Solve the first two equations respectively for x and y in terms of z to obtain

$$x = -\frac{9}{23}z + \frac{5}{23}$$

and

$$y = \frac{10}{23}z - \frac{3}{23}.$$

The solution of the system is

$$\left(-\frac{9}{23}z + \frac{5}{23}, \frac{10}{23}z - \frac{3}{23}, z\right).$$

37. $2x + 3y + z = 9$

$4x - y + 3z = -1$

$6x + 2y - 4z = -8$

Write the augmented matrix and use row operations.

$$\begin{bmatrix} 2 & 3 & 1 & 9 \\ 4 & -1 & 3 & -1 \\ 6 & 2 & -4 & -8 \end{bmatrix}$$

$$\begin{matrix} -2R_1 + R_2 \to R_2 \\ -3R_1 + R_3 \to R_3 \end{matrix} \begin{bmatrix} 2 & 3 & 1 & 9 \\ 0 & -7 & 1 & -19 \\ 0 & -7 & -7 & -35 \end{bmatrix}$$

$$\begin{matrix} 7R_1 + 3R_2 \to R_1 \\ R_2 + (-1)R_3 \to R_3 \end{matrix} \begin{bmatrix} 14 & 0 & 10 & 6 \\ 0 & -7 & 1 & -19 \\ 0 & 0 & 8 & 16 \end{bmatrix}$$

$$\begin{matrix} 8R_1 + (-10)R_3 \to R_1 \\ -8R_2 + R_3 \to R_2 \end{matrix} \begin{bmatrix} 112 & 0 & 0 & -112 \\ 0 & 56 & 0 & 168 \\ 0 & 0 & 8 & 16 \end{bmatrix}$$

$$\begin{matrix} \frac{1}{112}R_1 \to R_1 \\ \frac{1}{56}R_2 \to R_2 \\ \frac{1}{8}R_3 \to R_3 \end{matrix} \begin{bmatrix} 1 & 0 & 0 & -1 \\ 0 & 1 & 0 & 3 \\ 0 & 0 & 1 & 2 \end{bmatrix}.$$

The solution of the system is $(-1, 3, 2)$.

39. $5x - 4y + 2z = 4$

$5x + 3y - z = 17$

$15x - 5y + 3z = 25$

Write the augmented matrix and use row operations.

$$\begin{bmatrix} 5 & -4 & 2 & 4 \\ 5 & 3 & -1 & 17 \\ 15 & -5 & 3 & 25 \end{bmatrix}$$

$$\begin{matrix} -1R_1 + R_2 \to R_2 \\ -3R_1 + R_3 \to R_3 \end{matrix} \begin{bmatrix} 5 & -4 & 2 & 4 \\ 0 & 7 & -3 & 13 \\ 0 & 7 & -3 & 13 \end{bmatrix}$$

$$\begin{matrix} 7R_1 + 4R_2 \to R_1 \\ -1R_2 + R_3 \to R_3 \end{matrix} \begin{bmatrix} 35 & 0 & 2 & 80 \\ 0 & 7 & -3 & 13 \\ 0 & 0 & 0 & 0 \end{bmatrix}$$

$$\begin{matrix} \frac{1}{35}R_1 \to R_1 \\ \frac{1}{7}R_2 \to R_2 \end{matrix} \begin{bmatrix} 1 & 0 & \frac{2}{35} & \frac{16}{7} \\ 0 & 1 & -\frac{3}{7} & \frac{13}{7} \\ 0 & 0 & 0 & 0 \end{bmatrix}$$

The row of zeros indicates dependent equations. Solve the first two equations respectively for x and y in terms of z to obtain

$x = -\dfrac{2}{35}z + \dfrac{16}{7}$ and $y = \dfrac{3}{7}z + \dfrac{13}{7}$.

The solution of the system is

$\left(-\dfrac{2}{35}z + \dfrac{16}{7}, \ \dfrac{3}{7}z + \dfrac{13}{7}, \ z\right)$.

41. $\begin{aligned} x + 2y \quad\ \ - w &= 3 \\ 2x \quad\ + 4z + 2w &= -6 \\ x + 2y - z \quad\ &= 6 \\ 2x - y + z + w &= -3 \end{aligned}$

Write the augmented matrix and use row operations.

$$\begin{bmatrix} 1 & 2 & 0 & -1 & | & 3 \\ 2 & 0 & 4 & 2 & | & -6 \\ 1 & 2 & -1 & 0 & | & 6 \\ 2 & -1 & 1 & 1 & | & -3 \end{bmatrix}$$

$\begin{matrix} \\ -2R_1 + R_2 \rightarrow R_2 \\ -1R_1 + R_3 \rightarrow R_3 \\ -2R_1 + R_4 \rightarrow R_4 \end{matrix} \begin{bmatrix} 1 & 2 & 0 & -1 & | & 3 \\ 0 & -4 & 4 & 4 & | & -12 \\ 0 & 0 & -1 & 1 & | & 3 \\ 0 & -5 & 1 & 3 & | & -9 \end{bmatrix}$

$\begin{matrix} 2R_1 + R_2 \rightarrow R_1 \\ \\ \\ -5R_2 + 4R_4 \rightarrow R_4 \end{matrix} \begin{bmatrix} 2 & 0 & 4 & 2 & | & -6 \\ 0 & -4 & 4 & 4 & | & -12 \\ 0 & 0 & -1 & 1 & | & 3 \\ 0 & 0 & -16 & -8 & | & 24 \end{bmatrix}$

$\begin{matrix} 4R_3 + R_1 \rightarrow R_1 \\ 4R_3 + R_2 \rightarrow R_2 \\ \\ 16R_3 + (-1)R_4 \rightarrow R_4 \end{matrix} \begin{bmatrix} 2 & 0 & 0 & 6 & | & 6 \\ 0 & -4 & 0 & 8 & | & 0 \\ 0 & 0 & -1 & 1 & | & 3 \\ 0 & 0 & 0 & 24 & | & 24 \end{bmatrix}$

$\begin{matrix} -4R_1 + R_4 \rightarrow R_1 \\ -3R_2 + R_4 \rightarrow R_2 \\ -24R_3 + R_4 \rightarrow R_3 \\ \\ \end{matrix} \begin{bmatrix} -8 & 0 & 0 & 0 & | & 0 \\ 0 & 12 & 0 & 0 & | & 24 \\ 0 & 0 & 24 & 0 & | & -48 \\ 0 & 0 & 0 & 24 & | & 24 \end{bmatrix}$

$\begin{matrix} -\frac{1}{8}R_1 \rightarrow R_1 \\ \frac{1}{12}R_2 \rightarrow R_2 \\ \frac{1}{24}R_3 \rightarrow R_3 \\ \frac{1}{24}R_4 \rightarrow R_4 \end{matrix} \begin{bmatrix} 1 & 0 & 0 & 0 & | & 0 \\ 0 & 1 & 0 & 0 & | & 2 \\ 0 & 0 & 1 & 0 & | & -2 \\ 0 & 0 & 0 & 1 & | & 1 \end{bmatrix}$

The solution of the system is $x = 0$, $y = 2$, $z = -2$, $w = 1$, or $(0, 2, -2, 1)$.

43. Let x_1 = the number of cars sent from I to A;

x_2 = the number of cars sent from II to A;

x_3 = the number of cars sent from I to B; and

x_4 = the number of cars sent from II to B.

	A	B
I	x_1	x_3
II	x_2	x_4

Plant I has 28 cars, so

$$x_1 + x_3 = 28. \quad (1)$$

Plant II has 8 cars, so

$$x_2 + x_4 = 8. \quad (2)$$

Dealer A needs 20 cars, so

$$x_1 + x_2 = 20. \quad (3)$$

Dealer B needs 16 cars, so

$$x_3 + x_4 = 16. \quad (4)$$

The total transportation cost is $10,640, so

$$220x_1 + 400x_2 + 300x_3 + 180x_4$$
$$= 10,640. \quad (5)$$

The system to be solved is

$$\begin{aligned} x_1 \quad\ + x_3 \quad\ &= 28 \\ x_2 \quad\ + x_4 &= 8 \\ x_1 + x_2 \quad\ &= 20 \\ x_3 + x_4 &= 16 \\ 220x_1 + 400x_2 + 300x_3 + 180x_4 &= 10,640. \end{aligned}$$

Write the augmented matrix and use row operations.

$$\begin{bmatrix} 1 & 0 & 1 & 0 & | & 28 \\ 0 & 1 & 0 & 1 & | & 8 \\ 1 & 1 & 0 & 0 & | & 20 \\ 0 & 0 & 1 & 1 & | & 16 \\ 220 & 400 & 300 & 180 & | & 10,640 \end{bmatrix}$$

$$\begin{array}{c} \\ \\ -1R_1 + R_3 \to R_3 \\ \\ -220R_1 + R_5 \to R_5 \end{array} \left[\begin{array}{cccc|c} 1 & 0 & 1 & 0 & 28 \\ 0 & 1 & 0 & 1 & 8 \\ 0 & 1 & -1 & 0 & -8 \\ 0 & 0 & 1 & 1 & 16 \\ 0 & 400 & 80 & 180 & 4480 \end{array}\right]$$

$$\begin{array}{c} \\ \\ -1R_2 + R_3 \to R_3 \\ \\ -400R_2 + R_5 \to R_5 \end{array} \left[\begin{array}{cccc|c} 1 & 0 & 1 & 0 & 28 \\ 0 & 1 & 0 & 1 & 8 \\ 0 & 0 & -1 & -1 & -16 \\ 0 & 0 & 1 & 1 & 16 \\ 0 & 0 & 80 & -220 & 1280 \end{array}\right]$$

$$\begin{array}{c} R_1 + R_3 \to R_1 \\ \\ \\ R_3 + R_4 \to R_4 \\ 80R_3 + R_5 \to R_5 \end{array} \left[\begin{array}{cccc|c} 1 & 0 & 0 & -1 & 12 \\ 0 & 1 & 0 & 1 & 8 \\ 0 & 0 & -1 & -1 & -16 \\ 0 & 0 & 0 & 0 & 0 \\ 0 & 0 & 0 & -300 & 0 \end{array}\right]$$

There is a 0 now in row 4, column 4, where we would like to get a 1. To proceed, interchange the fourth and fifth rows.

$$\left[\begin{array}{cccc|c} 1 & 0 & 0 & -1 & 12 \\ 0 & 1 & 0 & 1 & 8 \\ 0 & 0 & -1 & -1 & -16 \\ 0 & 0 & 0 & -300 & 0 \\ 0 & 0 & 0 & 0 & 0 \end{array}\right]$$

$$\begin{array}{c} 300R_1 + (-1)R_4 \to R_1 \\ 300R_2 + R_4 \to R_2 \\ 300R_3 + (-1)R_4 \to R_3 \\ \\ \\ \end{array} \left[\begin{array}{cccc|c} 300 & 0 & 0 & 0 & 3600 \\ 0 & 300 & 0 & 0 & 2400 \\ 0 & 0 & -300 & 0 & -4800 \\ 0 & 0 & 0 & -300 & 0 \\ 0 & 0 & 0 & 0 & 0 \end{array}\right]$$

$$\begin{array}{c} \frac{1}{300}R_1 \to R_1 \\ \frac{1}{300}R_2 \to R_2 \\ -\frac{1}{300}R_3 \to R_3 \\ -\frac{1}{300}R_4 \to R_4 \\ \\ \end{array} \left[\begin{array}{cccc|c} 1 & 0 & 0 & 0 & 12 \\ 0 & 1 & 0 & 0 & 8 \\ 0 & 0 & 1 & 0 & 16 \\ 0 & 0 & 0 & 1 & 0 \\ 0 & 0 & 0 & 0 & 0 \end{array}\right]$$

Each of the original variables has a value, so the last row of all zeros may be ignored.

The solution of the system is

$x_1 = 12$, $x_2 = 8$, $x_3 = 16$, $x_4 = 0$.

12 cars should be sent from I to A, 8 cars from II to A, 16 cars from I to B, and no cars from II to B.

45.   **(a)**   Let $x_1 =$ the number of units from first supplier for Roseville;

$x_2 =$ the number of units from first supplier for Akron;

$x_3 =$ the number of units from second supplier for Roseville; and

$x_4 =$ the number of units from second supplier for Akron.

(b)
$$\begin{aligned} x_1 + x_2 && = 75 \\ x_3 + x_4 &= 40 \\ x_1 + x_3 && = 40 \\ x_2 + x_4 &= 75 \\ 70x_1 + 90x_2 + 80x_3 + 120x_4 &= 10{,}750 \end{aligned}$$

(c)   Write the augmented matrix and use row operations.

$$\left[\begin{array}{cccc|c} 1 & 1 & 0 & 0 & 75 \\ 0 & 0 & 1 & 1 & 40 \\ 1 & 0 & 1 & 0 & 40 \\ 0 & 1 & 0 & 1 & 75 \\ 70 & 90 & 80 & 120 & 10{,}750 \end{array}\right]$$

$$\begin{array}{c} \\ \\ -1R_1 + R_3 \to R_3 \\ \\ -70R_1 + R_5 \to R_5 \end{array} \left[\begin{array}{cccc|c} 1 & 1 & 0 & 0 & 75 \\ 0 & 0 & 1 & 1 & 40 \\ 0 & -1 & 1 & 0 & -35 \\ 0 & 1 & 0 & 1 & 75 \\ 0 & 20 & 80 & 120 & 5500 \end{array}\right]$$

Interchange rows 2 and 4.

$$\left[\begin{array}{cccc|c} 1 & 1 & 0 & 0 & 75 \\ 0 & 1 & 0 & 1 & 75 \\ 0 & -1 & 1 & 0 & -35 \\ 0 & 0 & 1 & 1 & 40 \\ 0 & 20 & 80 & 120 & 5500 \end{array}\right]$$

$$\begin{array}{c} -1R_2 + R_1 \to R_1 \\ \\ R_2 + R_3 \to R_3 \\ \\ -20R_2 + R_5 \to R_5 \end{array} \left[\begin{array}{cccc|c} 1 & 0 & 0 & -1 & 0 \\ 0 & 1 & 0 & 1 & 75 \\ 0 & 0 & 1 & 1 & 40 \\ 0 & 0 & 1 & 1 & 40 \\ 0 & 0 & 80 & 100 & 4000 \end{array}\right]$$

$$\begin{array}{c} \\ \\ \\ -1R_3 + R_4 \to R_4 \\ -80R_3 + R_5 \to R_5 \end{array} \left[\begin{array}{cccc|c} 1 & 0 & 0 & -1 & 0 \\ 0 & 1 & 0 & 1 & 75 \\ 0 & 0 & 1 & 1 & 40 \\ 0 & 0 & 0 & 0 & 0 \\ 0 & 0 & 0 & 20 & 800 \end{array}\right]$$

Interchange rows 4 and 5.

$$\begin{bmatrix} 1 & 0 & 0 & -1 & \bigm| & 0 \\ 0 & 1 & 0 & 1 & \bigm| & 75 \\ 0 & 0 & 1 & 1 & \bigm| & 40 \\ 0 & 0 & 0 & 20 & \bigm| & 800 \\ 0 & 0 & 0 & 0 & \bigm| & 0 \end{bmatrix}$$

$$\frac{1}{20}R_4 \to R_4 \begin{bmatrix} 1 & 0 & 0 & -1 & \bigm| & 0 \\ 0 & 1 & 0 & 1 & \bigm| & 75 \\ 0 & 0 & 1 & 1 & \bigm| & 40 \\ 0 & 0 & 0 & 1 & \bigm| & 40 \\ 0 & 0 & 0 & 0 & \bigm| & 0 \end{bmatrix}$$

$$\begin{matrix} R_1 + R_4 \to R_1 \\ R_2 + (-1)R_4 \to R_2 \\ R_3 + (-1)R_4 \to R_3 \end{matrix} \begin{bmatrix} 1 & 0 & 0 & 0 & \bigm| & 40 \\ 0 & 1 & 0 & 0 & \bigm| & 35 \\ 0 & 0 & 1 & 0 & \bigm| & 0 \\ 0 & 0 & 0 & 1 & \bigm| & 40 \\ 0 & 0 & 0 & 0 & \bigm| & 0 \end{bmatrix}$$

Each of the original variables has a value, so the last row of all zeros may be ignored. The solution of the system is $x_1 = 40$, $x_2 = 35$, $x_3 = 0$, $x_4 = 40$, or $(40, 35, 0, 40)$.
The manufacturer should purchase 40 units for Roseville from the first supplier, 35 units for Akron from the first supplier, 0 units for Roseville from the second supplier, and 40 units for Akron from the second supplier.

47. This exercise should be solved by computer methods. The solution may vary based on the computer and software that are used.
The solution is $x \approx 30.7209$, $y \approx 39.6513$, $z \approx 31.386$, $w \approx 50.3966$.

49. Let x = the number of kilograms of the first chemical;

    y = the number of kilograms of the second chemical; and

    z = the number of kilograms of the third chemical.

The system to be solved is

$$x + y + z = 750$$
$$x = .108(750)$$
$$\frac{y}{z} = \frac{4}{3}.$$

Rewrite this system as

$$x + y + z = 750$$
$$x \qquad\quad = 81$$
$$3y - 4z = 0.$$

Use computer methods to solve this system. The solution is that 81 kg of the first chemical, about 382.286 kg of the second chemical, and about 286.714 kg of the third chemical should be used.

51. Let x = the number of species A;
    y = the number of species B; and
    z = the number of species C.

Use a chart to organize the information.

		A	B	C	Totals
	I	1.32	2.1	.86	490
Food	II	2.9	.95	1.52	897
	III	1.75	.6	2.01	653

Species

The system to be solved is

$$1.32x + 2.1y + .86z = 490$$
$$2.9x + .95y + 1.52z = 897$$
$$1.75x + .6y + 2.01z = 653.$$

Use computer methods to solve this system. The solution is that about 243 fish of species A, 38 fish of species B, and 101 fish of species C should be stocked in the lake.

Section 2.3

1. $\begin{bmatrix} 1 & 3 \\ 5 & 7 \end{bmatrix} = \begin{bmatrix} 1 & 5 \\ 3 & 7 \end{bmatrix}$

This statement is false, since not all corresponding elements are equal.

3. $\begin{bmatrix} x \\ y \end{bmatrix} = \begin{bmatrix} 3 \\ 5 \end{bmatrix}$ if x = 3 and y = 5.

This statement is true. The matrices are the same size and corresponding elements are equal.

5. $\begin{bmatrix} 1 & 9 & -4 \\ 3 & 7 & 2 \\ -1 & 1 & 0 \end{bmatrix}$ is a square matrix.

This statement is true. The matrix has 3 rows and 3 columns.

7. $\begin{bmatrix} -4 & 8 \\ 2 & 3 \end{bmatrix}$ is a 2 × 2 square matrix.

Its additive inverse is $\begin{bmatrix} 4 & -8 \\ -2 & -3 \end{bmatrix}$.

9. $\begin{bmatrix} -6 & 8 & 0 & 0 \\ 4 & 1 & 9 & 2 \\ 3 & -5 & 7 & 1 \end{bmatrix}$ is a 3 × 4 matrix.

Its additive inverse is

$\begin{bmatrix} 6 & -8 & 0 & 0 \\ -4 & -1 & -9 & -2 \\ -3 & 5 & -7 & -1 \end{bmatrix}$.

11. $\begin{bmatrix} 2 \\ 4 \end{bmatrix}$ is a 2 × 1 column matrix.

Its additive inverse is $\begin{bmatrix} -2 \\ -4 \end{bmatrix}$.

13. The sum of an n × m matrix and its additive inverse is the n × m zero matrix.

15. $\begin{bmatrix} 2 & 1 \\ 4 & 8 \end{bmatrix} = \begin{bmatrix} x & 1 \\ y & z \end{bmatrix}$

Corresponding elements must be equal for the matrices to be equal. Therefore, x = 2, y = 4, and z = 8.

17. $\begin{bmatrix} x+6 & y+2 \\ 8 & 3 \end{bmatrix} = \begin{bmatrix} -9 & 7 \\ 8 & k \end{bmatrix}$

Corresponding elements must be equal.

x + 6 = −9   y + 2 = 7   k = 3
x = −15   y = 5

19. $\begin{bmatrix} -7+z & 4r & 8s \\ 6p & 2 & 5 \end{bmatrix} + \begin{bmatrix} -9 & 8r & 3 \\ 2 & 5 & 4 \end{bmatrix}$

$= \begin{bmatrix} 2 & 36 & 27 \\ 20 & 7 & 12a \end{bmatrix}$

Add the two matrices on the left side of this equation to obtain

$\begin{bmatrix} -7+z & 4r & 8s \\ 6p & 2 & 5 \end{bmatrix} + \begin{bmatrix} -9 & 8r & 3 \\ 2 & 5 & 4 \end{bmatrix}$

$= \begin{bmatrix} (-7+z)+(-9) & 4r+8r & 8s+3 \\ 6p+2 & 7 & 9 \end{bmatrix}$

$= \begin{bmatrix} -16+z & 12r & 8s+3 \\ 6p+2 & 7 & 9 \end{bmatrix}$.

Corresponding elements of this matrix and the matrix on the right side of the original equation must be equal.

−16 + z = 2   12r = 36   8s + 3 = 27
z = 18   r = 3   s = 3

6p + 2 = 30   9 = 12a
p = 3 $a = \dfrac{3}{4}$

21. $\begin{bmatrix} 1 & 2 & 5 & -1 \\ 3 & 0 & 2 & -4 \end{bmatrix} + \begin{bmatrix} 8 & 10 & -5 & 3 \\ -2 & -1 & 0 & 0 \end{bmatrix}$

$= \begin{bmatrix} 1+8 & 2+10 & 5+(-5) & -1+3 \\ 3+(-2) & 0+(-1) & 2+0 & -4+0 \end{bmatrix}$

$= \begin{bmatrix} 9 & 12 & 0 & 2 \\ 1 & -1 & 2 & -4 \end{bmatrix}$

23. $\begin{bmatrix} 1 & 3 & -2 \\ 4 & 7 & 1 \end{bmatrix} + \begin{bmatrix} 3 & 0 \\ 6 & 4 \\ -5 & 2 \end{bmatrix}$

These matrices cannot be added since the first matrix has size 2 × 3, while the second has size 3 × 2. Only matrices that are the same size can be added.

25.   The matrices have the same size, so the subtraction can be done. Let A and B represent the given matrices. Using the definition of subtraction, we have

$A - B = A + (-B)$

$= \begin{bmatrix} 2 & 8 & 12 & 0 \\ 7 & 4 & -1 & 5 \\ 1 & 2 & 0 & 10 \end{bmatrix} + \begin{bmatrix} -1 & -3 & -6 & -9 \\ -2 & 3 & 3 & -4 \\ -8 & 0 & 2 & -17 \end{bmatrix}$

$= \begin{bmatrix} 1 & 5 & 6 & -9 \\ 5 & 7 & 2 & 1 \\ -7 & 2 & 2 & -7 \end{bmatrix}.$

27. $\begin{bmatrix} 2 & 3 \\ -2 & 4 \end{bmatrix} + \begin{bmatrix} 4 & 3 \\ 7 & 8 \end{bmatrix} - \begin{bmatrix} 3 & 2 \\ 1 & 4 \end{bmatrix}$

$= \begin{bmatrix} 2+4-3 & 3+3-2 \\ -2+7-1 & 4+8-4 \end{bmatrix} = \begin{bmatrix} 3 & 4 \\ 4 & 8 \end{bmatrix}$

29. $\begin{bmatrix} 1 & 5 \\ -3 & 7 \end{bmatrix} - \begin{bmatrix} 6 & 3 \\ 2 & 4 \end{bmatrix} + \begin{bmatrix} 8 & 10 \\ -1 & 0 \end{bmatrix}$

$= \begin{bmatrix} 1-6+8 & 5-3+10 \\ -3-2+(-1) & 7-4+0 \end{bmatrix}$

$= \begin{bmatrix} 3 & 12 \\ -6 & 3 \end{bmatrix}$

31. $\begin{bmatrix} -4x+2y & -3y+y \\ 6x-3y & 2x-5y \end{bmatrix} + \begin{bmatrix} -8x+6y & 2x \\ 3y-5x & 6x+4y \end{bmatrix}$

$= \begin{bmatrix} (-4x+2y)+(-8x+6y) & (-3x+y)+2x \\ (6x-3y)+(3y-5x) & (2x-5y)+(6x+4y) \end{bmatrix}$

$= \begin{bmatrix} -12x+8y & -x+y \\ x & 8x-y \end{bmatrix}$

33.   The additive inverse of

$$X = \begin{bmatrix} x & y \\ z & w \end{bmatrix}$$

is

$$-X = \begin{bmatrix} -x & -y \\ -z & -w \end{bmatrix}.$$

35.   Show that $X + (T + P) = (X + T) + P$.

On the left side, the sum $T + P$ is obtained first, and then

$$X + (T + P).$$

This gives the matrix

$$\begin{bmatrix} x+(r+m) & y+(s+n) \\ z+(t+p) & w+(u+q) \end{bmatrix}.$$

For the right side, first the sum $X + T$ is obtained, and then

$$(X + T) + P.$$

This gives the matrix

$$\begin{bmatrix} (x+r)+m & (y+s)+n \\ (z+t)+p & (w+u)+q \end{bmatrix}.$$

Comparing corresponding elements, we see that they are equal by the associative property of addition of real numbers. Thus,

$$X + (T + P) = (X + T) + P.$$

37.   Show that $P + O = P$.

$$P + O = \begin{bmatrix} m & n \\ p & q \end{bmatrix} + \begin{bmatrix} 0 & 0 \\ 0 & 0 \end{bmatrix}$$

$$= \begin{bmatrix} m+0 & n+0 \\ p+0 & q+0 \end{bmatrix}$$

$$= \begin{bmatrix} m & n \\ p & q \end{bmatrix}$$

$$= P$$

Thus, $P + O = P$.

39. **(a)** The production cost matrix for Chicago is

	Phones	Calculators
Material	4.05	7.01
Labor	3.27	3.51

The production cost matrix for Seattle is

	Phones	Calculators
Material	4.40	6.90
Labor	3.54	3.76

(b) $\dfrac{4.27 + 4.05 + 4.40}{3} = 4.24$

$\dfrac{3.45 + 3.27 + 3.54}{3} = 3.42$

$\dfrac{6.94 + 7.01 + 6.90}{3} = 6.95$

$\dfrac{3.65 + 3.51 + 3.76}{3} = 3.64$

The average production cost matrix is

	Phones	Calculators
Material	4.24	6.95
Labor	3.42	3.64

(c) The new production cost matrix for Chicago is

	Phones	Calculators
Material	4.05 + .37	7.01 + .42
Labor	3.27 + .11	3.51 + .11

or $\begin{bmatrix} 4.42 & 7.43 \\ 3.38 & 3.62 \end{bmatrix}$.

(d) $\dfrac{4.42 + 4.40}{2} = 4.41$

$\dfrac{3.38 + 3.54}{2} = 3.46$

$\dfrac{7.43 + 6.90}{2} \approx 7.17$

$\dfrac{3.62 + 3.76}{2} = 3.69$

The new average production cost matrix is

	Phones	Calculators
Material	4.41	7.17
Labor	3.46	3.69

41. **(a)** There are four food groups and three meals. To represent the data by a 3 × 4 matrix, we must use the rows to correspond to the meals: breakfast, lunch, and dinner, and the columns to correspond to the four food groups. Thus, we obtain the matrix

$$\begin{bmatrix} 2 & 1 & 2 & 1 \\ 3 & 2 & 2 & 1 \\ 4 & 3 & 2 & 1 \end{bmatrix}.$$

(b) There are four food groups. These will correspond to the four rows. There are three components in each food group: fat, carbohydrates, and protein. These will correspond to the three columns. The matrix is

$$\begin{bmatrix} 5 & 0 & 7 \\ 0 & 10 & 1 \\ 0 & 15 & 2 \\ 10 & 12 & 8 \end{bmatrix}.$$

(c) The matrix is

$$\begin{bmatrix} 8 \\ 4 \\ 5 \end{bmatrix}.$$

Obtained Pain Relief

43.

	Yes	No
Patient Took Painfree	22	3
Patient Took Placebo	8	17

(a) Of the 25 patients who took the placebo, 8 got relief.

(b) Of the 25 patients who took Painfree, 3 got no relief.

(c) $\begin{bmatrix} 22 & 3 \\ 8 & 17 \end{bmatrix} + \begin{bmatrix} 21 & 4 \\ 6 & 19 \end{bmatrix} + \begin{bmatrix} 19 & 6 \\ 10 & 15 \end{bmatrix}$

$+ \begin{bmatrix} 23 & 2 \\ 3 & 22 \end{bmatrix}$

$= \begin{bmatrix} 85 & 15 \\ 27 & 73 \end{bmatrix}$

(d) Yes, it appears that Painfree is effective. Of the 100 patients who took the medication, 85% got relief.

Matrix B
Size
3 × 4

Matrix A
Size
4 × 2

must match
Size of BA
is 3 × 2.

Section 2.4

1. Since A is 2 × 2 and B is 2 × 2, we have the following diagram:

Matrix A
Size
2 × 2

Matrix B
Size
2 × 2

must match
Size of AB
is 2 × 2.

The product AB exists because A has two columns and B has two rows. The size of the product AB is 2 × 2. A similar diagram with B and A interchanged shows that the product BA exists and its size is also 2 × 2.

3.
Matrix A
Size
3 × 5

Matrix B
Size
5 × 2

must match
Size of AB
is 3 × 2.

Matrix B
Size
5 × 2

Matrix A
Size
3 × 5

Do not match
BA does not exist.

5.
Matrix A
Size
4 × 2

Matrix B
Size
3 × 4

Do not match
AB does not exist.

7. $2A = 2 \begin{bmatrix} -2 & 4 \\ 0 & 3 \end{bmatrix} = \begin{bmatrix} -4 & 8 \\ 0 & 6 \end{bmatrix}$

9. $-4B = -4 \begin{bmatrix} -6 & 2 \\ 4 & 0 \end{bmatrix} = \begin{bmatrix} 24 & -8 \\ -16 & 0 \end{bmatrix}$

11. $-4A + 5B = -4 \begin{bmatrix} -2 & 4 \\ 0 & 3 \end{bmatrix} + 5 \begin{bmatrix} -6 & 2 \\ 4 & 0 \end{bmatrix}$

$= \begin{bmatrix} 8 & -16 \\ 0 & -12 \end{bmatrix} + \begin{bmatrix} -30 & 10 \\ 20 & 0 \end{bmatrix}$

$= \begin{bmatrix} -22 & -6 \\ 20 & -12 \end{bmatrix}$

13. To find the product matrix AB, the number of columns of A must be the same as the number of rows of B.

15. Call the first matrix A and the second matrix B. The product matrix AB will have size 2 × 1.

Step 1: Multiply the elements of the first row of A by the corresponding elements of the column of B and add.

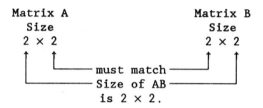

$1(-1) + 2(7) = 13$

Therefore, 13 is the first row entry of the product matrix AB.

Step 2: Multiply the elements of the second row of A by the corresponding elements of the column of B and add.

$$\begin{bmatrix} 1 & 2 \\ \boxed{3 \quad 4} \end{bmatrix}\begin{bmatrix} \boxed{\begin{matrix} -1 \\ 7 \end{matrix}} \end{bmatrix} \quad 3(-1) + 4(7) = 25$$

The second row entry of the product
is 25.

Step 3: Write the product using
the two entries found above.

$$AB = \begin{bmatrix} 1 & 2 \\ 3 & 4 \end{bmatrix}\begin{bmatrix} -1 \\ 7 \end{bmatrix} = \begin{bmatrix} 13 \\ 25 \end{bmatrix}$$

17. $\begin{bmatrix} 1 & 5 & 3 \\ -1 & 2 & 7 \end{bmatrix}\begin{bmatrix} 4 \\ 2 \\ -3 \end{bmatrix}$

$$= \begin{bmatrix} 1 \cdot 4 + 5 \cdot 2 + 3(-3) \\ -1(4) + 2 \cdot 2 + 7(-3) \end{bmatrix}$$

$$= \begin{bmatrix} 5 \\ -21 \end{bmatrix}$$

19. $\begin{bmatrix} 5 & 1 \\ 2 & 3 \end{bmatrix}\begin{bmatrix} 3 & -1 & 0 \\ 1 & 0 & 2 \end{bmatrix}$

$$= \begin{bmatrix} 5 \cdot 3 + 1 \cdot 1 & 5(-1) + 1 \cdot 0 & 5 \cdot 0 + 1 \cdot 2 \\ 2 \cdot 3 + 3 \cdot 1 & 2(-1) + 3 \cdot 0 & 2 \cdot 0 + 3 \cdot 2 \end{bmatrix}$$

$$= \begin{bmatrix} 16 & -5 & 2 \\ 9 & -2 & 6 \end{bmatrix}$$

21. $\begin{bmatrix} 2 & 2 & -1 \\ 3 & 0 & 1 \end{bmatrix}\begin{bmatrix} 0 & 2 \\ -1 & 4 \\ 0 & 2 \end{bmatrix}$

$$= \begin{bmatrix} 2 \cdot 0 + 2(-1) + (-1)(0) & 2 \cdot 2 + 2 \cdot 4 + (-1)(2) \\ 3 \cdot 0 + 0(-1) + 1(0) & 3 \cdot 2 + 0 \cdot 4 + 1 \cdot 2 \end{bmatrix}$$

$$= \begin{bmatrix} -2 & 10 \\ 0 & 8 \end{bmatrix}$$

23. $\begin{bmatrix} 1 & 2 \\ 3 & 4 \end{bmatrix}\begin{bmatrix} -1 & 5 \\ 7 & 0 \end{bmatrix}$

$$= \begin{bmatrix} 1(-1) + 2 \cdot 7 & 1 \cdot 5 + 2 \cdot 0 \\ 3(-1) + 4 \cdot 7 & 3 \cdot 5 + 4 \cdot 0 \end{bmatrix}$$

$$= \begin{bmatrix} 13 & 5 \\ 25 & 15 \end{bmatrix}$$

25. $\begin{bmatrix} -2 & -3 & 7 \\ 1 & 5 & 6 \end{bmatrix}\begin{bmatrix} 1 \\ 2 \\ 3 \end{bmatrix}$

$$= \begin{bmatrix} -2(1) + (-3)(2) + 7 \cdot 3 \\ 1 \cdot 1 + 5 \cdot 2 + 6 \cdot 3 \end{bmatrix}$$

$$= \begin{bmatrix} 13 \\ 29 \end{bmatrix}$$

27. $\left(\begin{bmatrix} 4 & 3 \\ 0 & 2 \\ 0 & -5 \end{bmatrix}\begin{bmatrix} 2 & -2 \\ 1 & -1 \end{bmatrix}\right)\begin{bmatrix} 10 \\ 0 \end{bmatrix}$

$$= \begin{bmatrix} 11 & -11 \\ 4 & -4 \\ -5 & 5 \end{bmatrix}\begin{bmatrix} 10 \\ 0 \end{bmatrix} = \begin{bmatrix} 110 \\ 40 \\ -50 \end{bmatrix}$$

29. $\begin{bmatrix} 2 & -2 \\ 1 & -1 \end{bmatrix}\left(\begin{bmatrix} 4 & 3 \\ 1 & 2 \end{bmatrix} + \begin{bmatrix} 7 & 0 \\ -1 & 5 \end{bmatrix}\right)$

$$= \begin{bmatrix} 2 & -2 \\ 1 & -1 \end{bmatrix}\begin{bmatrix} 11 & 3 \\ 0 & 7 \end{bmatrix}$$

$$= \begin{bmatrix} 22 & -8 \\ 11 & -4 \end{bmatrix}$$

31. (a) $AB = \begin{bmatrix} -2 & 4 \\ 1 & 3 \end{bmatrix}\begin{bmatrix} -2 & 1 \\ 3 & 6 \end{bmatrix} = \begin{bmatrix} 16 & 22 \\ 7 & 19 \end{bmatrix}$

(b) $BA = \begin{bmatrix} -2 & 1 \\ 3 & 6 \end{bmatrix}\begin{bmatrix} -2 & 4 \\ 1 & 3 \end{bmatrix} = \begin{bmatrix} 5 & -5 \\ 0 & 30 \end{bmatrix}$

(c) No, AB and BA are not equal here.

(d) No, AB does not always equal BA.

33. Verify that P(X + T) = PX + PT.
Find P(X + T) and PX + PT separately
and compare their values to see if
they are the same.

P(X + T)

$$= \begin{bmatrix} m & n \\ p & q \end{bmatrix}\left(\begin{bmatrix} x & y \\ z & w \end{bmatrix} + \begin{bmatrix} r & s \\ t & u \end{bmatrix}\right)$$

$$= \begin{bmatrix} m & n \\ p & q \end{bmatrix}\left(\begin{bmatrix} x+r & y+s \\ z+t & w+u \end{bmatrix}\right)$$

$$= \begin{bmatrix} m(x+r) + n(z+t) & m(y+s) + n(w+u) \\ p(x+r) + q(z+t) & p(y+s) + q(w+u) \end{bmatrix}$$

$$= \begin{bmatrix} mx + mr + nz + nt & my + ms + nw + nu \\ px + pr + qz + qt & py + ps + qw + qu \end{bmatrix}$$

PX + PT

$$= \begin{bmatrix} m & n \\ p & q \end{bmatrix} \begin{bmatrix} x & y \\ z & w \end{bmatrix} + \begin{bmatrix} m & n \\ p & q \end{bmatrix} \begin{bmatrix} r & s \\ t & u \end{bmatrix}$$

$$= \begin{bmatrix} mx + nz & my + nw \\ px + qz & py + qw \end{bmatrix} + \begin{bmatrix} mr + nt & ms + nu \\ pr + qt & ps + qu \end{bmatrix}$$

$$= \begin{bmatrix} (mx+nz)+(mr+nt) & (my+nw)+(ms+nu) \\ (px+qz)+(pr+qt) & (py+qw)+(ps+qu) \end{bmatrix}$$

$$= \begin{bmatrix} mx + nz + mr + nt & my + nw + ms + nu \\ px + qz + pr + qt & py + qw + ps + qu \end{bmatrix}$$

$$= \begin{bmatrix} mx + mr + nz + nt & my + ms + nw + nu \\ px + pr + qz + qt & py + ps + qw + qu \end{bmatrix}$$

Observe that the two results are identical. Thus, $P(X + T) = PX + PT$.

35. Verify that $(k + h)P = kP + hP$ for any real numbers k and h.

$$(k + h)P = (k + h) \begin{bmatrix} m & n \\ p & q \end{bmatrix}$$

$$= \begin{bmatrix} (k+h)m & (k+h)n \\ (k+h)p & (k+h)q \end{bmatrix}$$

$$= \begin{bmatrix} km + hm & kn + hn \\ kp + hp & kq + hq \end{bmatrix}$$

$$= \begin{bmatrix} km & kn \\ kp & kq \end{bmatrix} + \begin{bmatrix} hm & hn \\ hp & hq \end{bmatrix}$$

$$= k \begin{bmatrix} m & n \\ p & q \end{bmatrix} + h \begin{bmatrix} m & n \\ p & q \end{bmatrix}$$

$$= kP + hP$$

Thus, $(k + h)P = kP + hP$ for any real numbers k and h.

37. $\begin{bmatrix} 2 & 3 & 1 \\ 1 & -4 & 5 \end{bmatrix} \begin{bmatrix} x_1 \\ x_2 \\ x_3 \end{bmatrix} = \begin{bmatrix} 2x_1 + 3x_2 + x_3 \\ x_1 - 4x_2 + 5x_3 \end{bmatrix}$,

and $\begin{bmatrix} 2x_1 + 3x_2 + x_3 \\ x_1 - 4x_2 + 5x_3 \end{bmatrix} = \begin{bmatrix} 5 \\ 8 \end{bmatrix}$.

This is equivalent to

$$2x_1 + 3x_2 + x_3 = 5$$
$$x_1 - 4x_2 + 5x_3 = 8$$

since corresponding elements of equal matrices must be equal. Reversing this, observe that the given system of linear equations can be written as the matrix equation

$$\begin{bmatrix} 2 & 3 & 1 \\ 1 & -4 & 5 \end{bmatrix} \begin{bmatrix} x_1 \\ x_2 \\ x_3 \end{bmatrix} = \begin{bmatrix} 5 \\ 8 \end{bmatrix}.$$

39. (a) $\begin{bmatrix} 10 & 4 & 3 & 5 & 6 \\ 7 & 2 & 2 & 3 & 8 \\ 4 & 5 & 1 & 0 & 10 \\ 0 & 3 & 4 & 5 & 5 \end{bmatrix} \begin{bmatrix} 2 & 3 \\ 1 & 1 \\ 4 & 3 \\ 3 & 3 \\ 1 & 2 \end{bmatrix}$

	A	B
Dept. 1	57	70
Dept. 2	41	54
Dept. 3	27	40
Dept. 4	39	40

(with an $=$ sign to the left of the matrix)

(b) The total cost to buy from supplier A is $57 + 41 + 27 + 39 = \$164$, and the total cost to buy from supplier B is $70 + 54 + 40 + 40 = \$204$. The company should make the purchase from supplier A, since \$164 is a lower total cost than \$204.

41. (a) The matrices are

$$S = \begin{bmatrix} .027 & .009 \\ .030 & .007 \\ .015 & .009 \\ .013 & .011 \\ .019 & .011 \end{bmatrix} \text{ and}$$

$$P = \begin{bmatrix} 1596 & 218 & 199 & 425 & 214 \\ 1996 & 286 & 226 & 460 & 243 \\ 2440 & 365 & 252 & 484 & 266 \\ 2906 & 455 & 277 & 499 & 291 \end{bmatrix}.$$

	Births	Deaths
1960	62.208	24.710
1970	76.459	29.733
1980	91.956	35.033
1990	108.28	40.522

(b)   PS = (above matrix)

This product matrix gives the total number of births and deaths in each year (in millions).

43.   (a)   Use a computer to find the product matrix. The answer is

$$AC = \begin{bmatrix} 6 & 106 & 158 & 222 & 28 \\ 120 & 139 & 64 & 75 & 115 \\ -146 & -2 & 184 & 144 & -129 \\ 106 & 94 & 24 & 116 & 110 \end{bmatrix}$$

(b)   CA does not exist.

(c)   AC and CA are clearly not equal, since CA does not even exist.

45.   Use a computer to find the matrix products and sums.
The answers are as follows.

(a) $C + D = \begin{bmatrix} -1 & 5 & 9 & 13 & -1 \\ 7 & 17 & 2 & -10 & 6 \\ 18 & 9 & -12 & 12 & 22 \\ 9 & 4 & 18 & 10 & -3 \\ 1 & 6 & 10 & 28 & 5 \end{bmatrix}$

(b) $(C + D)B = \begin{bmatrix} -2 & -9 & 90 & 77 \\ -42 & -63 & 127 & 62 \\ 413 & 76 & 180 & -56 \\ -29 & -44 & 198 & 85 \\ 137 & 20 & 162 & 103 \end{bmatrix}$

(c) $CB = \begin{bmatrix} -56 & -1 & 1 & 45 \\ -156 & -119 & 76 & 122 \\ 315 & 86 & 118 & -91 \\ -17 & -17 & 116 & 51 \\ 118 & 19 & 125 & 77 \end{bmatrix}$

(d) $DB = \begin{bmatrix} 54 & -8 & 89 & 32 \\ 114 & 56 & 51 & -60 \\ 98 & -10 & 62 & 35 \\ -12 & -27 & 82 & 34 \\ 19 & 1 & 37 & 26 \end{bmatrix}$

(e) $CB + DB = \begin{bmatrix} -2 & -9 & 90 & 77 \\ -42 & -63 & 127 & 62 \\ 413 & 76 & 180 & -56 \\ -29 & -44 & 198 & 85 \\ 137 & 20 & 162 & 103 \end{bmatrix}$

(f)   Yes, (C + D)B and CB + DB are equal, as can be seen by observing that the answers to parts (b) and (e) are identical.

Section 2.5

1. $\begin{bmatrix} 2 & 3 \\ 1 & 1 \end{bmatrix}\begin{bmatrix} -1 & 3 \\ 1 & -2 \end{bmatrix} = \begin{bmatrix} 1 & 0 \\ 0 & 1 \end{bmatrix} = I$

$\begin{bmatrix} -1 & 3 \\ 1 & -2 \end{bmatrix}\begin{bmatrix} 2 & 3 \\ 1 & 1 \end{bmatrix} = \begin{bmatrix} 1 & 0 \\ 0 & 1 \end{bmatrix} = I$

Yes, these matrices are inverses of each other since their product matrix (both ways) is I.

3. $\begin{bmatrix} 2 & 1 \\ 3 & 2 \end{bmatrix}\begin{bmatrix} 2 & 1 \\ -3 & 2 \end{bmatrix} = \begin{bmatrix} 1 & 4 \\ 0 & 7 \end{bmatrix} \neq I$

No, these matrices are not inverses of each other since their product matrix is not I.

5. $\begin{bmatrix} 1 & 2 & 0 \\ 0 & 1 & 0 \\ 0 & 1 & 0 \end{bmatrix}\begin{bmatrix} 1 & -2 & 0 \\ 0 & 1 & 0 \\ 0 & -1 & 1 \end{bmatrix}$

$= \begin{bmatrix} 1 & 0 & 0 \\ 0 & 1 & 0 \\ 0 & 1 & 0 \end{bmatrix} \neq I$

No, these matrices are not inverses of each other.

7. $\begin{bmatrix} 1 & 3 & 3 \\ 1 & 4 & 3 \\ 1 & 3 & 4 \end{bmatrix} \begin{bmatrix} 7 & -3 & -3 \\ -1 & 1 & 0 \\ -1 & 0 & 1 \end{bmatrix}$

$= \begin{bmatrix} 1 & 0 & 0 \\ 0 & 1 & 0 \\ 0 & 0 & 1 \end{bmatrix} = I$

$\begin{bmatrix} 7 & -3 & -3 \\ -1 & 1 & 0 \\ -1 & 0 & 1 \end{bmatrix} \begin{bmatrix} 1 & 3 & 3 \\ 1 & 4 & 3 \\ 1 & 3 & 4 \end{bmatrix}$

$= \begin{bmatrix} 1 & 0 & 0 \\ 0 & 1 & 0 \\ 0 & 0 & 1 \end{bmatrix} = I$

Yes, these matrices are inverses of each other.

9. No, a matrix with a row of all zeros does not have an inverse; the row of all zeros makes it impossible to get all the 1's in the main diagonal of the identity matrix.

11. Let $A = \begin{bmatrix} 1 & -1 \\ 2 & 0 \end{bmatrix}$.

Form the augmented matrix $[A \mid I]$.

$[A \mid I] = \begin{bmatrix} 1 & -1 & 1 & 0 \\ 2 & 0 & 0 & 1 \end{bmatrix}$

Perform row operations on $[A \mid I]$ to get a matrix of the form $[I \mid B]$.

$\begin{bmatrix} 1 & -1 & 1 & 0 \\ 2 & 0 & 0 & 1 \end{bmatrix}$

$-2R_1 + R_2 \rightarrow R_2 \begin{bmatrix} 1 & -1 & 1 & 0 \\ 0 & 2 & -2 & 1 \end{bmatrix}$

$2R_1 + R_2 \rightarrow R_1 \begin{bmatrix} 2 & 0 & 0 & 1 \\ 0 & 2 & -2 & 1 \end{bmatrix}$

$\begin{array}{c} \frac{1}{2}R_1 \rightarrow R_1 \\ \frac{1}{2}R_2 \rightarrow R_2 \end{array} \begin{bmatrix} 1 & 0 & 0 & \frac{1}{2} \\ 0 & 1 & -1 & \frac{1}{2} \end{bmatrix} = [I \mid B]$

The matrix B in the last transformation is the desired multiplicative inverse.

$$A^{-1} = \begin{bmatrix} 0 & \frac{1}{2} \\ -1 & \frac{1}{2} \end{bmatrix}$$

This answer may be checked by showing that $AA^{-1} = I$ and $A^{-1}A = I$.

13. Let $A = \begin{bmatrix} 3 & -1 \\ -5 & 2 \end{bmatrix}$.

$[A \mid I] = \begin{bmatrix} 3 & -1 & 1 & 0 \\ -5 & 2 & 0 & 1 \end{bmatrix}$

$5R_1 + 3R_2 \rightarrow R_2 \begin{bmatrix} 3 & -1 & 1 & 0 \\ 0 & 1 & 5 & 3 \end{bmatrix}$

$R_1 + R_2 \rightarrow R_1 \begin{bmatrix} 3 & 0 & 6 & 3 \\ 0 & 1 & 5 & 3 \end{bmatrix}$

$\frac{1}{3}R_1 \rightarrow R_1 \begin{bmatrix} 1 & 0 & 2 & 1 \\ 0 & 1 & 5 & 3 \end{bmatrix} = [I \mid B]$

The desired inverse is

$$A^{-1} = \begin{bmatrix} 2 & 1 \\ 5 & 3 \end{bmatrix}.$$

15. Let $A = \begin{bmatrix} -6 & 4 \\ -3 & 2 \end{bmatrix}$.

$[A \mid I] = \begin{bmatrix} -6 & 4 & 1 & 0 \\ -3 & 2 & 0 & 1 \end{bmatrix}$

$R_1 + (-2)R_2 \rightarrow R_2 \begin{bmatrix} -6 & 4 & 1 & 0 \\ 0 & 0 & 1 & -2 \end{bmatrix}$

Because the last row has all zeros to the left of the vertical bar, there is no way to complete the desired transformation. A has no inverse.

17. Let $A = \begin{bmatrix} 1 & 0 & 0 \\ 0 & -1 & 0 \\ 1 & 0 & 1 \end{bmatrix}$.

$[A \mid I] = \begin{bmatrix} 1 & 0 & 0 & 1 & 0 & 0 \\ 0 & -1 & 0 & 0 & 1 & 0 \\ 1 & 0 & 1 & 0 & 0 & 1 \end{bmatrix}$

$$-1R_1 + R_3 \rightarrow R_3 \begin{bmatrix} 1 & 0 & 0 & 1 & 0 & 0 \\ 0 & -1 & 0 & 0 & 1 & 0 \\ 0 & 0 & 1 & -1 & 0 & 1 \end{bmatrix}$$

$$-1R_2 \rightarrow R_2 \begin{bmatrix} 1 & 0 & 0 & 1 & 0 & 0 \\ 0 & 1 & 0 & 0 & -1 & 0 \\ 0 & 0 & 1 & -1 & 0 & 1 \end{bmatrix}$$

$$A^{-1} = \begin{bmatrix} 1 & 0 & 0 \\ 0 & -1 & 0 \\ -1 & 0 & 1 \end{bmatrix}$$

Because the last row has all zeros to the left of the vertical bar, there is no way to complete the desired transformation. A has no inverse.

19. Let $A = \begin{bmatrix} -1 & -1 & -1 \\ 4 & 5 & 0 \\ 0 & 1 & -3 \end{bmatrix}$.

$$[A \mid I] = \begin{bmatrix} -1 & -1 & -1 & 1 & 0 & 0 \\ 4 & 5 & 0 & 0 & 1 & 0 \\ 0 & 1 & -3 & 0 & 0 & 1 \end{bmatrix}$$

$$4R_1 + R_2 \rightarrow R_2 \begin{bmatrix} -1 & -1 & -1 & 1 & 0 & 0 \\ 0 & 1 & -4 & 4 & 1 & 0 \\ 0 & 1 & -3 & 0 & 0 & 1 \end{bmatrix}$$

$$\begin{array}{l} R_1 + R_2 \rightarrow R_1 \\ \\ -1R_2 + R_3 \rightarrow R_3 \end{array} \begin{bmatrix} -1 & 0 & -5 & 5 & 1 & 0 \\ 0 & 1 & -4 & 4 & 1 & 0 \\ 0 & 0 & 1 & -4 & -1 & 1 \end{bmatrix}$$

$$\begin{array}{l} 5R_3 + R_1 \rightarrow R_1 \\ 4R_3 + R_2 \rightarrow R_2 \end{array} \begin{bmatrix} -1 & 0 & 0 & -15 & -4 & 5 \\ 0 & 1 & 0 & -12 & -3 & 4 \\ 0 & 0 & 1 & -4 & -1 & 1 \end{bmatrix}$$

$$-1R_1 \rightarrow R_1 \begin{bmatrix} -1 & 0 & 0 & -15 & 4 & -5 \\ 0 & 1 & 0 & -12 & -3 & 4 \\ 0 & 0 & 1 & -4 & -1 & 1 \end{bmatrix}$$

$$A^{-1} = \begin{bmatrix} 15 & 4 & -5 \\ -12 & -3 & 4 \\ -4 & -1 & 1 \end{bmatrix}$$

21. Let $A = \begin{bmatrix} 1 & 2 & 3 \\ -3 & -2 & -1 \\ -1 & 0 & 1 \end{bmatrix}$.

$$[A \mid I] = \begin{bmatrix} 1 & 2 & 3 & 1 & 0 & 0 \\ -3 & -2 & -1 & 0 & 1 & 0 \\ -1 & 0 & 1 & 0 & 0 & 1 \end{bmatrix}$$

$$\begin{array}{l} 3R_1 + R_2 \rightarrow R_2 \\ R_1 + R_3 \rightarrow R_3 \end{array} \begin{bmatrix} 1 & 2 & 3 & 1 & 0 & 0 \\ 0 & 4 & 8 & 3 & 1 & 0 \\ 0 & 2 & 4 & 1 & 0 & 1 \end{bmatrix}$$

$$\begin{array}{l} -2R_1 + R_2 \rightarrow R_1 \\ \\ -2R_3 + R_2 \rightarrow R_3 \end{array} \begin{bmatrix} -2 & 0 & 2 & 1 & 1 & 0 \\ 0 & 4 & 8 & 3 & 1 & 0 \\ 0 & 0 & 0 & 1 & 1 & -2 \end{bmatrix}$$

23. Let $A = \begin{bmatrix} 2 & 4 & 6 \\ -1 & -4 & -3 \\ 0 & 1 & -1 \end{bmatrix}$.

$$[A \mid I] = \begin{bmatrix} 2 & 4 & 6 & 1 & 0 & 0 \\ -1 & -4 & -3 & 0 & 1 & 0 \\ 0 & 1 & -1 & 0 & 0 & 1 \end{bmatrix}$$

$$R_1 + 2R_2 \rightarrow R_2 \begin{bmatrix} 2 & 4 & 6 & 1 & 0 & 0 \\ 0 & -4 & 0 & 1 & 2 & 0 \\ 0 & 1 & -1 & 0 & 0 & 1 \end{bmatrix}$$

$$\begin{array}{l} R_1 + R_2 \rightarrow R_1 \\ \\ R_2 + 4R_3 \rightarrow R_3 \end{array} \begin{bmatrix} 2 & 0 & 6 & 2 & 2 & 0 \\ 0 & -4 & 0 & 1 & 2 & 0 \\ 0 & 0 & -4 & 1 & 2 & 4 \end{bmatrix}$$

$$4R_1 + 6R_3 \rightarrow R_1 \begin{bmatrix} 8 & 0 & 0 & 14 & 20 & 24 \\ 0 & -4 & 0 & 1 & 2 & 0 \\ 0 & 0 & -4 & 1 & 2 & 4 \end{bmatrix}$$

$$\begin{array}{l} \frac{1}{8}R_1 \rightarrow R_1 \\ \\ -\frac{1}{4}R_2 \rightarrow R_2 \\ \\ -\frac{1}{4}R_3 \rightarrow R_3 \end{array} \begin{bmatrix} 1 & 0 & 0 & \frac{7}{4} & \frac{5}{2} & 3 \\ 0 & 1 & 0 & -\frac{1}{4} & -\frac{1}{2} & 0 \\ 0 & 0 & 1 & -\frac{1}{4} & -\frac{1}{2} & -1 \end{bmatrix}$$

$$A^{-1} = \begin{bmatrix} \frac{7}{4} & \frac{5}{2} & 3 \\ -\frac{1}{4} & -\frac{1}{2} & 0 \\ -\frac{1}{4} & -\frac{1}{2} & -1 \end{bmatrix}$$

25. Let $A = \begin{bmatrix} 1 & -2 & 3 & 0 \\ 0 & 1 & -1 & 1 \\ -2 & 2 & -2 & 4 \\ 0 & 2 & -3 & 1 \end{bmatrix}$

$$[A \mid I] = \begin{bmatrix} 1 & -2 & 3 & 0 & 1 & 0 & 0 & 0 \\ 0 & 1 & -1 & 1 & 0 & 1 & 0 & 0 \\ -2 & 2 & -2 & 4 & 0 & 0 & 1 & 0 \\ 0 & 2 & -3 & 1 & 0 & 0 & 0 & 1 \end{bmatrix}$$

$$2R_1 + R_3 \rightarrow R_3 \begin{bmatrix} 1 & -2 & 3 & 0 & 1 & 0 & 0 & 0 \\ 0 & 1 & -1 & 1 & 0 & 1 & 0 & 0 \\ 0 & -2 & 4 & 4 & 2 & 0 & 1 & 0 \\ 0 & 2 & -3 & 1 & 0 & 0 & 0 & 1 \end{bmatrix}$$

$$\begin{matrix}2R_2+R_1\to R_1\\ \\2R_2+R_3\to R_3\\-2R_2+R_4\to R_4\end{matrix}\left[\begin{array}{cccc|cccc}1&0&1&2&1&2&0&0\\0&1&-1&1&0&1&0&0\\0&0&2&6&2&2&1&0\\0&0&-1&-1&0&-2&0&1\end{array}\right]$$

$$\begin{matrix}R_3+(-2)R_1\to R_1\\R_3+2R_2\to R_2\\ \\R_3+2R_4\to R_4\end{matrix}\left[\begin{array}{cccc|cccc}-2&0&0&2&0&-2&1&0\\0&2&0&8&2&4&1&0\\0&0&2&6&2&2&1&0\\0&0&0&4&2&-2&1&2\end{array}\right]$$

$$\begin{matrix}-2R_1+R_4\to R_1\\R_2+(-2)R_4\to R_2\\2R_3+(-3)R_4\to R_3\\ \end{matrix}\left[\begin{array}{cccc|cccc}4&0&0&0&2&2&-1&2\\0&2&0&0&-2&8&-1&-4\\0&0&4&0&-2&10&-1&-6\\0&0&0&4&2&-2&1&2\end{array}\right]$$

$$\begin{matrix}\frac{1}{4}R_1\to R_1\\[4pt]\frac{1}{2}R_2\to R_2\\[4pt]\frac{1}{4}R_3\to R_3\\[4pt]\frac{1}{4}R_4\to R_4\end{matrix}\left[\begin{array}{cccc|cccc}1&0&0&0&\frac{1}{2}&\frac{1}{2}&-\frac{1}{4}&\frac{1}{2}\\[4pt]0&1&0&0&-1&4&-\frac{1}{2}&-2\\[4pt]0&0&1&0&-\frac{1}{2}&\frac{5}{2}&-\frac{1}{4}&-\frac{3}{2}\\[4pt]0&0&0&1&\frac{1}{2}&-\frac{1}{2}&\frac{1}{4}&\frac{1}{2}\end{array}\right]$$

$$A^{-1}=\begin{bmatrix}\frac{1}{2}&\frac{1}{2}&-\frac{1}{4}&\frac{1}{2}\\[4pt]-1&4&-\frac{1}{2}&-2\\[4pt]-\frac{1}{2}&\frac{5}{2}&-\frac{1}{4}&-\frac{3}{2}\\[4pt]\frac{1}{2}&-\frac{1}{2}&\frac{1}{4}&\frac{1}{2}\end{bmatrix}$$

27. $2x+3y=10$

$\qquad x-\ y=-5$

First write the system in matrix form.

$$\begin{bmatrix}2&3\\1&-1\end{bmatrix}\begin{bmatrix}x\\y\end{bmatrix}=\begin{bmatrix}10\\-5\end{bmatrix}$$

Let $A=\begin{bmatrix}2&3\\1&-1\end{bmatrix}$, $X=\begin{bmatrix}x\\y\end{bmatrix}$, $B=\begin{bmatrix}10\\-5\end{bmatrix}$.

The system is (in matrix form) $AX=B$.
Now use row operations to find A^{-1}.

$$[A\mid I]=\begin{bmatrix}2&3&1&0\\1&-1&0&1\end{bmatrix}$$

$$R_1+(-2)R_2\to R_2\begin{bmatrix}2&3&1&0\\0&5&1&-2\end{bmatrix}$$

$$5R_1+(-3)R_2\to R_1\begin{bmatrix}10&0&2&6\\0&5&1&-2\end{bmatrix}$$

$$\begin{matrix}\frac{1}{10}R_1\to R_1\\[4pt]\frac{1}{5}R_2\to R_2\end{matrix}\left[\begin{array}{cc|cc}1&0&\frac{1}{5}&\frac{3}{5}\\[4pt]0&1&\frac{1}{5}&-\frac{2}{5}\end{array}\right]$$

$$A^{-1}=\begin{bmatrix}\frac{1}{5}&\frac{3}{5}\\[4pt]\frac{1}{5}&-\frac{2}{5}\end{bmatrix}.$$

Next, find the product $A^{-1}B$.

$$A^{-1}B=\begin{bmatrix}\frac{1}{5}&\frac{3}{5}\\[4pt]\frac{1}{5}&-\frac{2}{5}\end{bmatrix}\begin{bmatrix}10\\-5\end{bmatrix}=\begin{bmatrix}-1\\4\end{bmatrix}$$

Since $X=A^{-1}B$,

$$X=\begin{bmatrix}x\\y\end{bmatrix}=\begin{bmatrix}-1\\4\end{bmatrix}.$$

Thus, the solution of the system is $(-1,4)$.

29. $2x+\ y=\ 5$

$\qquad 5x+3y=13$

Let $A=\begin{bmatrix}2&1\\5&3\end{bmatrix}$, $X=\begin{bmatrix}x\\y\end{bmatrix}$, $B=\begin{bmatrix}5\\13\end{bmatrix}$.

Use row operations to obtain

$$A^{-1}=\begin{bmatrix}3&-1\\-5&2\end{bmatrix}.$$

$$X=A^{-1}B=\begin{bmatrix}3&-1\\-5&2\end{bmatrix}\begin{bmatrix}5\\13\end{bmatrix}=\begin{bmatrix}2\\1\end{bmatrix}$$

The solution of the system is $(2,1)$.

31. $-x+y=1$

$\qquad 2x-y=1$

Let $A=\begin{bmatrix}-1&1\\2&-1\end{bmatrix}$, $X=\begin{bmatrix}x\\y\end{bmatrix}$, $B=\begin{bmatrix}1\\1\end{bmatrix}$.

Use row operations to obtain

$$A^{-1}=\begin{bmatrix}1&1\\2&1\end{bmatrix}.$$

$$X=A^{-1}B=\begin{bmatrix}1&1\\2&1\end{bmatrix}\begin{bmatrix}1\\1\end{bmatrix}=\begin{bmatrix}2\\3\end{bmatrix}$$

The solution of the system is $(2,3)$.

33. $-x - 8y = 12$

$3x + 24y = -36$

Let $A = \begin{bmatrix} -1 & -8 \\ 3 & 24 \end{bmatrix}$, $X = \begin{bmatrix} x \\ y \end{bmatrix}$, $B = \begin{bmatrix} 12 \\ -36 \end{bmatrix}$.

Using row operations on $[A \mid I]$ leads to the matrix

$$\begin{bmatrix} 1 & -8 & -1 & 0 \\ 0 & 0 & 3 & 1 \end{bmatrix},$$

but the zeros in the second row indicate that matrix A does not have an inverse. We cannot complete the solution by this method.

Since the second equation is a multiple of the first, the equations are dependent.

Solve the first equation of the system for x.

$$-x - 8y = 12$$
$$-x = 8y + 12$$
$$x = -8y - 12$$

The solution of the system is $(-8y - 12, y)$.

35. $-x - y - z = 1$

$4x + 5y = -2$

$ y - 3z = 3$

has coefficient matrix

$$A = \begin{bmatrix} -1 & -1 & -1 \\ 4 & 5 & 0 \\ 0 & 1 & 3 \end{bmatrix}.$$

In Exercise 19, it was found that

$$A^{-1} = \begin{bmatrix} -1 & -1 & -1 \\ 4 & 5 & 0 \\ 0 & 1 & 3 \end{bmatrix}^{-1}$$

$$= \begin{bmatrix} 15 & 4 & -5 \\ -12 & -3 & 4 \\ -4 & -1 & 1 \end{bmatrix}.$$

Since $X = A^{-1}B$,

$$\begin{bmatrix} x \\ y \\ z \end{bmatrix} = \begin{bmatrix} 15 & 4 & -5 \\ -12 & -3 & 4 \\ -4 & -1 & 1 \end{bmatrix} \begin{bmatrix} 1 \\ -2 \\ 3 \end{bmatrix} = \begin{bmatrix} -8 \\ 6 \\ 1 \end{bmatrix}.$$

The solution of the system is $(-8, 6, 1)$.

37. $2x + 4y + 6z = 4$

$-x - 4y - 3z = 8$

$ y - z = -4$

has coefficient matrix

$$A = \begin{bmatrix} 2 & 4 & 6 \\ -1 & -4 & -3 \\ 0 & 1 & -1 \end{bmatrix}.$$

In Exercise 23, it was found that

$$A^{-1} = \begin{bmatrix} 2 & 4 & 6 \\ -1 & -4 & -3 \\ 0 & 1 & -1 \end{bmatrix}^{-1}$$

$$= \begin{bmatrix} \frac{7}{4} & \frac{5}{2} & 3 \\ -\frac{1}{4} & -\frac{1}{2} & 0 \\ -\frac{1}{4} & -\frac{1}{2} & -1 \end{bmatrix}.$$

Since $X = A^{-1}B$,

$$\begin{bmatrix} x \\ y \\ z \end{bmatrix} = \begin{bmatrix} \frac{7}{4} & \frac{5}{2} & 3 \\ -\frac{1}{4} & -\frac{1}{2} & 0 \\ -\frac{1}{4} & -\frac{1}{2} & -1 \end{bmatrix} \begin{bmatrix} 4 \\ 8 \\ -4 \end{bmatrix} = \begin{bmatrix} 15 \\ -5 \\ -1 \end{bmatrix}.$$

The solution of the system is $(15, -5, -1)$.

39. $2x - 2y = 5$

$ 4y + 8z = 7$

$x + 2z = 1$

has coefficient matrix

$$A = \begin{bmatrix} 2 & -2 & 0 \\ 0 & 4 & 8 \\ 1 & 0 & 2 \end{bmatrix}.$$

However, using row operations on $[A \mid I]$ shows that A does not have an inverse, so another method must be used.

Try the Gauss–Jordon method. The augmented matrix is

$$\begin{bmatrix} 2 & -2 & 0 & 5 \\ 0 & 4 & 8 & 7 \\ 1 & 0 & 2 & 1 \end{bmatrix}.$$

After several row operations, we obtain the matrix

$$\begin{bmatrix} 1 & 0 & 2 & \frac{17}{4} \\ 0 & 1 & 2 & \frac{7}{4} \\ 0 & 0 & 0 & 13 \end{bmatrix}.$$

The bottom row of this matrix shows that the system has no solution, since $0 = 13$ is a false statement.

41. $\begin{aligned} x - 2y + 3z + &= 4 \\ y - z + w &= -8 \\ -2x + 2y - 2z + 4w &= 12 \\ 2y - 3z + w &= -4 \end{aligned}$

has coefficient matrix

$$A = \begin{bmatrix} 1 & -2 & 3 & 0 \\ 0 & -1 & -1 & 1 \\ -2 & 2 & -2 & 4 \\ 0 & 2 & -3 & 1 \end{bmatrix}.$$

In Exercise 25, it was found that

$$A^{-1} = \begin{bmatrix} \frac{1}{2} & \frac{1}{2} & -\frac{1}{4} & \frac{1}{2} \\ -1 & 4 & -\frac{1}{2} & -2 \\ -\frac{1}{2} & \frac{5}{2} & -\frac{1}{4} & -\frac{3}{2} \\ \frac{1}{2} & -\frac{1}{2} & \frac{1}{4} & \frac{1}{2} \end{bmatrix}.$$

Since $X = A^{-1}B$,

$$\begin{bmatrix} x \\ y \\ z \\ w \end{bmatrix} = \begin{bmatrix} \frac{1}{2} & \frac{1}{2} & -\frac{1}{4} & \frac{1}{2} \\ -1 & 4 & -\frac{1}{2} & -2 \\ -\frac{1}{2} & \frac{5}{2} & -\frac{1}{4} & -\frac{3}{2} \\ \frac{1}{2} & -\frac{1}{2} & \frac{1}{4} & \frac{1}{2} \end{bmatrix} \begin{bmatrix} 4 \\ -8 \\ 12 \\ -4 \end{bmatrix} = \begin{bmatrix} -7 \\ -34 \\ -19 \\ 7 \end{bmatrix}.$$

The solution of the system is $(-7, -34, -19, 7)$.

43. $A = \begin{bmatrix} a & b \\ c & d \end{bmatrix}$

$IA = \begin{bmatrix} 1 & 0 \\ 0 & 1 \end{bmatrix}\begin{bmatrix} a & b \\ c & d \end{bmatrix} = \begin{bmatrix} a & b \\ c & d \end{bmatrix} = A$

Thus, $IA = A$.

45. $A = \begin{bmatrix} a & b \\ c & d \end{bmatrix}$, $0 = \begin{bmatrix} 0 & 0 \\ 0 & 0 \end{bmatrix}$

$A \cdot 0 = \begin{bmatrix} a & b \\ c & d \end{bmatrix}\begin{bmatrix} 0 & 0 \\ 0 & 0 \end{bmatrix} = \begin{bmatrix} 0 & 0 \\ 0 & 0 \end{bmatrix} = 0$

Thus, $A \cdot 0 = 0$

47. $A = \begin{bmatrix} a & b \\ c & d \end{bmatrix}$

In Exercise 46, it was found that

$$A^{-1} = \frac{1}{ad - bc}\begin{bmatrix} d & -b \\ -c & a \end{bmatrix}.$$

$$A^{-1}A = \left(\frac{1}{ad - bc}\begin{bmatrix} d & -b \\ -c & a \end{bmatrix}\right)\begin{bmatrix} a & b \\ c & d \end{bmatrix}$$

$$= \frac{1}{ad - bc}\left(\begin{bmatrix} d & -b \\ -c & a \end{bmatrix}\begin{bmatrix} a & b \\ c & d \end{bmatrix}\right)$$

$$= \frac{1}{ad - bc}\begin{bmatrix} ad - bc & 0 \\ 0 & ad - bc \end{bmatrix}$$

$$= \begin{bmatrix} 1 & 0 \\ 0 & 1 \end{bmatrix} = I$$

Thus, $A^{-1}A = I$.

49.

$$AB = 0$$

$$A^{-1}(AB) = A^{-1} \cdot 0$$

$$(A^{-1}A)B = 0$$

$$I \cdot B = 0$$

$$B = 0$$

Thus, if $AB = 0$ and A^{-1} exists, then $B = 0$.

51. (a) The matrix is $B = \begin{bmatrix} 72 \\ 48 \\ 60 \end{bmatrix}$.

(b) The matrix equation is

$$\begin{bmatrix} 2 & 4 & 2 \\ 2 & 1 & 2 \\ 2 & 1 & 3 \end{bmatrix} \begin{bmatrix} x_1 \\ x_2 \\ x_3 \end{bmatrix} = \begin{bmatrix} 72 \\ 48 \\ 60 \end{bmatrix}.$$

(c) To solve the system, begin by using row operations to find A^{-1}.

$$[A \mid I] = \begin{bmatrix} 2 & 4 & 2 & | & 1 & 0 & 0 \\ 2 & 1 & 2 & | & 0 & 1 & 0 \\ 2 & 1 & 3 & | & 0 & 0 & 1 \end{bmatrix}$$

$$\begin{matrix} \\ R_1 - 1R_2 \to R_2 \\ R_1 - 1R_3 \to R_3 \end{matrix} \begin{bmatrix} 2 & 4 & 2 & | & 1 & 0 & 0 \\ 0 & 3 & 0 & | & 1 & -1 & 0 \\ 0 & 3 & -1 & | & 1 & 0 & -1 \end{bmatrix}$$

$$\begin{matrix} 3R_1 - 4R_2 \to R_1 \\ \\ R_2 - 1R_3 \to R_3 \end{matrix} \begin{bmatrix} 6 & 0 & 6 & | & -1 & 4 & 0 \\ 0 & 3 & 0 & | & 1 & -1 & 0 \\ 0 & 0 & 1 & | & 0 & -1 & 1 \end{bmatrix}$$

$$R_1 - 6R_3 \to R_1 \begin{bmatrix} 6 & 0 & 0 & | & -1 & 10 & -6 \\ 0 & 3 & 0 & | & 1 & -1 & 0 \\ 0 & 0 & 1 & | & 0 & -1 & 1 \end{bmatrix}$$

$$\begin{matrix} \frac{1}{6}R_1 \to R_1 \\ \frac{1}{3}R_2 \to R_2 \\ \\ \end{matrix} \begin{bmatrix} 1 & 0 & 0 & | & -\frac{1}{6} & \frac{5}{3} & -1 \\ 0 & 1 & 0 & | & \frac{1}{3} & -\frac{1}{3} & 0 \\ 0 & 0 & 1 & | & 0 & -1 & 1 \end{bmatrix}$$

The inverse matrix is

$$A^{-1} = \begin{bmatrix} -\frac{1}{6} & \frac{5}{3} & -1 \\ \frac{1}{3} & -\frac{1}{3} & 0 \\ 0 & -1 & 1 \end{bmatrix}.$$

Since $X = A^{-1}B$,

$$\begin{bmatrix} x_1 \\ x_2 \\ x_3 \end{bmatrix} = \begin{bmatrix} -\frac{1}{6} & \frac{5}{3} & -1 \\ \frac{1}{3} & -\frac{1}{3} & 0 \\ 0 & -1 & 1 \end{bmatrix} \begin{bmatrix} 72 \\ 48 \\ 60 \end{bmatrix} = \begin{bmatrix} 8 \\ 8 \\ 12 \end{bmatrix}.$$

There are 8 daily orders for type I, 8 for type II, and 12 for type III.

53. Let x_1 = amount invested in AAA bonds;

x_2 = amount invested in A bonds; and

x_3 = amount invested in B bonds.

The total investment is

$$x_1 + x_2 + x_3.$$

The annual return is

$$.06x_1 + .07x_2 + .10x_3.$$

Since twice as much is invested in AAA bonds as in B bonds,

$$x_1 = 2x_3.$$

(a) The system to be solved is

$$x_1 + x_2 + x_3 = 25{,}000$$

$$.06x_1 + .07x_2 + .10x_3 = 1810$$

$$x_1 \qquad\quad - 2x_3 = 0.$$

Let $A = \begin{bmatrix} 1 & 1 & 1 \\ .06 & .07 & .10 \\ 1 & 0 & -2 \end{bmatrix}$, $X = \begin{bmatrix} x_1 \\ x_2 \\ x_3 \end{bmatrix}$,

$$B = \begin{bmatrix} 25{,}000 \\ 1800 \\ 0 \end{bmatrix}.$$

Use row operations to obtain

$$A^{-1} = \begin{bmatrix} -14 & 200 & 3 \\ 22 & -300 & -4 \\ -7 & 100 & 1 \end{bmatrix}.$$

Since $X = A^{-1}B$,

$$\begin{bmatrix} x_1 \\ x_2 \\ x_3 \end{bmatrix} = \begin{bmatrix} -14 & 200 & 3 \\ 22 & -300 & -4 \\ -7 & 100 & 1 \end{bmatrix} \begin{bmatrix} 25{,}000 \\ 1810 \\ 0 \end{bmatrix} = \begin{bmatrix} 12{,}000 \\ 7000 \\ 6000 \end{bmatrix}.$$

$12{,}000 should be invested in AAA bonds at 6%, $7000 in A bonds at 7%, and $6000 in B bonds at 10%.

(b) The matrix of constants is changed to

$$B = \begin{bmatrix} 30,000 \\ 2150 \\ 0 \end{bmatrix}.$$

$$X = A^{-1}B = \begin{bmatrix} -14 & 200 & 3 \\ 22 & -300 & -4 \\ -7 & 100 & 1 \end{bmatrix} \begin{bmatrix} 30,000 \\ 2150 \\ 0 \end{bmatrix}$$

$$= \begin{bmatrix} 10,000 \\ 15,000 \\ 5000 \end{bmatrix}$$

$10,000 should be invested in AAA bonds at 6%, $15,000 in A bonds at 7%, and $5000 in B bonds at 10%.

(c) The matrix of constants is changed to

$$B = \begin{bmatrix} 40,000 \\ 2900 \\ 0 \end{bmatrix}.$$

$$X = A^{-1}B = \begin{bmatrix} -14 & 200 & 3 \\ 22 & -300 & -4 \\ -7 & 100 & 1 \end{bmatrix} \begin{bmatrix} 40,000 \\ 2900 \\ 0 \end{bmatrix}$$

$$= \begin{bmatrix} 20,000 \\ 10,000 \\ 10,000 \end{bmatrix}$$

$20,000 should be invested in AAA bonds at 6%, $10,000 in A bonds at 7%, and $10,000 in B bonds at 10%.

55. Use a computer to perform this calculation. With entries rounded to 6 places, the answer is

$$(CD)^{-1} = \begin{bmatrix} .010146 & -.011883 & .002772 & .020724 & -.012273 \\ .006353 & .014233 & -.001861 & -.029146 & .019225 \\ -.000638 & .006782 & -.004823 & -.022658 & .019344 \\ -.005261 & .003781 & .006192 & .004837 & -.006910 \\ -.012252 & -.001177 & -.006126 & .006744 & .002792 \end{bmatrix}$$

57. Use a computer to determine that no, $C^{-1}D^{-1}$ and $(CD)^{-1}$ are not equal.

59. Use a computer to obtain $X = \begin{bmatrix} .62963 \\ .148148 \\ .259259 \end{bmatrix}.$

61. Use a computer to obtain $X = \begin{bmatrix} .489558 \\ 1.00104 \\ 2.11853 \\ -1.20793 \\ -.961346 \end{bmatrix}.$

Section 2.6

1. $A = \begin{bmatrix} .5 & .4 \\ .25 & .2 \end{bmatrix}$, $D = \begin{bmatrix} 2 \\ 4 \end{bmatrix}$

To find the production matrix, we first calculate I - A.

$$I - A = \begin{bmatrix} 1 & 0 \\ 0 & 1 \end{bmatrix} - \begin{bmatrix} .5 & .4 \\ .25 & .2 \end{bmatrix}$$

$$= \begin{bmatrix} .5 & -.4 \\ -.25 & .8 \end{bmatrix}$$

Using row operations, find the inverse of I - A.

$$[\,I - A \mid I\,] = \begin{bmatrix} .5 & -.4 & 1 & 0 \\ -.25 & .8 & 0 & 1 \end{bmatrix}$$

$$\begin{matrix} 10R_1 \to R_1 \\ 100R_2 \to R_2 \end{matrix} \begin{bmatrix} 5 & -4 & 10 & 0 \\ -25 & 80 & 0 & 100 \end{bmatrix}$$

$$5R_1 + R_2 \to R_2 \begin{bmatrix} 5 & -4 & 10 & 0 \\ 0 & 60 & 50 & 100 \end{bmatrix}$$

$$15R_1 + R_2 \to R_1 \begin{bmatrix} 75 & 0 & 200 & 100 \\ 0 & 60 & 50 & 100 \end{bmatrix}$$

$$\begin{matrix} \frac{1}{75}R_1 \to R_1 \\ \frac{1}{60}R_2 \to R_2 \end{matrix} \begin{bmatrix} 1 & 0 & \frac{8}{3} & \frac{4}{3} \\ 0 & 1 & \frac{5}{6} & \frac{5}{3} \end{bmatrix}$$

$$(I - A)^{-1} = \begin{bmatrix} \frac{8}{3} & \frac{4}{3} \\ \frac{5}{6} & \frac{5}{3} \end{bmatrix} \approx \begin{bmatrix} 2.67 & 1.33 \\ .83 & 1.67 \end{bmatrix}$$

Since X = (I - A)⁻¹D, the production matrix is

$$X = \begin{bmatrix} 2.67 & 1.33 \\ .83 & 1.67 \end{bmatrix} \begin{bmatrix} 2 \\ 4 \end{bmatrix} = \begin{bmatrix} 10.67 \\ 8.33 \end{bmatrix}.$$

3. $A = \begin{bmatrix} .1 & .03 \\ .07 & .6 \end{bmatrix}$, $D = \begin{bmatrix} 5 \\ 10 \end{bmatrix}$

First, calculate I - A.

$$I - A = \begin{bmatrix} .9 & -.03 \\ -.07 & .4 \end{bmatrix}$$

Use row operations to find the inverse of I - A, which is

$$(I - A)^{-1} \approx \begin{bmatrix} 1.118 & .084 \\ .195 & 2.515 \end{bmatrix}.$$

Since X = (I - A)⁻¹D, the production matrix is

$$X = \begin{bmatrix} 1.118 & .084 \\ .195 & 2.515 \end{bmatrix} \begin{bmatrix} 5 \\ 10 \end{bmatrix} = \begin{bmatrix} 6.43 \\ 26.12 \end{bmatrix}.$$

5. $A = \begin{bmatrix} .4 & 0 & .3 \\ 0 & .8 & .1 \\ 0 & .2 & .4 \end{bmatrix}$, $D = \begin{bmatrix} 1 \\ 3 \\ 2 \end{bmatrix}$

First, calculate I - A.

$$I - A = \begin{bmatrix} 1 & 0 & 0 \\ 0 & 1 & 0 \\ 0 & 0 & 1 \end{bmatrix} - \begin{bmatrix} .4 & 0 & .3 \\ 0 & .8 & .1 \\ 0 & .2 & .4 \end{bmatrix}$$

$$= \begin{bmatrix} .6 & 0 & -.3 \\ 0 & .2 & -.1 \\ 0 & -.2 & .6 \end{bmatrix}$$

Now use row operations to find (I - A)⁻¹.

$$[\,I - A \mid I\,] = \begin{bmatrix} .6 & 0 & -.3 & 1 & 0 & 0 \\ 0 & .2 & -.1 & 0 & 1 & 0 \\ 0 & -.2 & .6 & 0 & 0 & 1 \end{bmatrix}$$

$$R_2 + R_3 \to R_3 \begin{bmatrix} .6 & 0 & -.3 & 1 & 0 & 0 \\ 0 & .2 & -.1 & 0 & 1 & 0 \\ 0 & 0 & .5 & 0 & 1 & 1 \end{bmatrix}$$

$$\begin{matrix} 5R_1 + 3R_3 \to R_1 \\ 5R_2 + R_3 \to R_2 \end{matrix} \begin{bmatrix} 3 & 0 & 0 & 5 & 3 & 3 \\ 0 & 1 & 0 & 0 & 6 & 1 \\ 0 & 0 & .5 & 0 & 1 & 1 \end{bmatrix}$$

$$\begin{matrix} \frac{1}{3}R_1 \to R_1 \\ \\ 2R_3 \to R_3 \end{matrix} \begin{bmatrix} 1 & 0 & 0 & \frac{5}{3} & 1 & 1 \\ 0 & 1 & 0 & 0 & 6 & 1 \\ 0 & 0 & 1 & 0 & 2 & 2 \end{bmatrix}$$

$$(I - A)^{-1} \approx \begin{bmatrix} 1.67 & 1 & 1 \\ 0 & 6 & 1 \\ 0 & 2 & 2 \end{bmatrix}.$$

Since X = (I - A)⁻¹D, the production matrix is

$$X = \begin{bmatrix} 1.67 & 1 & 1 \\ 0 & 6 & 1 \\ 0 & 2 & 2 \end{bmatrix} \begin{bmatrix} 1 \\ 3 \\ 2 \end{bmatrix} = \begin{bmatrix} 6.67 \\ 20 \\ 10 \end{bmatrix}.$$

7.
$$\begin{array}{c} \\ A \\ B \\ C \end{array}\begin{array}{ccc} A & B & C \\ \left[\begin{array}{ccc} .3 & .1 & .8 \\ .5 & .6 & .1 \\ .2 & .3 & .1 \end{array}\right] \end{array} = A$$

$$I - A = \left[\begin{array}{ccc} .7 & -.1 & -.8 \\ -.5 & .4 & -.1 \\ -.2 & -.3 & .9 \end{array}\right]$$

Set $(I - A)X = 0$ to obtain the following.

$$\left[\begin{array}{ccc} .7 & -.1 & -.8 \\ -.5 & .4 & -.1 \\ -.2 & -.3 & .9 \end{array}\right]\left[\begin{array}{c} x_1 \\ x_2 \\ x_3 \end{array}\right] = \left[\begin{array}{c} 0 \\ 0 \\ 0 \end{array}\right]$$

$$\left[\begin{array}{c} .7x_1 - .1x_2 - .8x_3 \\ -.5x_1 + .4x_2 - .1x_3 \\ -.2x_1 - .3x_2 + .9x_3 \end{array}\right]\left[\begin{array}{c} x_1 \\ x_2 \\ x_3 \end{array}\right] = \left[\begin{array}{c} 0 \\ 0 \\ 0 \end{array}\right]$$

Rewrite this matrix equation as a system of equations.

$$.7x_1 - .1x_2 - .8x_3 = 0$$
$$-.5x_1 + .4x_2 - .1x_3 = 0$$
$$-.2x_1 - .3x_2 + .9x_3 = 0$$

Rewrite the equations without decimals.

$$7x_1 - x_2 - 8x_3 = 0 \quad (1)$$
$$-5x_1 + 4x_2 - x_3 = 0 \quad (2)$$
$$-2x_1 - 3x_2 + 9x_3 = 0 \quad (3)$$

Use row operations to solve this system of equations. Begin by eliminating x_1 in equations (2) and (3).

$$7x_1 - x_2 - 8x_3 = 0 \quad (1)$$
$$5R_1 + 7R_2 \rightarrow R_2 \qquad 23x_2 - 47x_3 = 0 \quad (4)$$
$$2R_1 + 7R_3 \rightarrow R_3 \qquad -23x_2 + 47x_3 = 0 \quad (5)$$

Eliminate x_2 in equations (1) and (5).

$$23R_1 + R_2 \rightarrow R_1 \quad 161x_1 \qquad - 231x_3 = 0 \quad (6)$$
$$23x_2 - 47x_3 = 0 \quad (4)$$
$$0 = 0 \quad (7)$$

The true statement in equation (7) indicates that the equations are dependent. Solve equation (6) for x_1 and equation (4) for x_2, each in terms of x_3.

$$x_1 = \frac{231}{161}x_3 = \frac{33}{23}x_3$$

$$x_2 = \frac{47}{23}x_3$$

The solution of the system is $\left(\frac{33}{23}x_3, \frac{47}{23}x_3, x_3\right)$.

If $x_3 = 23$, then $x_1 = 33$ and $x_2 = 47$, so the production of the three commodities should be in the ratio $33:47:23$.

9. In Example 4, it was found that

$$(I - A)^{-1} \approx \left[\begin{array}{cc} 1.39 & .13 \\ .51 & 1.17 \end{array}\right].$$

Since $X = (I - A)^{-1}D$, the production matrix is

$$X = \left[\begin{array}{cc} 1.39 & .13 \\ .51 & 1.17 \end{array}\right]\left[\begin{array}{c} 690 \\ 920 \end{array}\right] = \left[\begin{array}{c} 1078.7 \\ 1428.3 \end{array}\right].$$

Thus, about 1079 metric tons of wheat and 1428 metric tons of oil should be produced.

11. In Example 3, it was found that

$$(I - A)^{-1} \approx \left[\begin{array}{ccc} 1.40 & .50 & .59 \\ .84 & 1.36 & .62 \\ .56 & .47 & 1.30 \end{array}\right].$$

Since $X = (I - A)^{-1}D$, the production matrix is

$$X = \left[\begin{array}{ccc} 1.40 & .50 & .59 \\ .84 & 1.36 & .62 \\ .56 & .47 & 1.30 \end{array}\right]\left[\begin{array}{c} 516 \\ 516 \\ 516 \end{array}\right]$$

$$= \left[\begin{array}{c} 1284.8 \\ 1455.1 \\ 1202.3 \end{array}\right].$$

Thus, about 1285 units of agriculture, 1455 units of manufacturing,
and 1202 units of transportation
should be produced.

13. From the given data, we get the
input–output matrix

$$A = \begin{bmatrix} 0 & \frac{1}{2} & \frac{1}{4} \\ \frac{1}{4} & 0 & \frac{1}{4} \\ \frac{1}{2} & \frac{1}{4} & 0 \end{bmatrix}.$$

$$I - A = \begin{bmatrix} 1 & -\frac{1}{2} & -\frac{1}{4} \\ -\frac{1}{4} & 1 & -\frac{1}{4} \\ -\frac{1}{2} & -\frac{1}{4} & 1 \end{bmatrix}$$

Use row operations to find the
inverse of $I - A$, which is

$$(I - A)^{-1} \approx \begin{bmatrix} 1.538 & .923 & .615 \\ .615 & 1.436 & .513 \\ .923 & .821 & 1.436 \end{bmatrix}.$$

Since $X = (I - A)^{-1}D$, the production
matrix is

$$X = \begin{bmatrix} 1.538 & .923 & .615 \\ .615 & 1.436 & .513 \\ .923 & .821 & 1.436 \end{bmatrix} \begin{bmatrix} 1000 \\ 1000 \\ 1000 \end{bmatrix}$$

$$\approx \begin{bmatrix} 3077 \\ 2564 \\ 3179 \end{bmatrix}.$$

Thus, the production should be about
3077 units of agriculture, 2564 units
of manufacturing, and 3179 units of
transportation.

15. From the given data, we get the
input–output matrix

$$A = \begin{bmatrix} \frac{1}{4} & \frac{1}{6} \\ \frac{1}{2} & 0 \end{bmatrix}.$$

$$I - A = \begin{bmatrix} \frac{3}{4} & -\frac{1}{6} \\ -\frac{1}{2} & 1 \end{bmatrix}$$

Use row operations to find the
inverse of $I - A$, which is

$$(I - A)^{-1} = \begin{bmatrix} \frac{3}{2} & \frac{1}{4} \\ \frac{3}{4} & \frac{9}{8} \end{bmatrix}.$$

(a) The production matrix is

$$X = (I - A)^{-1}D = \begin{bmatrix} \frac{3}{2} & \frac{1}{4} \\ \frac{3}{4} & \frac{9}{8} \end{bmatrix} \begin{bmatrix} 1 \\ 1 \end{bmatrix} = \begin{bmatrix} \frac{7}{4} \\ \frac{15}{8} \end{bmatrix}.$$

Thus, $\frac{7}{4}$ bushels of yams and $\frac{15}{8} \approx 2$
pigs should be produced.

(b) The production matrix is

$$X = (I - A)^{-1}D = \begin{bmatrix} \frac{3}{2} & \frac{1}{4} \\ \frac{3}{4} & \frac{9}{8} \end{bmatrix} \begin{bmatrix} 100 \\ 70 \end{bmatrix} = \begin{bmatrix} 167.5 \\ 153.75 \end{bmatrix}.$$

Thus, 167.5 bushels of yams and
$153.75 \approx 154$ pigs should be produced.

17. Use a computer to find the production
matrix $X = (I - A)^{-1}D$. The answer is

$$X = \begin{bmatrix} 2930 \\ 3570 \\ 2300 \\ 580 \end{bmatrix}.$$

19. $A = \begin{bmatrix} .1 & .2 & .1 \\ .2 & .1 & .05 \\ 0 & .05 & .1 \end{bmatrix}$, $D = \begin{bmatrix} 1000 \\ 1000 \\ 1000 \end{bmatrix}$

Use a computer to find the production
matrix $X = (I - A)^{-1}D$. The answer is

$$X = \begin{bmatrix} 1583.91 \\ 1529.54 \\ 1196.09 \end{bmatrix}.$$

Chapter 2 Review Exercises

3. $2x + 3y = 10$ *(1)*
     $-3x + y = 18$ *(2)*

   Eliminate x in equation (2).

   $$2x + 3y = 10 \quad (1)$$
   $3R_2 + 2R_1 \to R_2$ $\qquad 11y = 66 \quad (3)$

   Make each leading coefficient equal 1.

   $\frac{1}{2}R_1 \to R_1$ $\qquad x + \frac{3}{2}y = 5 \quad (4)$

   $\frac{1}{11}R_2 \to R_2$ $\qquad y = 6 \quad (5)$

   Substitute 6 for y in equation (4) to
   get $x = -4$.
   The solution of the system is $(-4, 6)$.

5. $2x - 3y + z = -5$ *(1)*
     $x + 4y + 2z = 13$ *(2)*
     $5x + 5y + 3z = 14$ *(3)*

   Eliminate x in equations (2) and (3).

   $$2x - 3y + z = -5 \quad (1)$$
   $-2R_2 + R_1 \to R_2$ $\qquad -11y - 3z = -31 \quad (4)$
   $5R_1 + (-2)R_3 \to R_3$ $\qquad -25y - z = -53 \quad (5)$

   Eliminate y in equation (5).

   $$2x - 3y + z = -5 \quad (1)$$
   $$-11y - 3z = -31 \quad (6)$$
   $-25R_2 + R_3 \to R_3$ $\qquad 64z = 192 \quad (7)$

   Make each leading coefficient equal 1.

   $\frac{1}{2}R_1 \to R_1$ $\quad x - \frac{3}{2}y + \frac{1}{2}z = -\frac{5}{2} \quad (8)$

   $-\frac{1}{11}R_2 \to R_2$ $\qquad y + \frac{3}{11}z = \frac{31}{11} \quad (9)$

   $\frac{1}{64}R_3 \to R_3$ $\qquad z = 3 \quad (10)$

   Substitute 3 for z in equation (9) to
   get $y = 2$. Substitute 3 for z and 2 for
   y in equation (8) to get $x = -1$. The
   solution of the system is $(-1, 2, 3)$.

7. $2x + 4y = -6$
     $-3x - 5y = 12$

   Write the augmented matrix and use row
   operations.

   $$\begin{bmatrix} 2 & 4 & | & -6 \\ -3 & -5 & | & 12 \end{bmatrix}$$

   $3R_1 + 2R_2 \to R_2$ $\begin{bmatrix} 2 & 4 & | & -6 \\ 0 & 2 & | & 6 \end{bmatrix}$

   $-2R_2 + R_1 \to R_1$ $\begin{bmatrix} 2 & 0 & | & -18 \\ 0 & 2 & | & 6 \end{bmatrix}$

   $\frac{1}{2}R_1 \to R_1$ $\begin{bmatrix} 1 & 0 & | & -9 \\ 0 & 1 & | & 3 \end{bmatrix}$
   $\frac{1}{2}R_2 \to R_2$

   The solution of the system is $(-9, 3)$.

9. $x - y + 3z = 13$
     $4x + y + 2z = 17$
     $3x + 2y + 2z = 1$

   Write the augmented matrix and use row
   operations.

   $$\begin{bmatrix} 1 & -1 & 3 & | & 13 \\ 4 & 1 & 2 & | & 17 \\ 3 & 2 & 2 & | & 1 \end{bmatrix}$$

   $-4R_1 + R_2 \to R_2$ $\begin{bmatrix} 1 & -1 & 3 & | & 13 \\ 0 & 5 & -10 & | & -35 \\ 0 & 5 & -7 & | & -38 \end{bmatrix}$
   $-3R_1 + R_3 \to R_3$

   $5R_1 + R_2 \to R_1$ $\begin{bmatrix} 5 & 0 & 5 & | & 30 \\ 0 & 5 & -10 & | & -35 \\ 0 & 0 & 3 & | & -3 \end{bmatrix}$
   $-1R_2 + R_3 \to R_3$

   $-3R_1 + 5R_3 \to R_1$ $\begin{bmatrix} -15 & 0 & 0 & | & -105 \\ 0 & 15 & 0 & | & -135 \\ 0 & 0 & 3 & | & -3 \end{bmatrix}$
   $3R_2 + 10R_3 \to R_2$

   $-\frac{1}{15}R_1 \to R_1$ $\begin{bmatrix} 1 & 0 & 0 & | & 7 \\ 0 & 1 & 0 & | & -9 \\ 0 & 0 & 1 & | & -1 \end{bmatrix}$
   $\frac{1}{15}R_2 \to R_2$
   $\frac{1}{3}R_3 \to R_3$

   The solution of the system is
   $(7, -9, -1)$.

11. $3x - 6y + 9z = 12$

$-x + 2y - 3z = -4$

$x + y + 2z = 7$

Write the augmented matrix and use row operations.

$$\begin{bmatrix} 3 & -6 & 9 & | & 12 \\ -1 & 2 & -3 & | & -4 \\ 1 & 1 & 2 & | & 7 \end{bmatrix}$$

$$\begin{matrix} \\ R_1 + 3R_2 \rightarrow R_2 \\ -1R_1 + 3R_3 \rightarrow R_3 \end{matrix} \begin{bmatrix} 3 & -6 & 9 & | & 12 \\ 0 & 0 & 0 & | & 0 \\ 0 & 9 & -3 & | & 9 \end{bmatrix}$$

The zero in row 2, column 2 is an obstacle. To proceed, interchange the second and third rows.

$$\begin{bmatrix} 3 & -6 & 9 & | & 12 \\ 0 & 9 & -3 & | & 9 \\ 0 & 0 & 0 & | & 0 \end{bmatrix}$$

$$3R_1 + 2R_2 \rightarrow R_1 \begin{bmatrix} 9 & 0 & 21 & | & 54 \\ 0 & 9 & -3 & | & 9 \\ 0 & 0 & 0 & | & 0 \end{bmatrix}$$

$$\begin{matrix} \frac{1}{9}R_1 \rightarrow R_1 \\ \\ \frac{1}{9}R_2 \rightarrow R_2 \end{matrix} \begin{bmatrix} 1 & 0 & \frac{7}{3} & | & 6 \\ 0 & 1 & -\frac{1}{3} & | & 1 \\ 0 & 0 & 0 & | & 0 \end{bmatrix}$$

The row of zeros indicates dependent equations. Solve the first two equations respectively for x and y in terms of z to obtain

$$x = 6 - \frac{7}{3}z \quad \text{and} \quad y = 1 + \frac{1}{3}z.$$

The solution of the system is

$$\left(6 - \frac{7}{3}z, \ 1 + \frac{1}{3}z, \ z\right).$$

13. $$\begin{bmatrix} 2 & x \\ y & 6 \\ 5 & z \end{bmatrix} = \begin{bmatrix} a & -1 \\ 4 & 6 \\ p & 7 \end{bmatrix}$$

The size of these matrices is 3×2. For matrices to be equal, corresponding elements must be equal, so $a = 2$, $x = -1$, $y = 4$, $p = 5$, and $z = 7$.

15. $$\begin{bmatrix} a+5 & 3b & 6 \\ 4c & 2+d & -3 \\ -1 & 4p & q-1 \end{bmatrix} = \begin{bmatrix} -7 & b+2 & 2k-3 \\ 3 & 2d-1 & 4\ell \\ m & 12 & 8 \end{bmatrix}$$

These are 3×3 square matrices. Since corresponding elements must be equal,

$a + 5 = -7$, so $\quad a = -12$;

$3b = b + 2$, so $\quad b = 1$;

$6 = 2k - 3$, so $\quad k = \frac{9}{2}$;

$4c = 3$, so $\quad c = \frac{3}{4}$;

$2 + d = 2d - 1$, so $d = 3$;

$-3 = 4\ell$, so $\quad \ell = -\frac{3}{4}$;

$\quad m = -1$;

$4p = 12$, so $\quad p = 3$; and

$q - 1 = 8$, so $\quad q = 9$.

17. $2G - 4F = 2\begin{bmatrix} 2 & 5 \\ 1 & 6 \end{bmatrix} - 4\begin{bmatrix} -1 & 4 \\ 3 & 7 \end{bmatrix}$

$= \begin{bmatrix} 4 & 10 \\ 2 & 12 \end{bmatrix} + \begin{bmatrix} 4 & -16 \\ -12 & -28 \end{bmatrix}$

$= \begin{bmatrix} 8 & -6 \\ -10 & -16 \end{bmatrix}$

19.   Since B is a 3×3 matrix, and A is a 3×2 matrix, the calculation of $B - A$ is not possible.

21.   A has size 3×2

and

F has size 2×2,

so AF will have size 3×2.

$$AF = \begin{bmatrix} 4 & 10 \\ -2 & -3 \\ 6 & 9 \end{bmatrix} \begin{bmatrix} -1 & 4 \\ 3 & 7 \end{bmatrix}$$

$$= \begin{bmatrix} 26 & 86 \\ -7 & -29 \\ 21 & 87 \end{bmatrix}$$

23. D has size 3×1

and

E has size 1×3,

so DE will have size 3×3.

$$DE = \begin{bmatrix} 6 \\ 1 \\ 0 \end{bmatrix} \begin{bmatrix} 1 & 3 & -4 \end{bmatrix}$$

$$= \begin{bmatrix} 6 & 18 & -24 \\ 1 & 3 & -4 \\ 0 & 0 & 0 \end{bmatrix}$$

25. B has size 3×3

and

D has size 3×1,

so BD will have size 3×1.

$$BD = \begin{bmatrix} 2 & 3 & -2 \\ 2 & 4 & 0 \\ 0 & 1 & 2 \end{bmatrix} \begin{bmatrix} 6 \\ 1 \\ 0 \end{bmatrix} = \begin{bmatrix} 15 \\ 16 \\ 1 \end{bmatrix}$$

27. $F = \begin{bmatrix} -1 & 4 \\ 3 & 7 \end{bmatrix}$

$$[F \mid I] = \begin{bmatrix} -1 & 4 & 1 & 0 \\ 3 & 7 & 0 & 1 \end{bmatrix}$$

$$3R_1 + R_2 \rightarrow R_2 \begin{bmatrix} -1 & 4 & 1 & 0 \\ 0 & 19 & 3 & 1 \end{bmatrix}$$

$$-19R_1 + 4R_2 \rightarrow R_1 \begin{bmatrix} 19 & 0 & -7 & 4 \\ 0 & 19 & 3 & 1 \end{bmatrix}$$

$$\begin{array}{c} \frac{1}{19}R_1 \rightarrow R_1 \\ \frac{1}{19}R_2 \rightarrow R_2 \end{array} \begin{bmatrix} 1 & 0 & -\frac{7}{19} & \frac{4}{19} \\ 0 & 1 & \frac{3}{19} & \frac{1}{19} \end{bmatrix}$$

$$F^{-1} = \begin{bmatrix} -\frac{7}{19} & \frac{4}{19} \\ \frac{3}{19} & \frac{1}{19} \end{bmatrix}$$

29. A and C are 3×2 matrices, so their sum $A + C$ is a 3×2 matrix. Only square matrices have inverses. Therefore, $(A + C)^{-1}$ does not exist.

31. Let $A = \begin{bmatrix} -4 & 2 \\ 0 & 3 \end{bmatrix}$.

$$[A \mid I] = \begin{bmatrix} -4 & 2 & 1 & 0 \\ 0 & 3 & 0 & 1 \end{bmatrix}$$

$$-3R_1 + 2R_2 \rightarrow R_1 \begin{bmatrix} 12 & 0 & -3 & 2 \\ 0 & 3 & 0 & 1 \end{bmatrix}$$

$$\begin{array}{c} \frac{1}{12}R_1 \rightarrow R_1 \\ \frac{1}{3}R_2 \rightarrow R_2 \end{array} \begin{bmatrix} 1 & 0 & -\frac{1}{4} & \frac{1}{6} \\ 0 & 1 & 0 & \frac{1}{3} \end{bmatrix}$$

The inverse is $\begin{bmatrix} -\frac{1}{4} & \frac{1}{6} \\ 0 & \frac{1}{3} \end{bmatrix}$.

33. Let $A = \begin{bmatrix} 6 & 4 \\ 3 & 2 \end{bmatrix}$.

$$[A \mid I] = \begin{bmatrix} 6 & 4 & 1 & 0 \\ 3 & 2 & 0 & 1 \end{bmatrix}$$

$$R_1 + (-2)R_2 \rightarrow R_2 \begin{bmatrix} 6 & 4 & 1 & 0 \\ 0 & 0 & 1 & -2 \end{bmatrix}$$

The zeros in the second row indicate that the original matrix has no inverse.

35. Let $A = \begin{bmatrix} 2 & 0 & 4 \\ 1 & -1 & 0 \\ 0 & 1 & -2 \end{bmatrix}$.

$$[A \mid I] = \begin{bmatrix} 2 & 0 & 4 & 1 & 0 & 0 \\ 1 & -1 & 0 & 0 & 1 & 0 \\ 0 & 1 & -2 & 0 & 0 & 1 \end{bmatrix}$$

$$-2R_2 + R_1 \rightarrow R_2 \begin{bmatrix} 2 & 0 & 4 & 1 & 0 & 0 \\ 0 & 2 & 4 & 1 & -2 & 0 \\ 0 & 1 & -2 & 0 & 0 & 1 \end{bmatrix}$$

$$-2R_3 + R_2 \rightarrow R_3 \begin{bmatrix} 2 & 0 & 4 & 1 & 0 & 0 \\ 0 & 2 & 4 & 1 & -2 & 0 \\ 0 & 0 & 8 & 1 & -2 & -2 \end{bmatrix}$$

$$\begin{array}{c} -1R_3 + 2R_1 \rightarrow R_1 \\ -1R_3 + 2R_2 \rightarrow R_2 \end{array} \begin{bmatrix} 4 & 0 & 0 & 1 & 2 & 2 \\ 0 & 4 & 0 & 1 & -2 & 2 \\ 0 & 0 & 8 & 1 & -2 & -2 \end{bmatrix}$$

$\frac{1}{4}R_1 \rightarrow R_1$
$\frac{1}{4}R_2 \rightarrow R_2$
$\frac{1}{8}R_3 \rightarrow R_3$
$\begin{bmatrix} 1 & 0 & 0 & | & \frac{1}{4} & \frac{1}{2} & \frac{1}{2} \\ 0 & 1 & 0 & | & \frac{1}{4} & -\frac{1}{2} & \frac{1}{2} \\ 0 & 0 & 1 & | & \frac{1}{8} & -\frac{1}{4} & -\frac{1}{4} \end{bmatrix}$

The inverse is $\begin{bmatrix} \frac{1}{4} & \frac{1}{2} & \frac{1}{2} \\ \frac{1}{4} & -\frac{1}{2} & \frac{1}{2} \\ \frac{1}{8} & -\frac{1}{4} & -\frac{1}{4} \end{bmatrix}$.

37. Let $A = \begin{bmatrix} 2 & 3 & 5 \\ -2 & -3 & 5 \\ 1 & 4 & 2 \end{bmatrix}$.

$[A \mid I] = \begin{bmatrix} 2 & 3 & 5 & | & 1 & 0 & 0 \\ -2 & -3 & -5 & | & 0 & 1 & 0 \\ 1 & 4 & 2 & | & 0 & 0 & 1 \end{bmatrix}$

$\begin{matrix} R_1 + R_2 \rightarrow R_2 \\ -2R_3 + R_1 \rightarrow R_3 \end{matrix} \begin{bmatrix} 2 & 3 & 5 & | & 1 & 0 & 0 \\ 0 & 0 & 0 & | & 1 & 1 & 0 \\ 0 & -5 & 1 & | & 1 & 0 & -2 \end{bmatrix}$

The zeros in the second row to the left of the vertical bar indicate that the original matrix has no inverse.

39. $A = \begin{bmatrix} 1 & 2 \\ 2 & 4 \end{bmatrix}$, $B = \begin{bmatrix} 5 \\ 10 \end{bmatrix}$

Row operations may be used to see that matrix A has no inverse. The matrix equation AX = B may be written as the system of equations

$\begin{aligned} x + 2y &= 5 \quad (1) \\ 2x + 4y &= 10. \quad (2) \end{aligned}$

Use the elimination method to solve this system. Begin by eliminating x in equation (2).

$\begin{aligned} x + 2y &= 5 \quad (1) \\ -2R_1 + R_2 \rightarrow R_2 \qquad 0 &= 0 \quad (3) \end{aligned}$

The true statement in equation (3) indicates that the equations are dependent. Solve equation (1) for x in terms of y.

$$x = -2y + 5$$

The solution of the system is $(-2y + 5, y)$.

41. $A = \begin{bmatrix} 2 & 4 & 0 \\ 1 & -2 & 0 \\ 0 & 0 & 3 \end{bmatrix}$, $B = \begin{bmatrix} 72 \\ -24 \\ 48 \end{bmatrix}$

Use row operations to find the inverse of A, which is

$$A^{-1} = \begin{bmatrix} \frac{1}{4} & \frac{1}{2} & 0 \\ \frac{1}{8} & -\frac{1}{4} & 0 \\ 0 & 0 & \frac{1}{3} \end{bmatrix}.$$

Since $X = A^{-1}B$,

$$X = \begin{bmatrix} \frac{1}{4} & \frac{1}{2} & 0 \\ \frac{1}{8} & -\frac{1}{4} & 0 \\ 0 & 0 & \frac{1}{3} \end{bmatrix} \begin{bmatrix} 72 \\ -24 \\ 48 \end{bmatrix} = \begin{bmatrix} 6 \\ 15 \\ 16 \end{bmatrix}.$$

43. $\begin{aligned} 5x + 10y &= 80 \\ 3x - 2y &= 120 \end{aligned}$

Let $A = \begin{bmatrix} 5 & 10 \\ 3 & -2 \end{bmatrix}$, $X = \begin{bmatrix} x \\ y \end{bmatrix}$, $B = \begin{bmatrix} 80 \\ 120 \end{bmatrix}$.

Use row operations to find the inverse of A, which is

$$A^{-1} = \begin{bmatrix} \frac{1}{20} & \frac{1}{4} \\ \frac{3}{40} & -\frac{1}{8} \end{bmatrix}.$$

Since $X = A^{-1}B$,

$$\begin{bmatrix} x \\ y \end{bmatrix} = \begin{bmatrix} \frac{1}{20} & \frac{1}{4} \\ \frac{3}{40} & -\frac{1}{8} \end{bmatrix} \begin{bmatrix} 80 \\ 120 \end{bmatrix} = \begin{bmatrix} 34 \\ -9 \end{bmatrix}.$$

The solution of the system is $(34, -9)$.

45. $A = \begin{bmatrix} .01 & .05 \\ .04 & .03 \end{bmatrix}$, $D = \begin{bmatrix} 200 \\ 300 \end{bmatrix}$

$I - A = \begin{bmatrix} 1 & 0 \\ 0 & 1 \end{bmatrix} - \begin{bmatrix} .01 & .05 \\ .04 & .03 \end{bmatrix}$

$= \begin{bmatrix} .99 & -.05 \\ -.04 & .97 \end{bmatrix}$

Use row operations to find the inverse of $I - A$, which is

$(I - A)^{-1} \approx \begin{bmatrix} 1.0122 & .0522 \\ .0417 & 1.0331 \end{bmatrix}$.

Since $X = (I - A)^{-1}D$, the production matrix is

$X = \begin{bmatrix} 1.0122 & .0522 \\ .0417 & 1.0331 \end{bmatrix} \begin{bmatrix} 200 \\ 300 \end{bmatrix}$

$= \begin{bmatrix} 218.1 \\ 318.2 \end{bmatrix}$

47. Use a chart to organize the information.

	Standard	Extra Large	Time Available
Hours Cutting	$\frac{1}{4}$	$\frac{1}{3}$	4
Hours Shaping	$\frac{1}{2}$	$\frac{1}{3}$	6

Let x = the number of standard paper clips (in thousands); and

y = the number of extra large paper clips (in thousands).

The given information leads to the system

$\frac{1}{4}x + \frac{1}{3}y = 4$

$\frac{1}{2}x + \frac{1}{3}y = 6.$

Solve this system by any method to get x = 8, y = 6.

The manufacturer can make 8 thousand standard and 6 thousand extra large paper clips.

49. Let x_1 = the number of blankets;

x_2 = the number of rugs; and

x_3 = the number of skirts.

The given information leads to the system

$24x_1 + 30x_2 + 12x_3 = 306$ (1)

$4x_1 + 5x_2 + 3x_3 = 59$ (2)

$15x_1 + 18x_2 + 9x_3 = 201$ (3)

Simplify equations (1) and (3).

$\frac{1}{6}R_1 \rightarrow R_1$ $4x_1 + 5x_2 + 2x_3 = 51$ (4)

$4x_1 + 5x_2 + 3x_3 = 59$ (5)

$\frac{1}{3}R_3 \rightarrow R_3$ $5x_1 + 6x_2 + 3x_3 = 67$ (6)

Solve this system by the Gauss–Jordan method. Write the augmented matrix and use row operations.

$\begin{bmatrix} 4 & 5 & 2 & | & 51 \\ 4 & 5 & 3 & | & 59 \\ 5 & 6 & 3 & | & 67 \end{bmatrix}$

$\begin{matrix} -1R_1 + R_2 \rightarrow R_2 \\ -4R_3 + 5R_1 \rightarrow R_3 \end{matrix} \begin{bmatrix} 4 & 5 & 2 & | & 51 \\ 0 & 0 & 1 & | & 8 \\ 0 & 1 & -2 & | & -13 \end{bmatrix}$

Interchange the second and third rows.

$\begin{bmatrix} 4 & 5 & 2 & | & 51 \\ 0 & 1 & -2 & | & -13 \\ 0 & 0 & 1 & | & 8 \end{bmatrix}$

$-5R_2 + R_1 \rightarrow R_1 \begin{bmatrix} 4 & 0 & 12 & | & 116 \\ 0 & 1 & -2 & | & -13 \\ 0 & 0 & 1 & | & 8 \end{bmatrix}$

$\begin{matrix} -12R_3 + R_1 \rightarrow R_1 \\ R_2 + 2R_3 \rightarrow R_2 \end{matrix} \begin{bmatrix} 4 & 0 & 0 & | & 20 \\ 0 & 1 & 0 & | & 3 \\ 0 & 0 & 1 & | & 8 \end{bmatrix}$

$\frac{1}{4}R_1 \to R_1$ $\begin{bmatrix} 1 & 0 & 0 & | & 5 \\ 0 & 1 & 0 & | & 3 \\ 0 & 0 & 1 & | & 8 \end{bmatrix}$

The solution of the system is $x = 5$, $y = 3$, $z = 8$.
5 blankets, 3 rugs, and 8 skirts can be made.

51. The 4×5 matrix of stock reports is

$\begin{bmatrix} 5 & 7 & 2532 & 52\frac{3}{8} & -\frac{1}{4} \\ 3 & 9 & 1464 & 56 & \frac{1}{8} \\ 2.50 & 5 & 4974 & 41 & -1\frac{1}{2} \\ 1.36 & 10 & 1754 & 18\frac{7}{8} & \frac{1}{2} \end{bmatrix}$.

53. (a) The input–output matrix is

$$A = \begin{bmatrix} 0 & \frac{1}{2} \\ \frac{2}{3} & 0 \end{bmatrix} .$$

(b) $I - A = \begin{bmatrix} 1 & -\frac{1}{2} \\ -\frac{2}{3} & 1 \end{bmatrix}$, $D = \begin{bmatrix} 400 \\ 800 \end{bmatrix}$

Use row operations to find the inverse of $I - A$, which is

$$(I - A)^{-1} = \begin{bmatrix} \frac{3}{2} & \frac{3}{4} \\ 1 & \frac{3}{2} \end{bmatrix} .$$

Since $X = (I - A)^{-1}D$,

$$X = \begin{bmatrix} \frac{3}{2} & \frac{3}{4} \\ 1 & \frac{3}{2} \end{bmatrix} \begin{bmatrix} 400 \\ 800 \end{bmatrix} = \begin{bmatrix} 1200 \\ 1600 \end{bmatrix} .$$

The production required is 1200 units of cheese and 1600 units of goats.

55. $x + 2y + z = 7$ *(1)*
$2x - y - z = 2$ *(2)*
$3x - 3y + 2z = -5$ *(3)*

(a) To solve the system by elimination, begin by eliminating x in equations (2) and (3).

$x + 2y + z = 7$ *(1)*
$-2R_1 + R_2 \to R_2$ $\quad -5y - 3z = -12$ *(4)*
$-3R_1 + R_3 \to R_3$ $\quad -9y - z = -26$ *(5)*

Eliminate y in equation (5).

$x + 2y + z = 7$ *(1)*
$-5y - 3z = -12$ *(4)*
$-9R_2 + 5R_3 \to R_3$ $\quad 22z = -22$ *(6)*

Make each leading coefficient equal 1.

$x + 2y + z = 7$ *(1)*
$-\frac{1}{5}R_2 \to R_2$ $\quad y + \frac{3}{5}z = \frac{12}{5}$ *(7)*
$\frac{1}{22}R_3 \to R_3$ $\quad z = -1$ *(8)*

Substitute -1 for z in equation (7) to get $y = 3$. Substitute -1 for z and 3 for y in equation (1) to get $x = 2$. The solution of the system is $(2, 3, -1)$.

(b) The same system is to be solved using the Gaussian method. Write the augmented matrix and use row operations.

$\begin{bmatrix} 1 & 2 & 1 & | & 7 \\ 2 & -1 & -1 & | & 2 \\ 3 & -3 & 2 & | & -5 \end{bmatrix}$

$-2R_1 + R_2 \to R_2$
$-3R_1 + R_3 \to R_3$ $\begin{bmatrix} 1 & 2 & 1 & | & 7 \\ 0 & -5 & -3 & | & -12 \\ 0 & -9 & -1 & | & -26 \end{bmatrix}$

$-9R_2 + 5R_3 \to R_3$ $\begin{bmatrix} 1 & 2 & 1 & | & 7 \\ 0 & -5 & -3 & | & -12 \\ 0 & 0 & 22 & | & -22 \end{bmatrix}$

$$-\frac{1}{5}R_2 \rightarrow R_2 \quad \frac{1}{22}R_3 \rightarrow R_3 \quad \begin{bmatrix} 1 & 2 & 1 & | & 7 \\ 0 & 1 & \frac{3}{5} & | & \frac{12}{5} \\ 0 & 0 & 1 & | & -1 \end{bmatrix}$$

The corresponding system is

$$x + 2y + z = 7$$
$$y + \frac{3}{5}z = \frac{12}{5}$$
$$z = -1.$$

Use back–substitution to see again that the solution of the system is $(2, 3, -1)$.

(c) The same system is to be solved using the Gauss–Jordan method. Write the augmented matrix and use row operations.

$$\begin{bmatrix} 1 & 2 & 1 & | & 7 \\ 2 & -1 & -1 & | & 2 \\ 3 & -3 & 2 & | & -5 \end{bmatrix}$$

$$\begin{matrix} -2R_1 + R_2 \rightarrow R_2 \\ -3R_1 + R_3 \rightarrow R_3 \end{matrix} \begin{bmatrix} 1 & 2 & 1 & | & 7 \\ 0 & -5 & -3 & | & -12 \\ 0 & -9 & -1 & | & -26 \end{bmatrix}$$

$$\begin{matrix} 5R_1 + 2R_2 \rightarrow R_1 \\ \\ -9R_2 + 5R_3 \rightarrow R_3 \end{matrix} \begin{bmatrix} 5 & 0 & -1 & | & 11 \\ 0 & -5 & -3 & | & -12 \\ 0 & 0 & 22 & | & -22 \end{bmatrix}$$

$$\begin{matrix} 22R_1 + R_3 \rightarrow R_1 \\ 22R_2 + 3R_3 \rightarrow R_2 \end{matrix} \begin{bmatrix} 110 & 0 & 0 & | & 220 \\ 0 & -110 & 0 & | & -330 \\ 0 & 0 & 22 & | & -22 \end{bmatrix}$$

$$\begin{matrix} \frac{1}{110}R_1 \rightarrow R_1 \\ -\frac{1}{110}R_2 \rightarrow R_2 \\ \frac{1}{22}R_3 \rightarrow R_3 \end{matrix} \begin{bmatrix} 1 & 0 & 0 & | & 2 \\ 0 & 1 & 0 & | & 3 \\ 0 & 0 & 1 & | & -1 \end{bmatrix}$$

Once again, the solution of the system is $(2, 3, -1)$.

(d) The system can be written as a matrix equation $AX = B$ by writing

$$\begin{bmatrix} 1 & 2 & 1 \\ 2 & -1 & -1 \\ 3 & -3 & 2 \end{bmatrix} \begin{bmatrix} x \\ y \\ z \end{bmatrix} = \begin{bmatrix} 7 \\ 2 \\ -5 \end{bmatrix}.$$

(e) The inverse of the coefficient matrix A can be found by using row operations.

$$[A \mid I] = \begin{bmatrix} 1 & 2 & 1 & | & 1 & 0 & 0 \\ 2 & -1 & -1 & | & 0 & 1 & 0 \\ 3 & -3 & 2 & | & 0 & 0 & 1 \end{bmatrix}$$

$$\begin{matrix} -2R_1 + R_2 \rightarrow R_2 \\ -3R_1 + R_3 \rightarrow R_3 \end{matrix} \begin{bmatrix} 1 & 2 & 1 & | & 1 & 0 & 0 \\ 0 & -5 & -3 & | & -2 & 1 & 0 \\ 0 & -9 & -1 & | & -3 & 0 & 1 \end{bmatrix}$$

$$\begin{matrix} 5R_1 + 2R_2 \rightarrow R_1 \\ \\ -9R_2 + 5R_3 \rightarrow R_3 \end{matrix} \begin{bmatrix} 5 & 0 & -1 & | & 1 & 2 & 0 \\ 0 & -5 & -3 & | & -2 & 1 & 0 \\ 0 & 0 & 22 & | & 3 & -9 & 5 \end{bmatrix}$$

$$\begin{matrix} 22R_1 + R_3 \rightarrow R_1 \\ 22R_2 + 3R_3 \rightarrow R_2 \end{matrix} \begin{bmatrix} 110 & 0 & 0 & | & 25 & 35 & 5 \\ 0 & -110 & 0 & | & -35 & -5 & 15 \\ 0 & 0 & 22 & | & 3 & -9 & 5 \end{bmatrix}$$

$$\begin{matrix} \frac{1}{110}R_1 \rightarrow R_1 \\ -\frac{1}{110}R_2 \rightarrow R_2 \\ \frac{1}{22}R_3 \rightarrow R_3 \end{matrix} \begin{bmatrix} 1 & 0 & 0 & | & \frac{5}{22} & \frac{7}{22} & \frac{1}{22} \\ 0 & 1 & 0 & | & \frac{7}{22} & \frac{1}{22} & -\frac{3}{22} \\ 0 & 0 & 1 & | & \frac{3}{22} & -\frac{9}{22} & \frac{5}{22} \end{bmatrix}$$

The inverse of matrix A is

$$A^{-1} = \begin{bmatrix} \frac{5}{22} & \frac{7}{22} & \frac{1}{22} \\ \frac{7}{22} & \frac{1}{22} & -\frac{3}{22} \\ \frac{3}{22} & -\frac{9}{22} & \frac{5}{22} \end{bmatrix}$$

$$\approx \begin{bmatrix} .23 & .32 & .05 \\ .32 & .05 & -.14 \\ .14 & -.41 & .23 \end{bmatrix}.$$

(f) Since $X = A^{-1}B$,

$$\begin{bmatrix} x \\ y \\ z \end{bmatrix} = \begin{bmatrix} \frac{5}{22} & \frac{7}{22} & \frac{1}{22} \\ \frac{7}{22} & \frac{1}{22} & -\frac{3}{22} \\ \frac{3}{22} & -\frac{9}{22} & \frac{5}{22} \end{bmatrix} \begin{bmatrix} 7 \\ 2 \\ -5 \end{bmatrix} = \begin{bmatrix} 2 \\ 3 \\ -1 \end{bmatrix}.$$

Once again, the solution of the system is $(2, 3, -1)$.

Extended Application

1. **(a)** $A = \begin{bmatrix} .245 & .102 & .051 \\ .099 & .291 & .279 \\ .433 & .372 & .011 \end{bmatrix}$, $D = \begin{bmatrix} 2.88 \\ 31.45 \\ 30.91 \end{bmatrix}$, $X = \begin{bmatrix} x_1 \\ x_2 \\ x_3 \end{bmatrix}$

(b) $I - A = \begin{bmatrix} 1 & 0 & 0 \\ 0 & 1 & 0 \\ 0 & 0 & 1 \end{bmatrix} - \begin{bmatrix} .245 & .102 & .051 \\ .099 & .291 & .279 \\ .433 & .372 & .011 \end{bmatrix} = \begin{bmatrix} .755 & -.102 & -.051 \\ -.099 & .709 & -.279 \\ -.433 & -.372 & .989 \end{bmatrix}$

(c) $(I - A)^{-1}(I - A) = \begin{bmatrix} 1.454 & .291 & .157 \\ .533 & 1.763 & .525 \\ .837 & .791 & 1.278 \end{bmatrix}\begin{bmatrix} .755 & -.102 & -.051 \\ -.099 & .709 & -.279 \\ -.433 & -.372 & .989 \end{bmatrix}$

$= \begin{bmatrix} 1.00098 & .0004 & .00007 \\ .00055 & 1.0003 & .00017 \\ .00025 & .00003 & 1.00057 \end{bmatrix} \approx \begin{bmatrix} 1 & 0 & 0 \\ 0 & 1 & 0 \\ 0 & 0 & 1 \end{bmatrix}$

(d) $X = (I - A)^{-1}D = \begin{bmatrix} 1.454 & .291 & .157 \\ .533 & 1.763 & .525 \\ .837 & .791 & 1.278 \end{bmatrix}\begin{bmatrix} 2.88 \\ 31.45 \\ 30.91 \end{bmatrix} \approx \begin{bmatrix} 18.2 \\ 73.2 \\ 66.8 \end{bmatrix}$

(Each entry of X has been rounded to three significant digits.)

(e) $18.2 billion of agriculture, $73.2 billion of manufacturing, and $66.8 billion of household would be required to support a demand of $2.88 billion, $31.45 billion, and $30.91 billion, respectively.

2. **(a)** $A = \begin{bmatrix} .293 & 0 & 0 \\ .014 & .207 & .017 \\ .044 & .010 & .216 \end{bmatrix}$, $D = \begin{bmatrix} 138,213 \\ 17,597 \\ 1786 \end{bmatrix}$

(b) $I - A = \begin{bmatrix} 1 & 0 & 0 \\ 0 & 1 & 0 \\ 0 & 0 & 1 \end{bmatrix} - \begin{bmatrix} .293 & 0 & 0 \\ .014 & .207 & .017 \\ .044 & .010 & .216 \end{bmatrix} = \begin{bmatrix} .707 & 0 & 0 \\ -.014 & .793 & -.017 \\ -.044 & -.010 & .784 \end{bmatrix}$

(c) $(I - A)^{-1}(I - A) = \begin{bmatrix} 1.414 & 0 & 0 \\ .027 & 1.261 & .027 \\ .080 & .016 & 1.276 \end{bmatrix}\begin{bmatrix} .707 & 0 & 0 \\ -.014 & .793 & -.017 \\ -.044 & -.010 & .784 \end{bmatrix}$

$= \begin{bmatrix} .9997 & 0 & 0 \\ .0002 & .9997 & -.0003 \\ .0002 & -.00007 & 1.0001 \end{bmatrix} \approx \begin{bmatrix} 1 & 0 & 0 \\ 0 & 1 & 0 \\ 0 & 0 & 1 \end{bmatrix}$

(d) $X = (I - A)^{-1}D = \begin{bmatrix} 1.414 & 0 & 0 \\ .027 & 1.261 & .027 \\ .080 & .016 & 1.276 \end{bmatrix}\begin{bmatrix} 138,213 \\ 17,597 \\ 1786 \end{bmatrix} \approx \begin{bmatrix} 195,000 \\ 26,000 \\ 13,600 \end{bmatrix}$

Since the entries in D were in thousands of pounds, this means that 195 million pounds of agricultural products, 26 million pounds of manufactured goods, and 13.6 million pounds of energy are required.

CHAPTER 2   TEST

1. Solve the system below using the echelon method.

$$3x + y = 11$$
$$x - 2y = -8$$

2. Solve using the Gauss–Jordan method.

$$x + 2y + 3z = 5$$
$$2x - y + z = 5$$
$$x + y + z = 2$$

3. Find the values of the variables.

$$\begin{bmatrix} 6 - x & 2y + 1 \\ 3m & 5p + 2 \end{bmatrix} = \begin{bmatrix} 8 & 10 \\ -5 & 3p - 1 \end{bmatrix}$$

4. Given the matrices below, perform the indicated operations, if possible.

$$A = \begin{bmatrix} 1 & 2 & -1 \\ 0 & 1 & 1 \\ 1 & 0 & 1 \end{bmatrix} \qquad B = \begin{bmatrix} 1 & -2 \\ 1 & 1 \\ 0 & 1 \end{bmatrix} \qquad C = \begin{bmatrix} 2 & 1 & 3 \\ 0 & 4 & 1 \\ 1 & 1 & 1 \end{bmatrix}$$

   (a) AB   (b) 2A − C   (c) A + 2B

5. Find the inverse of each matrix which has an inverse.

$$A = \begin{bmatrix} 1 & 0 & -1 \\ 2 & 1 & 1 \\ 1 & 1 & 5 \end{bmatrix} \qquad B = \begin{bmatrix} 2 & -1 & 1 \\ 0 & 2 & 4 \\ 2 & 1 & 5 \end{bmatrix}$$

6. For $A = \begin{bmatrix} 1 & 0 & 1 \\ 1 & 1 & 1 \\ 2 & 1 & 3 \end{bmatrix}$, $A^{-1} = \begin{bmatrix} 2 & 1 & -1 \\ -1 & 1 & 0 \\ -1 & -1 & 1 \end{bmatrix}$.

   Use this inverse to solve the equation AX = B, where $B = \begin{bmatrix} 1 \\ 2 \\ -1 \end{bmatrix}$.

7. Solve the system below using the inverse of the coefficient matrix.

$$x - 2y = 3$$
$$x + 3y = 5$$

8. Find the production matrix, given the following input–output and demand matrices.

$$A = \begin{bmatrix} .2 & .1 & .3 \\ .1 & 0 & .2 \\ 0 & 0 & .4 \end{bmatrix} \qquad D = \begin{bmatrix} 1000 \\ 2000 \\ 5000 \end{bmatrix}$$

CHAPTER 2   TEST ANSWERS

1. (2, 5)

2. (1, -1, 2)

3. $x = -2$, $y = \dfrac{9}{2}$, $m = -\dfrac{5}{3}$, $p = -\dfrac{3}{2}$

4. (a) $\begin{bmatrix} 3 & -1 \\ 1 & 2 \\ 1 & -1 \end{bmatrix}$ (b) $\begin{bmatrix} 0 & 3 & -5 \\ 0 & -2 & 1 \\ 1 & -1 & 1 \end{bmatrix}$ (c) Not possible

5. $A^{-1} = \begin{bmatrix} \dfrac{4}{3} & -\dfrac{1}{3} & \dfrac{1}{3} \\ -3 & 2 & -1 \\ \dfrac{1}{3} & -\dfrac{1}{3} & \dfrac{1}{3} \end{bmatrix}$; B^{-1} does not exist.

6. $X = \begin{bmatrix} 5 \\ 1 \\ -4 \end{bmatrix}$

7. $\left(\dfrac{19}{5}, \dfrac{2}{5} \right)$

8. $\begin{bmatrix} 4895 \\ 4156 \\ 8333 \end{bmatrix}$

CHAPTER 3   LINEAR PROGRAMMING: THE GRAPHICAL METHOD

Section 3.1

For Exercises 1–39, see the answer graphs in the back of the textbook.

1. $x + y \leq 2$

First graph the boundary line $x + y = 2$ using the points $(2, 0)$ and $(0, 2)$. Since the points on this line satisfy $x + y \leq 2$, draw a solid line. To find the correct region to shade, choose any point not on the line. If $(0, 0)$ is used as the test point, we have

$$x + y \leq 2$$
$$0 + 0 \leq 2$$
$$0 \leq 2,$$

which is a true sttement. Shade the side of the line which includes $(0, 0)$, which is the region below the line.

3. $x \geq 3 + y$

First graph the boundary line $x = 3 + y$ using the points $(0, -3)$ and $(3, 0)$. This will be a solid line. Choose $(0, 0)$ as a test point, and we have

$$x \geq 3 + y$$
$$0 \geq 3 + 0$$
$$0 \geq 3,$$

which is a false statement.

Shade the side which does not include $(0, 0)$, which is the region below the line.

5. $4x - y < 6$

Graph $4x - y = 6$ as a dashed line, since the points on the line are not part of the solution; the line passes through the points $(0, -6)$ and $\left(\frac{3}{2}, 0\right)$. Using the test point $(0, 0)$, we get $0 - 0 < 6$ or $0 < 6$, a true statement. Shade the side of the boundary line that includes the origin $(0, 0)$, which is the region above the line.

7. $3x + y < 6$

Graph $3x + y = 6$ as a dashed line through $(2, 0)$ and $(0, 6)$. Using the test point $(0, 0)$, we get $3(0) + 0 < 6$ or $0 < 6$, a true statement. Shade the side that contains the origin, which is the region below the line.

9. $x + 3y \geq -2$

The graph includes the line $x + 3y = -2$, whose intercepts are the points $(0, -2/3)$ and $(-2, 0)$. Graph $x + 3y = -2$ as a solid line and use the origin as a test point. Since $0 + 3(0) \geq -2$ is true, shade the side which includes the origin, which is the region above the line.

11. $4x + 3y > -3$

Graph $4x + 3y = -3$ as a dashed line through the points $(0, -1)$ and $(-3/4, 0)$. Use the origin as a test point. Since $4(0) + 3(0) > -3$ is true, the origin will be included in the region, so we shade the half-plane above the line.

13. $2x - 4y < 3$

Graph $2x - 4y = 3$ as a dashed line through the points $(3/2, 0)$ and $(0, -3/4)$. Use the origin as a test point. $2(0) - 4(0) < 3$ is true, so the region above the line, which includes the origin, is the correct half-plane to shade.

15. $x \leq 5y$

Graph $x = 5y$ as a solid line through the points $(0, 0)$ and $(5, 1)$. Since this line contains the origin, some point other than $(0, 0)$ must be used as a test point. If we use the point $(1, 2)$, we obtain $1 \leq 5(2)$ or $1 \leq 10$, a true statement. Shade the side of the line containing $(1, 2)$, that is, the region above the line.

17. $-3x < y$

Graph $y = -3x$ as a dashed line through the points $(0, 0)$ and $(1, -3)$. Since this line contains the origin, use some point other than $(0, 0)$ as a test point. If $(1, 1)$ is used as a test point, we obtain $-3 < 3$, a true statement.

Shade the region containing $(1, 1)$, which is the half-plane above the line.

19. $x + y \leq 0$

Graph $x + y = 0$ as a solid line through the points $(0, 0)$ and $(1, -1)$. This line contains $(0, 0)$. If we use $(-1, 0)$ as a test point, we obtain $-1 + 0 \leq 0$ or $-1 \leq 0$, a true statement. Shade the region containing $(-1, 0)$, which is the half-plane below the line.

21. $y < x$

Graph $y = x$ as a dashed line through the points $(0, 0)$ and $(1, 1)$. Since this line contains the origin, choose a point other than $(0, 0)$ as a test point. If we use $(2, 3)$, we obtain $3 < 2$, which is false. Shade the region that does not contain the test point, which is the half-plane below the line.

23. $x < 4$

Graph $x = 4$ as a dashed line. This is the vertical line crossing the x-axis at the point $(4, 0)$. Using $(0, 0)$ as a test point, we obtain $0 < 4$, which is true. Shade the region containing the origin, which is the half-plane to the left of the line.

25. $y \le -2$

Graph $y = -2$ as a solid horizontal line through the point $(0, -2)$. Using the origin as a test point, we obtain $0 \le -2$, which is false. Shade the region which does not contain the origin, which is the half-plane below the line.

27. $x + y \le 1$
$x - y \ge 2$

Graph each inequality separately, using solid boundary lines. In both cases, the graph is the region below the line. Shade the overlapping part of these two half-planes, which is the region below both lines. The shaded region is the feasible region for this system.

29. $2x - y < 1$
$3x + y < 6$

Graph $2x - y < 1$ as the half-plane above the dashed line $2x - y = 1$. Graph $3x + y < 6$ as the half-plane below the dashed line $3x + y = 6$. Shade the overlapping part of these two half-planes to show the feasible region for this system.

31. $-x - y < 5$
$2x - y < 4$

Graph $-x - y < 5$ as the half-plane above the dashed line $-x - y = 5$.

Graph $2x - y < 4$ as the half-plane above the dashed line $2x - y = 4$. Shade the overlapping part of these two half-planes to show the feasible region.

33. $x + y \le 4$
$x - y \le 5$
$4x + y \le -4$

The graph of $x + y \le 4$ consists of the solid line $x + y = 4$ and all the points below it. The graph of $x - y \le 5$ consists of the solid line $x - y = 5$ and all the points above it. The graph of $4x + y \le -4$ consists of the solid line $4x + y = -4$ and all the points below it. The feasible region is the overlapping part of these three half-planes.

35. $-2 < x < 3$
$-1 \le y \le 5$
$2x + y < 6$

The graph of $-2 < x < 3$ is the region between the vertical lines $x = -2$ and $x = 3$, but not including the lines themselves (so the two vertical boundaries are drawn as dashed lines). The graph of $-1 \le y \le 5$ is the region between the horizontal lines $y = -1$ and $y = 5$, including the lines (so the two horizontal boundaries are drawn as solid lines). The graph of

2x + y < 6 is the region below the line 2x + y = 6 (so the boundary is drawn as a dashed line). Shade the region common to all three graphs to show the feasible region.

37.   2y + x ≥ -5
          y ≤ 3 + x
          x ≥ 0
          y ≥ 0

The graph of 2y + x ≥ -5 contains of the boundary line 2y + x = 5 and the region above it. The graph of y ≤ 3 + x consists of the boundary line y = 3 + x and the region below it. The inequalities x ≥ 0 and y ≤ 0 restrict the feasible region to the first quadrant. Shade the region in the first quadrant where the first two graphs overlap to show the feasible region.

39.   3x + 4y > 12
         2x - 3y < 6
          0 ≤ y ≤ 2
              x ≥ 0

3x + 4y > 12 is the set of points above the dashed line 3x + 4y = 12; 2x - 3y < 6 is the set of points above the dashed line 2x - 3y = 6; 0 ≤ y ≤ 2 is the rectangular strip of points lying on or between the horizontal lines y = 0 and y = 2; and x ≥ 0 consists of all the points on and to the right of the y-axis. Shade the feasible region, which is the triangular region satisfying all of the inequalities.

41.   (a)

	Number Made	Time on Wheel	Time in Kiln
Glazed	x	1/2	1
Unglazed	y	1	6
Maximum Time Available		8	20

(b) On the wheel, x glazed pots require $\frac{1}{2} \cdot x = \frac{1}{2}x$ hr and y unglazed pots require $1 \cdot y = y$ hr. Since the wheel is available for at most 8 hr per day,

$$\frac{1}{2}x + y \le 8.$$

In the kiln, x glazed pots require $1 \cdot x = x$ hr and y unglazed pots require $6 \cdot y = 6y$ hr. Since the kiln is available for at most 20 hr per day,

$$x + 6y \le 20.$$

Since it is not possible to produce a negative number of pots,

$$x \ge 0 \text{ and } y \ge 0.$$

Thus, we have the system

$$\frac{1}{2}x + y \le 8$$
$$x + 6y \le 20$$
$$x \ge 0$$
$$y \ge 0.$$

See the graph of the feasible region in the back of the textbook.

(c) Yes, 5 glazed and 2 unglazed planters can be made, since the point (5, 2) lies within the feasible region.

From the graph, it looks like the point (10, 2) might lie right on a boundary of the feasible region. However, (10, 2) does not satisfy the inequality $x + 6y \le 20$, so the point is definitely outside the feasible region. Therefore, no, 10 glazed and 2 unglazed planters cannot be made.

43. (a) $x \ge 3000$
            $y \ge 5000$
        $x + y \le 10,000$

(b) The first inequality gives the half-plane to the right of the vertical line $x = 3000$, including the points on the line. The second inequality gives the half-plane above the horizontal line $y = 5000$, including the points on the line. The third inequality gives the half-plane below the line $x + y \le 10,000$, including the points on the line. Shade the region where the three half-planes overlap to show the feasible region. (See the graph of the feasible region in the back of the textbook.)

45. (a) $x \ge 1000$
            $y \ge 800$
        $x + y \le 2400$

(b) The first inequality gives the set of points on and to the right of the vertical line $x = 1000$. The second inequality gives the set of points on and above the horizontal line $y = 800$. The third inequality gives the set of points on and below the line $x + y = 2400$. Shade the region where the three graphs overlap to show the feasible region. (See the graph of the feasible region in the back of the textbook.)

Section 3.2

1. Make a table indicating the value of the objective function $z = 3x + 5y$ at each corner point.

Corner Point	Value of $z = 3x + 5y$
(1, 1)	$3(1) + 5(1) = 8$ Minimum
(2, 7)	$3(2) + 5(7) = 41$
(5, 10)	$3(5) + 5(10) = 65$ Maximum
(6, 3)	$3(6) + 5(3) = 33$

The maximum value of 65 occurs at (5, 10). The minimum value of 8 occurs at (1, 1).

3.

Corner Point	Value of z = .40x + .75y
(0, 0)	.40(0) + .75(0) = 0
(0, 12)	.40(0) + .75(12) = 9
(4, 8)	.40(4) + .75(8) = 7.6
(7, 3)	.40(7) + .75(3) = 5.05
(8, 0)	.40(8) + .75(0) = 3.2

The maximum value is 9 at (0, 12);
the minimum value is 0 at (0, 0).

5. **(a)**

Corner Point	Value of z = 4x + 2y
(0, 8)	4(0) + 2(8) = 16
(3, 4)	4(3) + 2(4) = 20
(13/2, 2)	4(13/2) + 2(2) = 30
(12, 0)	4(12) + 2(0) = 48

The minimum value is 16 at (0, 8).
Since the feasible region is unbounded, there is no maximum value.

(b)

Corner Point	Value of z = 2x + 3y
(0, 8)	2(0) + 3(8) = 24
(3, 4)	2(3) + 3(4) = 18
(13/2, 2)	2(13/2) + 3(2) = 19
(12, 0)	2(12) + 3(0) = 24

The minimum value is 18 at (3, 4);
there is no maximum value.

(c)

Corner Point	Value of z = 2x + 4y
(0, 8)	2(0) + 4(8) = 32
(3, 4)	2(3) + 4(4) = 22
(13/2, 2)	2(13/2) + 4(2) = 21
(12, 0)	2(12) + 4(0) = 24

The minimum value is 21 at
(13/2, 2); there is no maximum
value.

(d)

Corner Point	Value of z = x + 4y
(0, 8)	0 + 4(8) = 32
(3, 4)	3 + 4(4) = 19
(13/2, 2)	13/2 + 4(2) = 29/2
(12, 0)	12 + 4(0) = 12

The minimum value is 12 at (12, 0);
there is no maximum value.

7. Maximize z = 5x + 2y

subject to: 2x + 3y ≤ 6
              4x + y ≥ 6
                   x ≥ 0
                   y ≥ 0.

Sketch the feasible region.

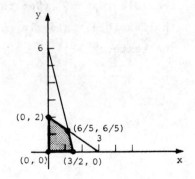

The graph shows that the feasible region is bounded.
The corner points are (0, 0), (0, 2), (3/2, 0), and (6/5, 6/5), which is the intersection of 2x + 3y = 6 and 4x + y = 6. Use the corner points to find the maximum value of the objective function.

Corner Point	Value of z = 5x + 2y
(0, 0)	$5(0) + 2(0) = 0$
(0, 2)	$5(0) + 2(2) = 4$
$\left(\frac{6}{5}, \frac{6}{5}\right)$	$5\left(\frac{6}{5}\right) + 2\left(\frac{6}{5}\right) = \frac{42}{5}$
$\left(\frac{3}{2}, 0\right)$	$5\left(\frac{3}{2}\right) + 2(0) = \frac{15}{2}$

The maximum value of z = 5x + 2y is 42/5 at the corner point (6/5, 6/5).

9. Maximize z = 2x + 2y

subject to: 3x − y ≤ 12

x + y ≤ 15

x ≥ 2

y ≥ 5.

Sketch the feasible region.

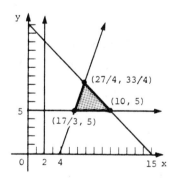

The graph shows that the feasible region is bounded. The corner points are: (17/3, 5), which is the intersection of y = 5 and 3x − y = 12; (27/4, 33/4), which is the intersection of 3x − y = 12 and x + y = 15; and (10, 5), which is the intersection of y = 5 and x + y = 15.
Use the corner points to find the maximum value of the objective function.

Corner Point	Value of z = 2x + 2y
$\left(\frac{17}{3}, 5\right)$	$2\left(\frac{17}{3}\right) + 2(5) = \frac{64}{3}$
$\left(\frac{27}{4}, \frac{33}{4}\right)$	$2\left(\frac{27}{4}\right) + 2\left(\frac{33}{4}\right) = 30$
(10, 5)	$2(10) + 2(5) = 30$

The maximum value of z = 2x + 2y is 30 at (27/4, 33/4) and it is also 30 at (10, 5), as well as at all points on the line segment joining (27/4, 33/4) and (10, 5).

11. Maximize z = 4x + 2y

subject to: x − y ≤ 10

5x + 3y ≤ 75

x + y ≤ 20

x ≥ 0

y ≥ 0.

Sketch the feasible region.

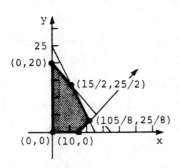

The region is bounded, with corner points (0, 0), (0, 20), (15/2, 25/2), (105/8, 25/8), and (10, 0).

Corner Point	Value of z = 4x + 2y
(0, 0)	$4(0) + 2(0) = 0$
(0, 20)	$4(0) + 2(20) = 40$
$\left(\frac{15}{2}, \frac{25}{2}\right)$	$4\left(\frac{15}{2}\right) + 2\left(\frac{25}{2}\right) = 55$
$\left(\frac{105}{8}, \frac{25}{8}\right)$	$4\left(\frac{105}{8}\right) + 2\left(\frac{25}{8}\right) = \frac{235}{4}$
(10, 0)	$4(10) + 2(0) = 40$

The maximum value of $z = 4x + 2y$ is 235/4 at (105/8, 25/8).

13. **(a)** $x + y \leq 20$
        $x + 3y \leq 24$

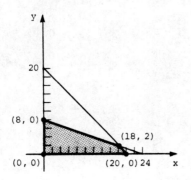

Corner Point	Value of z = 10x + 12y
(0, 0)	$10(0) + 12(0) = 0$
(0, 8)	$10(0) + 12(8) = 96$
(18, 2)	$10(18) + 12(2) = 204$
(20, 0)	$10(20) + 12(0) = 200$

The maximum value of 204 occurs 0 when x = 18 and y = 2.

(b) $3x + y \leq 15$
      $x + 2y \leq 18$

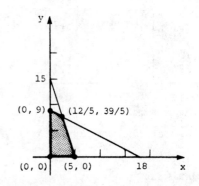

Corner Point	Value of z = 10x + 12y
(0, 0)	$10(0) + 12(0) = 0$
(0, 9)	$10(0) + 12(9) = 108$
$\left(\frac{12}{5}, \frac{39}{5}\right)$	$10\left(\frac{12}{5}\right) + 12\left(\frac{39}{5}\right) = \frac{588}{5}$
(5, 0)	$10(5) + 12(0) = 50$

The maximum value of 588/5 occurs when x = 12/5 and y = 39/5.

(c) $2x + 5y \geq 22$
      $4x + 3y \leq 28$
      $2x + 2y \leq 17$

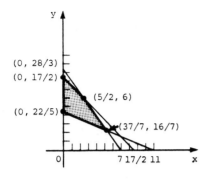

Corner Point	Value of z = 10x + 12y
$\left(0, \frac{22}{5}\right)$	$10(0) + 12\left(\frac{22}{5}\right) = \frac{264}{5}$
$\left(0, \frac{17}{2}\right)$	$10(0) + 12\left(\frac{17}{2}\right) = 102$
$\left(\frac{5}{2}, 6\right)$	$10\left(\frac{5}{2}\right) + 12(6) = 97$
$\left(\frac{37}{7}, \frac{16}{7}\right)$	$10\left(\frac{37}{7}\right) + 12\left(\frac{16}{7}\right) = \frac{562}{7}$

The minimum value of 102 occurs when $x = 0$ and $y = 17/2$.

15. Maximize $z = c_1 x_1 + c_2 x_2$

Subject to: $2x_1 + x_2 \le 11$

$-x_1 + 2x_2 \le 2$

$x_1 \ge 0,\ x_2 \ge 0$.

Sketch the feasible region.

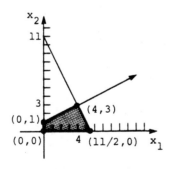

The region is bounded, with corner points $(0, 0)$ $(0, 1)$, $(4, 3)$, and $(11/2, 0)$.

Corner Point	Value of $z = c_1 x_1 + c_2 x_2$
$(0, 0)$	$c_1(0) + c_2(0) = 0$
$(0, 1)$	$c_1(0) + c_2(1) = c_2$
$(4, 3)$	$c_1(4) + c_2(3) = 4c_1 + 3c_2$
$\left(\frac{11}{2}, 0\right)$	$c_1\left(\frac{11}{2}\right) + c_2(0) = \frac{11}{2}c_1$

If we are to have $(x_1, x_2) = (4, 3)$ as an optimal solution, then it must be true that both $4c_1 + 3c_2 \ge c_2$ and $4c_1 + 3c_2 \ge \frac{11}{2}c_1$, because the value of z at $(4, 3)$ cannot be smaller than the other values of z in the table. Manipulate the symbols in these two inequalities in order to isolate $\frac{c_1}{c_2}$ in each; keep in mind the given information that $c_2 > 0$ when performing division by c_2. First,

$$4c_1 + 3c_2 \ge c_2$$

$$4c_1 \ge -2c_2$$

$$\frac{4c_1}{4c_2} \ge \frac{-2c_2}{4c_2}$$

$$\frac{c_1}{c_2} \ge -\frac{1}{2}.$$

Then,

$$4c_1 + 3c_2 \ge \frac{11}{2}c_1$$

$$-\frac{3}{2}c_1 + 3c_2 \ge 0$$

$$3c_1 - 6c_2 \le 0$$

$$3c_1 \le 6c_2$$

$$\frac{3c_1}{3c_2} \le \frac{6c_2}{3c_2}$$

$$\frac{c_1}{c_2} \le 2.$$

Since $\frac{c_1}{c_2} \geq -\frac{1}{2}$ and $\frac{c_1}{c_2} \leq 2$, the desired range for c_1/c_2 is $[-1/2, 2]$, which corresponds to choice (b).

Section 3.3

1. Let x represent the number of product A made, and y represent the number of product B. Each item of A uses 2 hr on the machine, so 2x represents the total hours required for x items of product A. Similarly, 3y represents the total hours used for product B. There are only 45 hr available, so

$$2x + 3y \leq 45.$$

3. Let x represent the number of green pills and y represent the number of red pills. Then 4x represents the number of vitamin units provided by the green pills, and y represents the vitamin units provided by the red ones. Since at least 25 units are needed per day,

$$4x + y \geq 25.$$

5. Let x represent the number of pounds of \$6 coffee and y represent the number of pounds of \$5 coffee. Since the mixture must weigh at least 50 lb,

$$x + y \geq 50.$$

(Notice that the price per pound is not used in setting up this inequality.)

For Exercises 7–21, the graphs of the feasible regions appear in the answer section of the textbook.

7. Let x represent the number of engines sent to plant I and y represent the number of engines sent to plant II.

    Minimize z = 20x + 35y

    subject to: x ≥ 50

    y ≥ 27

    x + y ≤ 85

    x ≥ 0

    y ≥ 0.

The corner points are found by solving the following systems of equations.

The solution of the system

$$x = 50$$
$$y = 27$$

is (50, 27).

The solution of the system

$$x = 50$$
$$x + y = 85$$

is (50, 35).

The solution of the system

$$y = 27$$
$$x + y = 85$$

is (58, 27).

The feasible region is the triangular region shown in the graph in the back of the textbook. The corner points are (50, 27), (50, 35), and (58, 27).

Use the corner points to find the
minimum value of the objective
function.

Corner Point	Value of z = 20z + 35y
(50, 27)	20(50) + 35(27) = 1945
(50, 35)	20(50) + 35(35) = 2225
(58, 27)	20(58) + 35(27) = 2105

The minimum value of z occurs at the
corner point (50, 27).
50 engines should be shipped to
plant I and 27 engines should be
shipped to plant II. The minimum
cost is $1945.

9. (a) Let x represent the number of
units of policy A to purchase and y
represent the number of units of
policy B to purchase.
Minimize z = 50x + 40y
subject to: 10x + 15y ≥ 100
80x + 120y ≥ 1000
x ≥ 0
y ≥ 0.

In the first quadrant, the graph of
80x + 120y = 100 lies entirely above
the graph of 10x + 15y = 100, so the
corner points are the x- and y-
intercepts of 80x + 120y = 1000.
These are (0, 25/3) and (25/2, 0).
Use the corner points to find the
minimum value of the objective
function.

Corner Point	Value of z = 50x + 40y
(0, 25/3)	1000/3 = 333 1/3 Minimum
(25/2, 0)	625

In order to minimize the premium
costs, 0 units of policy A and 25/3
(or 8 1/3) units of policy B should
be purchased for a minimum premium
cost of $333.33.

(b) The objective function to be
minimized is altered to
z = 25x + 40y, but the corner points
of the feasible region remain the
same since none of the constraints
are altered.

Corner Point	Value of z = 25x + 40y
(0, 25/3)	1000/3 = 333 1/3
(25/2, 0)	625/2 = 312 1/2 Minimum

In order to minimize the altered
premium costs, 25/2 (or 12 1/2)
units of policy A and 0 units of
policy B should be purchased for
a minimum premium cost of $312.50.

11. Let x represent the number of type 1
bolts and y represent the number of
type 2 bolts.
Maximize z = .10x + .12y
subject to: .1x + .1y ≤ 240
.1x + .4y ≤ 720
.1x + .5y ≤ 160
x ≥ 0
y ≥ 0.

The first two inequalities do not affect the solution. The feasible region is the triangular region bounded by the x- and y-axes and the line .1x + .5y = 160. The corner points are (0, 0), (0, 320), and (1600, 0).

Use the corner points to find the maximum value of the objective function.

Corner Point	Value of z = .10x + .12y
(0, 0)	0
(0, 320)	38.40
(1600, 0)	160.00 Maximum

1600 type 1 bolts and 0 type 2 bolts should be manufactured for a maximum revenue of $160 per day.

13. **(a)** Let x represent the number of kilograms of the half-and-half mix and y represent the number of kilograms of the other mix. The problem requests maximum revenue, so we need a revenue function: z = 6x + 4.8y. The constraints are on the available nuts and raisins that make up the mixes.

Maximize z = 6x + 4.8y

subject to: $\frac{1}{2}x + \frac{1}{3}y \le 100$

$\frac{1}{2}x + \frac{2}{3}y \le 125$

$x \ge 0$

$y \ge 0.$

Three of the corner points are (0, 0), (0, 187.5), (200, 0). The fourth corner point is the point of intersection of

$\frac{1}{2}x + \frac{1}{3}y = 100$ and $\frac{1}{2}x + \frac{2}{3}y = 125,$

which is (150, 75).

Corner Point	Value of z = 6x + 4.8y
(0, 0)	0
(0, 187.5)	900
(150, 75)	1260 Maximum
(200, 0)	1200

The company should prepare 150 kg of the half-and-half mix and 75 kg of the other mix for a maximum revenue of $1260.

(b) The objective function to be maximized is altered to

z = 8x + 4.8y,

but the corner points remain the same.

Corner Point	Value of z = 8x + 4.8y
(0, 0)	0
(0, 187.5)	900
(150, 75)	1560
(200, 0)	1600 Maximum

In order to maximize the revenue under the altered conditions, the company should prepare 200 kg of the half-and-half mix and 0 kg of the other mix for a maximum revenue of $1600.

15. Let x represent the number of gal-
    lons of milk from Dairy II and y
    represent the number of gallons of
    milk from Dairy I.

    Maximize z = .032x + .037y

    subject to: x ≤ 80

                 y ≤ 50

             x + y ≤ 100

                 x ≥ 0

                 y ≥ 0.

    The corner points are (0, 0),
    (80, 0), (80, 20), (50, 50),
    and (0, 50).

Corner Point	Value of z = .032x + .037y
(0, 0)	0
(80, 0)	2.56
(80, 20)	3.30
(50, 50)	3.45 Maximum
(0, 50)	1.85

    50 gal from Dairy I and 50 gal from
    Dairy II should be used to get milk
    with a maximum butterfat of 3.45
    gal.

17. Let x represent the amount invested
    in bonds and y represent the amount
    invested in mutual funds (in mil-
    lions of dollars).

    The amount of annual interest is

            .12x + .08y.

    Maximize z = .12x + .08y

    subject to: x ≥ 20

                 y ≥ 15

             x + y ≤ 40.

The corner points are (20, 15),
(20, 20), and (25, 15).

Corner Point	Value of z = .12x + .08y
(20, 15)	3.6
(20, 20)	4.0
(25, 15)	4.2 Maximum

He should invest $25 million in
bonds and $15 million in mutual
funds for maximum annual interest
of $4.20 million.

19. Let x represent the number of
    species I prey and y represent
    the number of species II prey.

		Protein	Fat
Species	I	5	2
	II	3	4

Minimize z = 2x + 3y

subject to: 5x + 3y ≥ 10

             2x + 4y ≥ 8

                  x ≥ 0

                  y ≥ 0.

The corner points are (0, 10/3),
(4, 0), and the intersection of
5x + 3y = 10 and 2x + 4y = 8,
which is (8/7, 10/7).

Corner Point	Value of z = 2x + 3y
(0, 10/3)	10
(4, 0)	8
(8/7, 10/7)	46/7 Minimum

The minimum value of z is $46/7 \approx 6.57$. $8/7$ units of species I and $10/7$ units of species II will meet the daily food requirements with the least expenditure of energy. However, a predator prob-ably can catch and digest only whole numbers of prey. This problem shows that it is important to consider whether a model produces a real-istic answer to a problem.

21. Let x represent the number of serv-ings of product A and y represent the number of servings of product B. The cost function (in dollars) is $z = .25x + .40y$.

Minimize $z = .25x + .40y$
subject to: $3x + 2y \geq 15$
              $2x + 4y \geq 15$
                     $x \geq 0$
                     $y \geq 0.$

The corner points are $(0, 15/2)$, $(15/4, 15/8)$, and $(15/2, 0)$.

Corner Point	Value of z = .25x + .40y
(0, 15/2)	3
(15/4, 15/8)	1.6875 Minimum
(15/2, 0)	1.875

$15/4$ (or $3\ 3/4$) servings of A and $15/8$ (or $1\ 7/8$) servings of B will satisfy the requirements at a mini-mum cost of $1.69 (rounded to the nearest cent).

25. Beta is limited to 400 units per day, so Beta ≤ 400. The correct answer is choice (a).

Chapter 3   Review Exercises

For Exercises 3–13, see the answer graphs in the back of the textbook.

3. $y \geq 2x + 3$

Graph $y = 2x + 3$ as a solid line, using the intercepts $(0, 3)$ and $(-3/2, 0)$. Using the origin as a test point, we get $0 \geq 2(0) + 3$ or $0 \geq 3$, which is false. Shade the region that does not contain the origin, that is, the half-plane above the line.

5. $3x + 4y \leq 12$

Graph $3x + 4y = 12$ as a solid line, using the intercepts $(0, 3)$ and $(4, 0)$. Using the origin as a test point, we get $0 \leq 12$, which is true. Shade the region that contains the origin, that is, the half-plane below the line.

7. $y \geq x$

Graph $y = x$ as a solid line. Since this line contains the origin, choose a point other than $(0, 0)$ as a test point. If we use $(1, 4)$, we get $4 \geq 1$, which is true. Shade the region that contains the test point, that is, the half-plane above the line.

9. $x + y \leq 6$
      $2x - y \geq 3$

$x + y \leq 6$ is the half-plane on or
below the line $x + y = 6$; $2x - y \geq 3$
is the half-plane on or below the
line $2x - y = 3$. Shade the overlapping part of these two half-
planes, which is the region below
both lines. The only corner point
is the intersection of the two
boundary lines, the point $(3, 3)$.

11. $-4 \leq x \leq 2$
      $-1 \leq y \leq 3$
      $x + y \leq 4$

$-4 \leq x \leq 2$ is the rectangular region
lying on or between the two vertical
lines, $x = -4$ and $x = 2$; $-1 \leq y \leq 3$
is the rectangular region lying on
or between the two horizontal lines,
$y = -1$ and $y = 3$; $x + y \leq 4$ is the
half-plane lying on or below the
line $x + y = 4$. Shade the overlapping part of these three regions.
The corner points are $(-4, -1)$,
$(-4, 3)$, $(1, 3)$, $(2, 2)$, and
$(2, -1)$.

13. $x + 3y \geq 6$
      $4x - 3y \leq 12$
      $x \geq 0$
      $y \geq 0$

$x + 3y \geq 6$ is the half-plane on or
above the line $x + 3y = 6$;
$4x - 3y \leq 12$ is the half-plane on or
above the line $4x - 3y = 12$; $x \geq 0$
and $y \geq 0$ together restrict the
graph to the first quadrant. Shade

the portion of the first quadrant
where the half-planes overlap.
The corner points are $(0, 2)$ and
$(18/5, 4/5)$.

15.

Corner Point	Value of $z = 2x + 4y$
$(1, 6)$	$2(1) + 4(6) = 26$
$(6, 7)$	$2(6) + 4(7) = 40$ Maximum
$(7, 3)$	$2(7) + 4(3) = 26$
$\left(1, 2\frac{1}{2}\right)$	$2(1) + 4\left(2\frac{1}{2}\right) = 12$
$(2, 1)$	$2(2) + 4(1) = 8$ Minimum

The maximum value of 40 occurs at
$(6, 7)$, and the minimum value of 8
occurs at $(2, 1)$.

17. Maximize $z = 2x + 4y$

subject to: $3x + 2y \leq 12$
             $5x + y \geq 5$
             $x \geq 0$
             $y \geq 0$.

Sketch the feasible region.

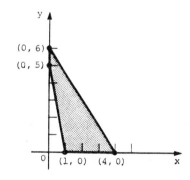

The corner points are $(0, 5)$,
$(0, 6)$, $(4, 0)$, and $(1, 0)$.

Corner Point	Value of z = 2x + 4y
(0, 5)	20
(0, 6)	24 Maximum
(4, 0)	8
(1, 0)	2

The maximum value of z = 2x + 4y is 24 at (0, 6).

19. Minimize z = 4x + 2y

subject to: x + y ≤ 50

2x + y ≥ 20

x + 2y ≥ 30

x ≥ 0

y ≥ 0.

Sketch the feasible region.

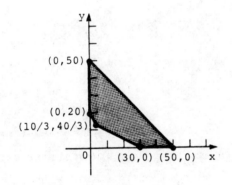

The corner points are (0, 20), (10/3, 40/3), (30, 0), (50, 0), and (0, 50).

Corner Point	Value of z = 4x + 2y
(0, 20)	40 Minimum
$(\frac{10}{3}, \frac{40}{3})$	40 Minimum
(30, 0)	120
(50, 0)	200
(0, 50)	100

Thus, the minimum value of 4x + 2y is 40 and occurs at every point on the line segment joining (0, 20) and (10/3, 40/3).

23. Let x represent the number of batches of cakes and y represent the number of batches of cookies. Then we have the following inequalities.

$$2x + \frac{3}{2}y \le 15 \quad \text{(oven time)}$$

$$3x + \frac{2}{3}y \le 13 \quad \text{(decorating)}$$

$$x \ge 0$$

$$y \ge 0$$

See the answer graph in the back of the textbook.

25. From the graph for Exercise 23, the corner points are (0, 10), (3, 6), (13/3, 0) and (0, 0). Since x was the number of batches of cakes and y the number of batches of cookies, the revenue function is

z = 30x + 20y.

Evaluate this objective function at each corner point.

Corner Point	Value of z = 30x + 20y
(0, 0)	200
(3, 6)	210 Maximum
$(\frac{13}{3}, 0)$	130
(0, 0)	0

Therefore, 3 batches of cakes and 6 batches of cookies should be made to produce a maximum profit of $210.

27. Let x represent the number of hours Charles should spend with his math tutor and y represent the number of hours he should spend with his accounting tutor. The number of points he expects to get on the two tests combined is z = 3x + 5y. The given information translates into the following problem.

Maximize z = 3x + 5y
subject to:

$$20x + 40y \leq 220 \quad \text{(finances)}$$
$$x + 3y \leq 16 \quad \text{(aspirin)}$$
$$x + 3y \leq 15 \quad \text{(sleep)}$$
$$x \geq 0$$
$$y \geq 0$$

Graph the feasible region.

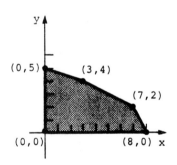

The corner points of the feasible region are (0, 0), (0, 5), (3, 4), (7, 2), and (8, 0). Evaluate the objective function at each corner point.

Corner Point	Value of z = 3x + 5y
(0, 0)	0
(0, 5)	25
(3, 4)	29
(7, 2)	31 Maximum
(8, 0)	24

Therefore, Charles should spend 7 hr with the math tutor and 2 hr with the accounting tutor in order to earn a maximum of 31 points.

29. Maximize z = 4x + 2y
subject to: x − y ≤ 10
5x + 3y ≤ 75
x + y ≤ 20
x ≥ 0
y ≥ 0.

The feasible region is the same as it was when we solved this problem previously. Add the line z = 40, which is the same as 4x + 2y = 40. (40 is chosen because the numbers work out simply and the line will show up clearly in our graph, but the chosen value of z is arbitrary.)

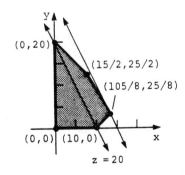

Also drawn in this graph is a line
parallel to the line z = 40 that is
as far from the origin as possible
but still touches the feasible region; this second line passes
through the corner point
(105/8, 25/8) and so, as before,
the objective function has a maximum value of 235/4 when x = 105/8,
y = 25/8.

CHAPTER 3   TEST

1.   Graph the inequality $2x + y \leq 4$.

2.   Graph the solution to the system of inequalities below.   Find all corner points.

$$x + y \leq 4$$
$$2x + y \geq 6$$
$$x \geq 0$$
$$y \geq 0$$

3.   Maximize the objective function $z = 3x + 2y$ for the region sketched below.

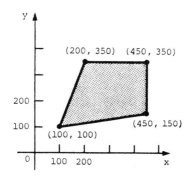

4.   Use the graphical method to solve the following linear programming problem.

$$\text{Maximize } z = 4x + 3y$$
$$\text{subject to: } 3x + y \leq 12$$
$$x + y \geq 3$$
$$x \geq 0$$
$$y \geq 0.$$

5.   The Gigantic Zipper Company manufactures two kinds of zippers.   Type I zippers require 2 min on machine I and 3 min on machine II.   Type II zippers require 1 min on machine I and 4 min on machine II.   Machine I is available for 20 min, while machine II is available for 12 min.   The profit on each type I zipper is $.30 and on each type II zipper is $.20. How many of each type of zipper should be manufactured to ensure the maximum profit?

CHAPTER 3   TEST ANSWERS

1.

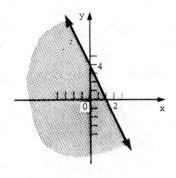

2.

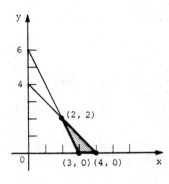

Corner points: (3, 0), (4, 0), (2, 2)

3.   Maximum of 2050 at (450, 350)

4.   Maximum of 36 at (0, 12)

5.   4 type I zippers and 0 type II zippers

CHAPTER 4 LINEAR PROGRAMMING: THE SIMPLEX METHOD

Section 4.1

1. $x_1 + 2x_2 \le 6$ becomes

   $x_1 + 2x_2 + x_3 = 6$.

3. $2x_1 + 4x_2 + 3x_3 \le 100$ becomes

   $2x_1 + 4x_2 + 3x_3 + x_4 = 100$.

5. **(a)** Since there are three constraints to be converted into equations, we need 3 slack variables.

   (b) The original problem uses x_1 and x_2, so we use x_3, x_4, x_5 for the slack variables.

   (c) The equations are

   $$4x_1 + 2x_2 + x_3 \qquad\qquad = 20$$
   $$5x_1 + x_2 \qquad + x_4 \qquad = 50$$
   $$2x_1 + 3x_2 \qquad\qquad + x_5 = 25.$$

7. **(a)** There are two constraints to be converted into equations, so we must introduce 2 slack variables.

   (b) x_1, x_2, and x_3 are already used in the problem, so call the slack variables x_4 and x_5.

   (c) The equations are

   $$7x_1 + 6x_2 + 8x_3 + x_4 \qquad = 118$$
   $$4x_1 + 5x_2 + 10x_3 \qquad + x_5 = 220.$$

9.

x_1	x_2	x_3	x_4	x_5	z	
2	2	0	3	1	0	15
3	4	1	6	0	0	20
-2	-1	0	1	0	1	10

The variables x_3 and x_5 are basic variables, because the columns for these variables have all zeros except for one nonzero entry. If the remaining variables x_1, x_2, and x_4 are zero, then $x_3 = 20$ and $x_5 = 15$. From the bottom row, $z = 10$. The basic feasible solution is

$x_1 = 0$, $x_2 = 0$, $x_3 = 20$, $x_4 = 0$, $x_5 = 15$, and $z = 10$.

11.

x_1	x_2	x_3	x_4	x_5	x_6	z	
6	2	2	3	0	0	0	16
2	2	0	1	0	5	0	35
2	1	0	3	1	0	0	6
-3	-2	0	2	0	0	3	36

The basic variables are x_3, x_5 and x_6. If x_1, x_2, and x_4 are zero, then $2x_3 = 16$, so $x_3 = 8$. Similarly, $x_5 = 6$ and $5x_6 = 35$, so $x_6 = 7$. From the bottom row, $3z = 36$, so $z = 12$. The basic feasible solution is

$x_1 = 0$, $x_2 = 0$, $x_3 = 8$, $x_4 = 0$, $x_5 = 6$, $x_6 = 7$, and $z = 12$.

13.

x_1	x_2	x_3	x_4	x_5	z	
1	2	4	1	0	0	56
2	②	1	0	1	0	40
-1	-3	-2	0	0	1	0

$-R_2 + 2R_1 \rightarrow R_1$
$$\begin{array}{c}\begin{array}{ccccccc}x_1 & x_2 & x_3 & x_4 & x_5 & z \end{array}\\\left[\begin{array}{cccccc|c}-1 & 0 & 3 & 1 & -1 & 0 & 16\\2 & ② & 1 & 0 & 1 & 0 & 40\\\hline -1 & -3 & -2 & 0 & 0 & 1 & 0\end{array}\right]\end{array}$$

$3R_2 + 2R_3 \rightarrow R_3$
$$\begin{array}{c}\begin{array}{cccccc}x_1 & x_2 & x_3 & x_4 & x_5 & z\end{array}\\\left[\begin{array}{cccccc|c}-1 & 0 & 3 & 1 & -1 & 0 & 16\\2 & 2 & 1 & 0 & 1 & 0 & 40\\\hline 4 & 0 & -1 & 0 & 3 & 2 & 120\end{array}\right]\end{array}$$

The solution is $x_1 = 0$, $x_2 = 20$, $x_3 = 0$, $x_4 = 16$, $x_5 = 0$, and $z = 60$.

15.
$$\begin{array}{c}\begin{array}{ccccccc}x_1 & x_2 & x_3 & x_4 & x_5 & x_6 & z\end{array}\\\left[\begin{array}{ccccccc|c}2 & 2 & ① & 1 & 0 & 0 & 0 & 12\\1 & 2 & 3 & 0 & 1 & 0 & 0 & 45\\3 & 1 & 1 & 0 & 0 & 1 & 0 & 20\\\hline -2 & -1 & -3 & 0 & 0 & 0 & 1 & 0\end{array}\right]\end{array}$$

$-3R_1 + R_2 \rightarrow R_2$
$-R_1 + R_3 \rightarrow R_3$
$3R_1 + R_4 \rightarrow R_4$
$$\begin{array}{c}\begin{array}{ccccccc}x_1 & x_2 & x_3 & x_4 & x_5 & x_6 & z\end{array}\\\left[\begin{array}{ccccccc|c}2 & 2 & 1 & 1 & 0 & 0 & 0 & 12\\-5 & -4 & 0 & -3 & 1 & 0 & 0 & 9\\1 & -1 & 0 & -1 & 0 & 1 & 0 & 8\\\hline 4 & 5 & 0 & 3 & 0 & 0 & 1 & 36\end{array}\right]\end{array}$$

The solution is $x_1 = 0$, $x_2 = 0$, $x_3 = 12$, $x_4 = 0$, $x_5 = 9$, $x_6 = 8$, and $z = 36$.

17.
$$\begin{array}{c}\begin{array}{ccccccc}x_1 & x_2 & x_3 & x_4 & x_5 & x_6 & z\end{array}\\\left[\begin{array}{ccccccc|c}1 & 1 & 1 & 1 & 0 & 0 & 0 & 60\\3 & 1 & ② & 0 & 1 & 0 & 0 & 100\\1 & 2 & 3 & 0 & 0 & 1 & 0 & 200\\\hline -1 & -1 & -2 & 0 & 0 & 0 & 1 & 0\end{array}\right]\end{array}$$

$2R_1 - R_2 \rightarrow R_1$
$-3R_2 + 2R_3 \rightarrow R_3$
$R_2 + R_4 \rightarrow R_4$
$$\begin{array}{c}\begin{array}{cccccccc}x_1 & x_2 & x_3 & x_4 & x_5 & x_6 & z\end{array}\\\left[\begin{array}{ccccccc|c}-1 & 1 & 0 & 2 & -1 & 0 & 0 & 20\\3 & 1 & 2 & 0 & 1 & 0 & 0 & 100\\-7 & 1 & 0 & 0 & -3 & 2 & 0 & 100\\\hline 2 & 0 & 0 & 0 & 1 & 1 & 1 & 100\end{array}\right]\end{array}$$

The solution is $x_1 = 0$, $x_2 = 0$, $x_3 = 50$, $x_4 = 10$, $x_5 = 0$, $x_6 = 50$, and $z = 100$.

19. Find $x_1 \geq 0$ and $x_2 \geq 0$ such that

$$2x_1 + 3x_2 \leq 6$$
$$4x_1 + x_2 \leq 6$$

and $z = 5x_1 + x_2$ is maximized. We need two slack variables, x_3 and x_4. Then the problem can be restated as: Find $x_1 \geq 0$, $x_2 \geq 0$, $x_3 \geq 0$, and $x_4 \geq 0$ such that

$$2x_1 + 3x_2 + x_3 \qquad = 6$$
$$4x_1 + x_2 \qquad + x_4 = 6$$

and $z = 5x_1 + x_2$ is maximized. Rewrite the objective function as $-5x_1 - x_2 + z = 0$. The initial simple tableau is

$$\begin{array}{c}\begin{array}{ccccc}x_1 & x_2 & x_3 & x_4 & z\end{array}\\\left[\begin{array}{cccc|c}2 & 3 & 1 & 0 & 0 & 6\\4 & 1 & 0 & 1 & 0 & 6\\\hline -5 & -1 & 0 & 0 & 1 & 0\end{array}\right]\end{array}.$$

21. Find $x_1 \geq 0$ and $x_2 \geq 0$ such that

$$x_1 + x_2 \leq 10$$
$$5x_1 + 2x_2 \leq 20$$
$$x_1 + 2x_2 \leq 36$$

and $z = x_1 + 3x_2$ is maximized.

Using slack variables x_3, x_4, and x_5, the problem can be restricted as:

Find $x_1 \geq 0$, $x_2 \geq 0$, $x_3 \geq 0$, $x_4 \geq 0$, and $x_5 \geq 0$ such that

$$
\begin{aligned}
x_1 + x_2 + x_3 &= 10 \\
5x_1 + 2x_2 \quad\quad + x_4 &= 20 \\
x_1 + 2x_2 \quad\quad\quad + x_5 &= 36
\end{aligned}
$$

and $z = x_1 + 3x_2$ is maximized. Rewrite the objective function as $-x_1 - 3x_2 + z = 0$.

The initial simplex tableau is

$$
\begin{array}{cccccc}
x_1 & x_2 & x_3 & x_4 & x_5 & z \\
\end{array}
$$
$$
\left[
\begin{array}{cccccc|c}
1 & 1 & 1 & 0 & 0 & 0 & 10 \\
5 & 2 & 0 & 1 & 0 & 0 & 20 \\
1 & 2 & 0 & 0 & 1 & 0 & 36 \\
\hline
-1 & -3 & 0 & 0 & 0 & 1 & 0
\end{array}
\right].
$$

23. Find $x_1 \geq 0$ and $x_2 \geq 0$ such that

$$
\begin{aligned}
3x_1 + x_2 &\leq 12 \\
x_1 + x_2 &\leq 15
\end{aligned}
$$

and $z = 2x_1 + x_2$ is maximized. Using slack variables x_3 and x_4, the problem can be restated as:
Find $x_1 \geq 0$, $x_2 \geq 0$, $x_3 \geq 0$, and $x_4 \geq 0$ such that

$$
\begin{aligned}
3x_1 + x_2 + x_3 &= 12 \\
x_1 + x_2 \quad\quad + x_4 &= 15
\end{aligned}
$$

and $z = 2x_1 + x_2$ is maximized. Rewrite the objective function as $-2x_1 - x_2 + z = 0$.
The initial simplex tableau is

$$
\begin{array}{ccccc}
x_1 & x_2 & x_3 & x_4 & z \\
\end{array}
$$
$$
\left[
\begin{array}{ccccc|c}
3 & 1 & 1 & 0 & 0 & 12 \\
1 & 1 & 0 & 1 & 0 & 15 \\
\hline
-2 & -1 & 0 & 0 & 1 & 0
\end{array}
\right].
$$

25. Let x_1 represent the number of prams, x_2 the number of runabouts, and x_3 the number of trimarans. Organize some of the information in a table.

	Pram	Run-about	Tri-maran	Available Time
Sec A	1	2	3	6240
Sec B	2	5	4	10,800
Profit	$75	$90	$100	

Using this information, together with the fact that the total number of boats cannot exceed 3000, the problem may be stated as:
Find $x_1 \geq 0$, $x_2 \geq 0$, and $x_3 \geq 0$ such that

$$
\begin{aligned}
x_1 + 2x_2 + 3x_3 &\leq 6240 \\
2x_1 + 5x_2 + 4x_3 &\leq 10,800 \\
x_1 + x_2 + x_3 &\leq 3000
\end{aligned}
$$

and $z = 75x_1 + 90x_2 + 100x_3$ is maximized.

Introduce slack variables x_4, x_5, and x_6, and the problem can be restated as:
Find $x_1 \geq 0$, $x_2 \geq 0$, $x_3 \geq 0$, $x_4 \geq 0$, $x_5 \geq 0$, and $x_6 \geq 0$ such that

$$
\begin{aligned}
x_1 + 2x_2 + 3x_3 + x_4 &= 6240 \\
2x_1 + 5x_2 + 4x_3 \quad\quad + x_5 &= 10,800 \\
x_1 + x_2 + x_3 \quad\quad\quad + x_6 &= 3000
\end{aligned}
$$

and $z = 75x_1 + 90x_2 + 100x_3$ is maximized.

Rewrite the objective as

$$-75x_1 - 90x_2 - 100x_3 + z = 0.$$

The initial simplex tableau is

$$
\begin{array}{c}
\begin{array}{ccccccc}
x_1 & x_2 & x_3 & x_4 & x_5 & x_6 & z
\end{array} \\
\left[
\begin{array}{ccccccc|c}
1 & 2 & 3 & 1 & 0 & 0 & 0 & 6240 \\
2 & 5 & 4 & 0 & 1 & 0 & 0 & 10{,}800 \\
1 & 1 & 1 & 0 & 0 & 1 & 0 & 3000 \\
\hline
-75 & -90 & -100 & 0 & 0 & 0 & 1 & 0
\end{array}
\right]
\end{array}
$$

27. Let x_1 represent the number of simple figures, x_2 the number of figures with additions, and x_3 the number of computer-drawn sketches. Organize some of the information in a table.

	Simple Figures	Figures with Additions	Computer-Drawn Sketches	Maximum Allowed
Cost	20	35	60	2200
Royalties	95	200	325	

The cost constraint is

$$20x_1 + 35x_2 + 60x_3 \leq 2200.$$

The limit of 400 figures leads to the constraint $x_1 + x_2 + x_3 \leq 400$. The other stated constraints are $x_3 \leq x_1 + x_2$ and $x_1 \geq 2x_2$, and these can be rewritten in standard form as $-x_1 - x_2 + x_3 \leq 0$ and $-x_1 + 2x_2 \leq 0$ respectively.

The problem may be stated as:

Find $x_1 \geq 0$, $x_2 \geq 0$, and $x_3 \geq 0$ such that

$$
\begin{aligned}
20x_1 + 35x_2 + 60x_3 &\leq 2200 \\
x_1 + x_2 + x_3 &\leq 400 \\
-x_1 - x_2 + x_3 &\leq 0 \\
-x_1 + 2x_2 &\leq 0
\end{aligned}
$$

and $z = 95x_1 + 200x_2 + 325x_3$ is maximized.

Introduce slack variables x_4, x_5, x_6, and x_7, and the problem can be restated as:

Find $x_1 \geq 0$, $x_2 \geq 0$, $x_3 \geq 0$, $x_4 \geq 0$, $x_5 \geq 0$, $x_6 \geq 0$, and $x_7 \geq 0$ such that

$$
\begin{aligned}
20x_1 + 35x_2 + 60x_3 + x_4 &&&&= 2200 \\
x_1 + x_2 + x_3 &+ x_5 &&&= 400 \\
-x_1 - x_2 + x_3 &&+ x_6 &&= 0 \\
-x_1 + 2x_2 &&&+ x_7 &= 0
\end{aligned}
$$

and $z = 95x_1 + 200x_2 + 325x_3$ is maximized.

Rewrite the objective function as

$$-95x_1 - 200x_2 - 325x_3 + z = 0.$$

The initial simplex tableau is

$$\begin{bmatrix}
x_1 & x_2 & x_3 & x_4 & x_5 & x_6 & x_7 & z & \\
20 & 35 & 60 & 1 & 0 & 0 & 0 & 0 & 2200 \\
1 & 1 & 1 & 0 & 1 & 0 & 0 & 0 & 400 \\
-1 & -1 & 1 & 0 & 0 & 1 & 0 & 0 & 0 \\
-1 & 2 & 0 & 0 & 0 & 0 & 1 & 0 & 0 \\
\hline
-95 & -200 & -325 & 0 & 0 & 0 & 0 & 1 & 0
\end{bmatrix}.$$

29. Let x_1 represent the number of one-speed bicycles, x_2 the number of three-speed bicycles, and x_3 the number of ten-speed bicycles.

Organize the information in a table.

	One-Speed	Three-Speed	Ten-Speed	Amount Available
Steel	17	27	34	91,800
Aluminum	12	21	15	42,000
Profit	$8	$12	$22	

Using this information, the problem may be stated as:

Find $x_1 \geq 0$, $x_2 \geq 0$, and $x_3 \geq 0$ such that

$$17x_1 + 27x_2 + 34x_3 \leq 91,800$$
$$12x_1 + 21x_2 + 15x_3 \leq 42,000$$

and $z = 8x_1 + 12x_2 + 22x_3$ is maximized.

Introduce slack variables x_4 and x_5 and the problem can be restated as:

Find $x_1 \geq 0$, $x_2 \geq 0$, $x_3 \geq 0$, $x_4 \geq 0$, and $x_5 \geq 0$ such that

$$17x_1 + 27x_2 + 34x_3 + x_4 = 91,800$$
$$12x_1 + 21x_2 + 15x_3 + x_5 = 42,000$$

and $z = 8x_1 + 12x_2 + 22x_3$ is maximized.

Rewrite the objective function as

$$-8x_1 - 12x_2 - 22x_3 + z = 0.$$

The initial simplex tableau is

$$\begin{bmatrix}
x_1 & x_2 & x_3 & x_4 & x_5 & z & \\
17 & 27 & 34 & 1 & 0 & 0 & 91,800 \\
12 & 21 & 15 & 0 & 1 & 0 & 42,000 \\
\hline
-8 & -12 & -22 & 0 & 0 & 1 & 0
\end{bmatrix}.$$

Section 4.2

1. The initial simplex tableau is as follows.

$$\begin{bmatrix}
x_1 & x_2 & x_3 & x_4 & x_5 & z & \\
1 & 2 & 4 & 1 & 0 & 0 & 8 \\
2 & 2 & 1 & 0 & 1 & 0 & 10 \\
\hline
-2 & -5 & -1 & 0 & 0 & 1 & 0
\end{bmatrix}$$

The most negative indicator is -5, in the second column. Find the quotients $8/2 = 4$ and $10/2 = 5$; since 4 is the smallest quotient, 2 in the first row, second column, is the pivot.

$$\begin{array}{c}
\\
8/2 = 4 \\
10/2 = 5 \\
\\
\end{array}
\begin{bmatrix}
x_1 & x_2 & x_3 & x_4 & x_5 & z & \\
1 & ② & 4 & 1 & 0 & 0 & 8 \\
2 & 2 & 1 & 0 & 1 & 0 & 10 \\
\hline
-2 & -5 & -1 & 0 & 0 & 1 & 0
\end{bmatrix}$$

Performing row transformations, we get the following tableau.

$$\begin{array}{c}
\\
-R_1 + R_2 \to R_2 \\
5R_1 + 2R_3 \to R_3
\end{array}
\begin{bmatrix}
x_1 & x_2 & x_3 & x_4 & x_5 & z & \\
1 & 2 & 4 & 1 & 0 & 0 & 8 \\
1 & 0 & -3 & -1 & 1 & 0 & 2 \\
1 & 0 & 18 & 5 & 0 & 2 & 40
\end{bmatrix}$$

All of the numbers in the last row are nonnegative, so we are finished pivoting.

Create a 1 in the columns corresponding to x_2 and z.

$$\begin{array}{c} \frac{1}{2}R_1 \to R_1 \\ \\ \\ \frac{1}{2}R_3 \to R_3 \end{array} \begin{array}{cccccc} x_1 & x_2 & x_3 & x_4 & x_5 & z \\ \left[\begin{array}{cccccc|c} \frac{1}{2} & 1 & 2 & \frac{1}{2} & 0 & 0 & 4 \\ 1 & 0 & -3 & -1 & 1 & 0 & 2 \\ \hline \frac{1}{2} & 0 & 9 & \frac{5}{2} & 0 & 1 & 20 \end{array}\right] \end{array}$$

The maximum value is 20 and occurs when $x_1 = 0$, $x_2 = 4$, $x_3 = 0$, $x_4 = 0$, and $x_5 = 2$.

3.

$$\begin{array}{cccccc} x_1 & x_2 & x_3 & x_4 & x_5 & z \\ \left[\begin{array}{cccccc|c} 1 & 3 & 1 & 0 & 0 & 0 & 12 \\ 2 & 1 & 0 & 1 & 0 & 0 & 10 \\ 1 & 1 & 0 & 0 & 1 & 0 & 4 \\ \hline -2 & -1 & 0 & 0 & 0 & 1 & 0 \end{array}\right] \end{array}$$

The most negative indicator is -2, in the first column. Find the quotients $12/1 = 12$, $10/2 = 5$, and $4/1 = 4$; since 4 is the smallest quotient, 1 in the third row, first column, is the pivot.

$$\begin{array}{cccccc} x_1 & x_2 & x_3 & x_4 & x_5 & z \\ \left[\begin{array}{cccccc|c} 1 & 3 & 1 & 0 & 0 & 0 & 12 \\ 2 & 1 & 0 & 1 & 0 & 0 & 10 \\ ① & 1 & 0 & 0 & 1 & 0 & 4 \\ \hline -2 & -1 & 0 & 0 & 0 & 1 & 0 \end{array}\right] \end{array}$$

$$\begin{array}{c} \\ -R_3+R_1 \to R_1 \\ -2R_3+R_2 \to R_2 \\ \\ 2R_3+R_4 \to R_4 \end{array} \begin{array}{cccccc} x_1 & x_2 & x_3 & x_4 & x_5 & z \\ \left[\begin{array}{cccccc|c} 0 & 2 & 1 & 0 & -1 & 0 & 8 \\ 0 & -1 & 0 & 1 & -2 & 0 & 2 \\ 1 & 1 & 0 & 0 & 1 & 0 & 4 \\ \hline 0 & 1 & 0 & 0 & 2 & 1 & 8 \end{array}\right] \end{array}$$

This is a final tableau, since all of the numbers in the last row are

nonnegative. The maximum value is 8 when $x_1 = 4$, $x_2 = 0$, $x_3 = 8$, $x_4 = 2$, and $x_5 = 0$.

5.

$$\begin{array}{ccccccc} x_1 & x_2 & x_3 & x_4 & x_5 & x_6 & z \\ \left[\begin{array}{ccccccc|c} 2 & 2 & 8 & 1 & 0 & 0 & 0 & 40 \\ 4 & -5 & 6 & 0 & 1 & 0 & 0 & 60 \\ 2 & -2 & 6 & 0 & 0 & 1 & 0 & 24 \\ \hline -14 & -10 & -12 & 0 & 0 & 0 & 1 & 0 \end{array}\right] \end{array}$$

The most negative indicator is -14, in the first column. Find the quotients $40/2 = 20$, $60/4 = 15$, and $24/2 = 12$; since 12 is the smallest quotient, 2 in the third row, first column, is the pivot.

$$\begin{array}{ccccccc} x_1 & x_2 & x_3 & x_4 & x_5 & x_6 & z \\ \left[\begin{array}{ccccccc|c} 2 & 2 & 8 & 1 & 0 & 0 & 0 & 40 \\ 4 & -5 & 6 & 0 & 1 & 0 & 0 & 60 \\ ② & -2 & 6 & 0 & 0 & 1 & 0 & 24 \\ \hline -14 & -10 & -12 & 0 & 0 & 0 & 1 & 0 \end{array}\right] \end{array}$$

Performing row transformations, we get the following tableau.

$$\begin{array}{c} -R_3+R_1 \to R_1 \\ -2R_3+R_2 \to R_2 \\ \\ 7R_3+R_4 \to R_4 \end{array} \begin{array}{ccccccc} x_1 & x_2 & x_3 & x_4 & x_5 & x_6 & z \\ \left[\begin{array}{ccccccc|c} 0 & ④ & 2 & 1 & 0 & -1 & 0 & 16 \\ 0 & -1 & -6 & 0 & 1 & -2 & 0 & 12 \\ 2 & -2 & 6 & 0 & 0 & 1 & 0 & 24 \\ \hline 0 & -24 & 30 & 0 & 0 & 7 & 1 & 168 \end{array}\right] \end{array}$$

Since there is still a negative indicator, we must repeat the process. The second pivot is the 4 in the second column, since $16/4$ is the only nonnegative quotient in the only column with a negative indicator.

Performing row transformations again, we get the following tableau.

$$
\begin{array}{c}
\\
R_1+4R_2\rightarrow R_2 \\
R_1+2R_3\rightarrow R_3 \\
\\
6R_1+R_4\rightarrow R_4
\end{array}
\begin{array}{ccccccc}
x_1 & x_2 & x_3 & x_4 & x_5 & x_6 & z \\
\end{array}
\left[
\begin{array}{ccccccc|c}
0 & 4 & 2 & 1 & 0 & -1 & 0 & 16 \\
0 & 0 & -22 & 1 & 4 & -9 & 0 & 64 \\
4 & 0 & 14 & 1 & 0 & 1 & 0 & 64 \\
\hline
0 & 0 & 42 & 6 & 0 & 1 & 1 & 264
\end{array}
\right]
$$

All of the numbers in the last row are nonnegative, so we are finished pivoting. Create a 1 in the columns corresponding to x_1, x_2, and x_5.

$$
\begin{array}{c}
\tfrac{1}{4}R_1\rightarrow R_1 \\[2mm]
\tfrac{1}{4}R_2\rightarrow R_2 \\[2mm]
\tfrac{1}{4}R_3\rightarrow R_3 \\[2mm]
\\
\end{array}
\begin{array}{ccccccc}
x_1 & x_2 & x_3 & x_4 & x_5 & x_6 & z \\
\end{array}
\left[
\begin{array}{ccccccc|c}
0 & 1 & \tfrac{1}{2} & \tfrac{1}{4} & 0 & -\tfrac{1}{4} & 0 & 4 \\[1mm]
0 & 0 & -\tfrac{11}{2} & \tfrac{1}{4} & 1 & -\tfrac{9}{4} & 0 & 16 \\[1mm]
1 & 0 & \tfrac{7}{2} & \tfrac{1}{4} & 0 & \tfrac{1}{4} & 0 & 16 \\[1mm]
\hline
0 & 0 & 42 & 6 & 0 & 1 & 1 & 264
\end{array}
\right]
$$

The maximum value is 264 and occurs when $x_1 = 16$, $x_2 = 4$, $x_3 = 0$, $x_4 = 0$, $x_5 = 16$, and $x_6 = 0$.

7. Maximize $z = 4x_1 + 3x_2$

subject to: $2x_1 + 3x_2 \le 11$

$x_1 + 2x_2 \le 6$

with $x_1 \ge 0,\ x_2 \ge 0.$

Two slack variables need to be introduced, x_3 and x_4.

The problem can be restated as:

Maximize $z = 4x_1 + 3x_2$

subject to: $2x_1 + 3x_2 + x_3 \qquad\quad = 11$

$x_1 + 2x_2 \qquad\quad + x_4 = 6$

with $x_1 \ge 0,\ x_2 \ge 0,$

$x_3 \ge 0,\ x_4 \ge 0.$

Rewrite the objective function as $-4x_1 - 3x_2 + z = 0.$

The initial simplex tableau follows.

$$
\begin{array}{ccccc}
x_1 & x_2 & x_3 & x_4 & z \\
\end{array}
\left[
\begin{array}{ccccc|c}
2 & 3 & 1 & 0 & 0 & 11 \\
1 & 2 & 0 & 1 & 0 & 6 \\
\hline
-4 & -3 & 0 & 0 & 1 & 0
\end{array}
\right]
$$

The most negative indicator is -4, in column 1; to select the pivot from column 1, find the quotients $11/2$ and $6/1$. The smallest is $11/2$, so 2 is the pivot.

$$
\begin{array}{ccccc}
x_1 & x_2 & x_3 & x_4 & z \\
\end{array}
\left[
\begin{array}{ccccc|c}
②\; & 3 & 1 & 0 & 0 & 11 \\
1 & 2 & 0 & 1 & 0 & 6 \\
\hline
-4 & -3 & 0 & 0 & 1 & 0
\end{array}
\right]
$$

$$
\begin{array}{c}
\\
-R_1 + 2R_2 \rightarrow R_2 \\
\\
2R_1 + R_3 \rightarrow R_3
\end{array}
\begin{array}{ccccc}
x_1 & x_2 & x_3 & x_4 & z \\
\end{array}
\left[
\begin{array}{ccccc|c}
2 & 3 & 1 & 0 & 0 & 11 \\
0 & 1 & -1 & 2 & 0 & 1 \\
\hline
0 & 3 & 2 & 0 & 1 & 22
\end{array}
\right]
$$

All of the indicators are nonnegative. Create a 1 in the columns corresponding to x_1 and x_4

$$
\begin{array}{c}
\tfrac{1}{2}R_1 \rightarrow R_1 \\[3mm]
\tfrac{1}{2}R_2 \rightarrow R_2 \\[3mm]
\\
\end{array}
\begin{array}{ccccc}
x_1 & x_2 & x_3 & x_4 & z \\
\end{array}
\left[
\begin{array}{ccccc|c}
1 & \tfrac{3}{2} & \tfrac{1}{2} & 0 & 0 & \tfrac{11}{2} \\[1mm]
0 & \tfrac{1}{2} & -\tfrac{1}{2} & 1 & 0 & \tfrac{1}{2} \\[1mm]
\hline
0 & 3 & 2 & 0 & 1 & 22
\end{array}
\right]
$$

The maximum value is 22 when $x_1 = 5.5$, $x_2 = 0$, $x_3 = 0$, and $x_4 = .5$.

9. Maximize $z = 10x_1 + 12x_2$

subject to: $4x_1 + 2x_2 \leq 20$

$5x_1 + x_2 \leq 50$

$2x_1 + 2x_2 \leq 24$

with $x_1 \geq 0, x_2 \geq 0$.

Three slack variables need to be introduced, x_3, x_4, and x_5. Then the initial tableau is as follows.

$$
\begin{array}{cccccc}
x_1 & x_2 & x_3 & x_4 & x_5 & z \\
\end{array}
$$

$$
\left[
\begin{array}{cccccc|c}
4 & 2 & 1 & 0 & 0 & 0 & 20 \\
5 & 1 & 0 & 1 & 0 & 0 & 50 \\
2 & 2 & 0 & 0 & 1 & 0 & 24 \\
\hline
-10 & -12 & 0 & 0 & 0 & 1 & 0 \\
\end{array}
\right]
$$

The most negative indicator is -12, in column 2. The quotients are $20/2 = 10$, $50/1 = 50$, and $24/2 = 12$; the smallest is 10, so 2 in the first row, second column, is the pivot.

$$
\begin{array}{cccccc}
x_1 & x_2 & x_3 & x_4 & x_5 & z \\
\end{array}
$$

$$
\left[
\begin{array}{cccccc|c}
4 & ② & 1 & 0 & 0 & 0 & 20 \\
5 & 1 & 0 & 1 & 0 & 0 & 50 \\
2 & 2 & 0 & 0 & 1 & 0 & 24 \\
\hline
-10 & -12 & 0 & 0 & 0 & 1 & 0 \\
\end{array}
\right]
$$

$$
\begin{array}{cccccc}
x_1 & x_2 & x_3 & x_4 & x_5 & z \\
\end{array}
$$

$$
\begin{array}{l}
\\
-R_1 + 2R_2 \rightarrow R_2 \\
\\
-R_1 + R_3 \rightarrow R_3 \\
\\
6R_1 + R_4 \rightarrow R_4
\end{array}
\left[
\begin{array}{cccccc|c}
4 & 2 & 1 & 0 & 0 & 0 & 20 \\
6 & 0 & -1 & 2 & 0 & 0 & 80 \\
-2 & 0 & -1 & 0 & 1 & 0 & 4 \\
14 & 0 & 6 & 0 & 0 & 1 & 120 \\
\end{array}
\right]
$$

All of the indicators are nonnegative, so we are finished pivoting. Create a 1 in the columns corresponding to x_2 and x_4.

$$
\begin{array}{cccccc}
x_1 & x_2 & x_3 & x_4 & x_5 & z \\
\end{array}
$$

$$
\begin{array}{l}
\frac{1}{2}R_1 \rightarrow R_1 \\
\\
\frac{1}{2}R_2 \rightarrow R_2 \\
\\
\\
\end{array}
\left[
\begin{array}{cccccc|c}
2 & 1 & \frac{1}{2} & 0 & 0 & 0 & 10 \\
3 & 0 & -\frac{1}{2} & 1 & 0 & 0 & 40 \\
-2 & 0 & -1 & 0 & 1 & 0 & 4 \\
\hline
14 & 0 & 6 & 0 & 0 & 1 & 120 \\
\end{array}
\right]
$$

The maximum value is 120 when $x_1 = 0$, $x_2 = 10$, $x_3 = 0$, $x_4 = 40$, and $x_5 = 4$.

11. Maximize $z = 8x_1 + 3x_2 + x_3$

subject to: $x_1 + 6x_2 + 8x_3 \leq 118$

$x_1 + 5x_2 + 10x_3 \leq 220$

with $x_1 \geq 0, x_2 \geq 0, x_3 \geq 0$.

Two slack variables need to be introduced, x_4 and x_5. The initial simplex tableau is as follows.

$$
\begin{array}{cccccc}
x_1 & x_2 & x_3 & x_4 & x_5 & z \\
\end{array}
$$

$$
\left[
\begin{array}{cccccc|c}
① & 6 & 8 & 1 & 0 & 0 & 118 \\
1 & 5 & 10 & 0 & 1 & 0 & 220 \\
\hline
-8 & -3 & -1 & 0 & 0 & 1 & 0 \\
\end{array}
\right]
$$

The most negative indicator is -8, in the first column. The quotients are $118/1 = 118$ and $220/1 = 220$; since 118 is the smallest, 1 in the first row, first column is the pivot. Performing row transformations, we get the following tableau.

$$
\begin{array}{cccccc}
x_1 & x_2 & x_3 & x_4 & x_5 & z \\
\end{array}
$$

$$
\begin{array}{l}
\\
-R_1 + R_2 \rightarrow R_2 \\
\\
8R_1 + R_3 \rightarrow R_3
\end{array}
\left[
\begin{array}{cccccc|c}
1 & 6 & 8 & 1 & 0 & 0 & 118 \\
0 & -1 & 2 & -1 & 1 & 0 & 102 \\
0 & 45 & 63 & 8 & 0 & 1 & 944 \\
\end{array}
\right]
$$

All of the indicators are nonnegative, so we are finished pivoting.

The maximum value is 944 when
$x_1 = 118$, $x_2 = 0$, $x_3 = 0$, $x_4 = 0$,
and $x_5 = 102$.

13. Maximize $z = x_1 + 2x_2 + x_3 + 5x_4$
subject to:

$$x_1 + 2x_2 + x_3 + x_4 \leq 50$$
$$3x_1 + x_2 + 2x_3 + x_4 \leq 100$$

with $x_1 \geq 0$, $x_2 \geq 0$,
$x_3 \geq 0$, $x_4 \geq 0$.

Two slack variables need to be
introduced, x_5 and x_6. Then the
initial simplex tableau is as
follows.

$$
\begin{array}{ccccccc}
x_1 & x_2 & x_3 & x_4 & x_5 & x_6 & z \\
\end{array}
$$
$$
\left[
\begin{array}{ccccccc|c}
1 & 2 & 1 & ① & 1 & 0 & 0 & 50 \\
3 & 1 & 2 & 1 & 0 & 1 & 0 & 100 \\
\hline
-1 & -2 & -1 & -5 & 0 & 0 & 1 & 0 \\
\end{array}
\right]
$$

In the column with the most negative indicator, -5, the quotients
are $50/1 = 50$ and $100/1 = 100$; the
smallest is 50, so 1 in the first
row, fourth column, is the pivot.

$$
\begin{array}{ccccccc}
 & x_1 & x_2 & x_3 & x_4 & x_5 & x_6 & z \\
\end{array}
$$
$$
\begin{array}{r}
 \\
-R_1+R_2\rightarrow R_2 \\
5R_1+R_3\rightarrow R_3
\end{array}
\left[
\begin{array}{ccccccc|c}
1 & 2 & 1 & 1 & 1 & 0 & 0 & 50 \\
2 & -1 & 1 & 0 & -1 & 1 & 0 & 50 \\
4 & 8 & 4 & 0 & 5 & 0 & 1 & 250 \\
\end{array}
\right]
$$

This is a final tableau, since all
of the indicators are nonnegative.
The maximum value is 250 when
$x_1 = 0$, $x_2 = 0$, $x_3 = 0$, $x_4 = 50$,
$x_5 = 0$, and $x_6 = 50$.

17. Organize the information in a table.

	Church group	Labor Union	Maximum Time Available
Letter Writing	2	2	16
Follow-up	1	3	12
Money Raised	$100	$200	

Let x_1 and x_2 be the number of church
groups and labor unions contacted
respectively. We need 2 slack variables, x_3 and x_4. We want to
maximize $z = 100x_1 + 200x_2$
subject to:

$$2x_1 + 2x_2 + x_3 \qquad = 16$$
$$x_1 + 3x_2 + \qquad + x_4 = 12$$

with $x_1 \geq 0$, $x_2 \geq 0$, $x_3 \geq 0$, $x_4 \geq 0$.

The initial simplex tableau is as
follows.

$$
\begin{array}{ccccc}
x_1 & x_2 & x_3 & x_4 & z \\
\end{array}
$$
$$
\left[
\begin{array}{ccccc|c}
2 & 2 & 1 & 0 & 0 & 16 \\
1 & ③ & 0 & 1 & 0 & 12 \\
\hline
-100 & -200 & 0 & 0 & 1 & 0 \\
\end{array}
\right]
$$

Pivot on the 3 in the second row,
second column.

$$
\begin{array}{ccccc}
 & x_1 & x_2 & x_3 & x_4 & z \\
\end{array}
$$
$$
\begin{array}{r}
3R_1-2R_2\rightarrow R_1 \\
 \\
200R_2+3R_3\rightarrow R_3
\end{array}
\left[
\begin{array}{ccccc|c}
④ & 0 & 3 & -2 & 0 & 24 \\
1 & 3 & 0 & 1 & 0 & 12 \\
\hline
-100 & 0 & 0 & 200 & 3 & 2400 \\
\end{array}
\right]
$$

Pivot on the 4 in the first row, first
column.

$$\begin{array}{c} \\ 4R_2 - R_1 \to R_2 \\ \\ 25R_1 + R_4 \to R_4 \end{array} \begin{array}{cccccc} x_1 & x_2 & x_3 & x_4 & z & \\ \left[\begin{array}{ccccc|c} 4 & 0 & 3 & -2 & 0 & 24 \\ 0 & 12 & -3 & 6 & 0 & 24 \\ \hline 0 & 0 & 75 & 150 & 3 & 3000 \end{array}\right] \end{array}$$

This is a final tableau, since all of the indicators are nonnegative. The maximum value is $\frac{3000}{3}$ = 1000 when $x_1 = \frac{24}{4}$ = 6, $x_2 = \frac{24}{12}$ = 2, x_3 = 0, and x_4 = 0.

She should contact 6 churches and 2 labor unions to raise a maximum of $1000 per month.

19.   Organize the information in a table.

	Recording	Mixing	Editing	Profit
Jazz	4	2	6	$.80
Blues	4	8	2	$.60
Reggae	10	4	6	$1.20
Maximum Available	80	52	54	

Let x_1, x_2, x_3 represent the number of jazz, blues and reggae albums respectively. We need 3 slack variables, x_4, x_5, and x_6.

We want to maximize $z = .8x_1 + .6x_2 + 1.2x_3$
subject to:
$$4x_1 + 4x_2 + 10x_3 + x_4 \qquad\qquad = 80$$
$$2x_1 + 8x_2 + 4x_3 \qquad + x_5 \qquad = 52$$
$$6x_1 + 2x_2 + 6x_3 \qquad\qquad + x_6 = 54$$
with $x_1 \geq 0$, $x_2 \geq 0$, $x_3 \geq 0$, $x_4 \geq 0$, $x_5 \geq 0$, and $x_6 \geq 0$.

The initial simplex tableau is as follows.

$$\begin{array}{ccccccc} x_1 & x_2 & x_3 & x_4 & x_5 & x_6 & z \\ \left[\begin{array}{ccccccc|c} 4 & 4 & \boxed{10} & 1 & 0 & 0 & 0 & 80 \\ 2 & 8 & 4 & 0 & 1 & 0 & 0 & 52 \\ 6 & 2 & 6 & 0 & 0 & 1 & 0 & 54 \\ \hline -.8 & -.6 & -1.2 & 0 & 0 & 0 & 1 & 0 \end{array}\right] \end{array}$$

Pivot on the 10 in the first row, third column.

$$
\begin{array}{c}
\\
-2R_1 + 5R_2 \rightarrow R_2 \\
-3R_1 + 5R_3 \rightarrow R_3 \\
\\
3R_1 + 25R_4 \rightarrow R_4
\end{array}
\begin{array}{cccccccc}
x_1 & x_2 & x_3 & x_4 & x_5 & x_6 & z & \\
\end{array}
$$

	x_1	x_2	x_3	x_4	x_5	x_6	z	
	4	4	10	1	0	0	0	80
$-2R_1 + 5R_2 \rightarrow R_2$	2	32	0	-2	5	0	0	100
$-3R_1 + 5R_3 \rightarrow R_3$	⑱	-2	0	-3	0	5	0	30
$3R_1 + 25R_4 \rightarrow R_4$	-8	-3	0	3	0	0	25	240

Pivot on the 18 in the third row, first column.

	x_1	x_2	x_3	x_4	x_5	x_6	z	
$9R_1 - 2R_3 \rightarrow R_1$	0	40	90	15	0	-10	0	660
$9R_2 - R_3 \rightarrow R_2$	0	㉙⓪	0	-15	45	-5	0	870
	18	-2	0	-3	0	5	0	30
$4R_3 + 9R_4 \rightarrow R_4$	0	-35	0	15	0	20	225	2280

Pivot on the 290 in the second row, second column.

	x_1	x_2	x_3	x_4	x_5	x_6	z	
$29R_1 - 4R_2 \rightarrow R_1$	0	0	2610	495	-180	-270	0	15,660
	0	290	0	-15	45	-5	0	870
$R_2 + 145R_3 \rightarrow R_3$	2610	0	0	-450	45	720	0	5220
$7R_2 + 58R_4 \rightarrow R_4$	0	0	0	765	315	1125	13,050	138,330

This is a final tableau, since all of the indicators are nonnegative. The maximum value is

$$
\frac{138,330}{13,050} = 10.60
$$

when $x_1 = \dfrac{5220}{2610} = 2$, $x_2 = \dfrac{870}{290} = 3$, $x_3 = \dfrac{15,660}{2610} = 6$, $x_4 = 0$, $x_5 = 0$, and $x_6 = 0$.

The company should produce 2 jazz albums, 3 blues albums, and 6 reggae albums for a maximum weekly profit of \$10.60.

21. Let x_1 represent the number of one- speed bicycles, x_2 the number of three-speed bicycles, and x_3 the number of ten-speed bicycles.

From Exercise 29 in Section 4.1, the initial simplex tableau is as follows.

x_1	x_2	x_3	x_4	x_5	z	
17	27	㉞	0	1	0	91,800
12	21	15	1	0	0	42,000
-8	-12	-22	0	0	1	0

Pivot on the 34 in the first row, third column.

$$
\begin{array}{c}
\\
-15R_1+34R_2 \rightarrow R_2 \\
\\
11R_1+17R_3 \rightarrow R_3
\end{array}
\begin{array}{c}
\begin{array}{cccccc}
x_1 & x_2 & x_3 & x_4 & x_5 & z
\end{array}\\
\left[\begin{array}{cccccc|c}
17 & 27 & 34 & 0 & 1 & 0 & 91,800 \\
153 & 309 & 0 & 34 & -15 & 0 & 51,000 \\
\hline
51 & 93 & 0 & 0 & 11 & 17 & 1,009,800
\end{array}\right]
\end{array}
$$

This is a final tableau, since all of the indicators are nonnegative. The maximum value is

$$\frac{1,009,800}{17} = 59,400 \text{ when}$$

$x_1 = 0$, $x_2 = 0$, $x_3 = \dfrac{91,800}{34} = 2700$,

$x_4 = \dfrac{51,000}{34} = 1500$, and $x_5 = 0$.

The company should make no one-speed or three-speed bicycles and 2700 ten-speed bicycles for a maximum profit of $59,400.

23. Let x_1 represent the amount of half nuts and half raisins mix, and x_2 be the amount of 1/3 nuts and 2/3 raisins mix. From Exercise 13 in Section 3.3, the problem is to Maximize $z = 6x_1 + 4.8x_2$ subject to:

$$\frac{1}{2}x_1 + \frac{1}{3}x_2 \le 100$$

$$\frac{1}{2}x_1 + \frac{2}{3}x_2 \le 125$$

with $x_1 \ge 0$ and $x_2 \ge 0$.

We need 2 slack variables, x_3 and x_4. Then the initial simplex tableau is as follows.

$$
\begin{array}{c}
\begin{array}{ccccc}
x_1 & x_2 & x_3 & x_4 & z
\end{array}\\
\left[\begin{array}{ccccc|c}
\boxed{\frac{1}{2}} & \frac{1}{3} & 1 & 0 & 0 & 100 \\
\frac{1}{2} & \frac{2}{3} & 0 & 1 & 0 & 125 \\
\hline
-6 & -4.8 & 0 & 0 & 1 & 0
\end{array}\right]
\end{array}
$$

Pivot on the $\frac{1}{2}$ in the first row, first column.

$$
\begin{array}{c}
\\
-R_1+R_2 \rightarrow R_2 \\
\\
12R_1+R_3 \rightarrow R_3
\end{array}
\begin{array}{c}
\begin{array}{ccccc}
x_1 & x_2 & x_3 & x_4 & z
\end{array}\\
\left[\begin{array}{ccccc|c}
\frac{1}{2} & \frac{1}{3} & 1 & 0 & 0 & 100 \\
0 & \boxed{\frac{1}{3}} & -1 & 1 & 0 & 25 \\
\hline
0 & -.8 & 12 & 0 & 1 & 1200
\end{array}\right]
\end{array}
$$

Pivot on the $\frac{1}{3}$ in the second row, second column.

$$
\begin{array}{c}
-R_2+R_1 \rightarrow R_1 \\
\\
12R_2+5R_3 \rightarrow R_3
\end{array}
\begin{array}{c}
\begin{array}{ccccc}
x_1 & x_2 & x_3 & x_4 & z
\end{array}\\
\left[\begin{array}{ccccc|c}
\frac{1}{2} & 0 & 2 & -1 & 0 & 75 \\
0 & \frac{1}{3} & -1 & 1 & 0 & 25 \\
\hline
0 & 0 & 48 & 12 & 5 & 6300
\end{array}\right]
\end{array}
$$

All of the indicators are non-negative, so we are finished pivoting. Create a 1 in the columns corresponding to x_1, x_2, and z.

$$
\begin{array}{c}
2R_1 \rightarrow R_1 \\
3R_2 \rightarrow R_2 \\
\\
\frac{1}{5}R_3 \rightarrow R_3
\end{array}
\begin{array}{c}
\begin{array}{ccccc}
x_1 & x_2 & x_3 & x_4 & z
\end{array}\\
\left[\begin{array}{ccccc|c}
1 & 0 & 4 & -2 & 0 & 150 \\
0 & 1 & -3 & 3 & 0 & 75 \\
\hline
0 & 0 & \frac{48}{5} & \frac{12}{5} & 1 & 1260
\end{array}\right]
\end{array}
$$

The maximum value is 1260 when $x_1 = 150$, $x_2 = 75$, $x_3 = 0$, and $x_4 = 0$.

The maximum revenue is $1260 when 150 kg of the half-and-half mix and 75 kg of the other mix are prepared.

25. (a) Look at the first chart, which has to do with the profits. The profit-maximization formula is

$2A + $5B + $4C = X$,

so the answer is choice 1.

(b) Look at the Painting row of the second chart.

The Painting constraint is

$$1A + 2B + 2C \leq 38,000,$$

so the answer is choice 3.

27. Let x_1 represent the number of minutes for the senator, x_2 the number of minutes for the congresswoman, and x_3 the number of minutes for the governor. Of the half-hour show's time, only $30 - 3 = 27$ min are available to be allotted to the politicans. The given information leads to the equation $x_1 + x_2 + x_3 = 27$ and the inequalities

$$x_1 \geq 2x_3 \quad \text{and} \quad x_1 + x_3 \geq 2x_2,$$

and we are to maximize the objective function $z = 40x_1 + 60x_2 + 50x_3$. Rewrite the equation as $x_3 = 27 - x_1 - x_2$ and the inequalities as

$$x_1 - 2x_3 \geq 0 \quad \text{and} \quad x_1 - 2x_2 + x_3 \geq 0.$$

Substitute $27 - x_1 - x_2$ for x_3 in the objective function and the inequalites, and the problem is as follows.

Maximize $z = 40x_1 + 60x_2 + 50(27 - x_1 - x_2)$ subject to:

$$x_1 \qquad - 2(27 - x_1 - x_2) \geq 0$$
$$x_1 - 2x_2 + (27 - x_1 - x_2) \geq 0$$

with $x_1 \geq 0$, $x_2 \geq 0$, and
$x_3 = 27 - x_1 - x_2 \geq 0$.
Simplify to obtain the following.

Maximize $z = 1350 - 10x_1 + 10x_2$ subject to: $3x_1 + 2x_2 \geq 54$
$$x_2 \leq 9$$
with $x_1 \geq 0$, $x_2 \geq 0$, and
$x_3 = 27 - x_1 - x_2 \geq 0$.

This linear programming problem is not in standard maximum form, so we will attempt to use reason rather than the simplex algorithm. (The graphical method of solution examined in Chapter 3 would also be acceptable.)
Decreasing x_2 would increase x_1 (since $x_1 + x_2 + x_3$ must remain equal to 27) and would decrease z (since $z = 1350 - 10x_1 + 10x_2$). Therefore, choose x_2 to be as large as possible, which means $x_2 = 9$. The constraint $3x_1 + 2x_2 \geq 54$ can be rewritten as $x_1 \geq \dfrac{54 - 2x_2}{3}$.
Since we have chosen $x_2 = 9$, this means $x_1 \geq \dfrac{54 - 2(9)}{3}$ or $x_1 \geq 12$.

Increasing x_1 would decrease z, so choose x_1 to be as small as possible. This means $x_1 = 12$.
With $x_1 = 12$ and $x_2 = 9$, we have $x_3 = 27 - 12 - 9$ or $x_3 = 6$, and $z = 1350 - 10(12) + 10(9)$ or $z = 1320$. For a maximum of 1,320,000 viewers, the time allotments should be 12 min for the senator, 9 min for the congresswoman, and 6 min for the governor.

29. **(a)** Let x_1 represent the number of species A,

x_2 represent the number of species B, and

x_3 represent the number of species C.

We are asked to

maximize $z = 1.62x_1 + 2.14x_2 + 3.01x_3$ subject to:

$$1.32x_1 + 2.1x_2 + .86x_3 \leq 490$$
$$2.9x_1 + .95x_2 + 1.52x_3 \leq 897$$
$$1.75x_1 + .6x_2 + 2.01x_3 \leq 653$$

Use a computer to solve this problem and find that the answer is to stock none of species A, 114 of species B, and 291 of species C for a maximum combined weight of 1119.72 kg.

(b) Many answers are possible. The idea is to choose average weights for species B and C that are considerably smaller than the average weight chosen for species A, so that species A dominates the objective function.

(c) Many answers are possible. The idea is to choose average weights for species A and B that are considerably smaller than the average weight chosen for species C.

Section 4.3

1. $2x_1 + 3x_2 \leq 8$

    $x_1 + 4x_2 \geq 7$

    Introduce the slack variable x_3 and the surplus variable x_4 to obtain the following equations.

    $$2x_1 + 3x_2 + x_3 \quad\quad = 8$$
    $$x_1 + 4x_2 \quad\quad - x_4 = 7$$

3. $x_1 + x_2 + x_3 \leq 100$

    $x_1 + x_2 + x_3 \geq 75$

    $x_1 + x_2 \quad\quad \geq 27$

    Introduce the slack variable x_4 and the surplus variables x_5 and x_6 to obtain the following equations.

    $$x_1 + x_2 + x_3 + x_4 \quad\quad = 100$$
    $$x_1 + x_2 + x_3 \quad\quad - x_5 \quad = 75$$
    $$x_1 + x_2 \quad\quad\quad - x_6 = 27$$

5. Minimize $w = 4y_1 + 3y_2 + 2y_3$

    subject to: $y_1 + y_2 + y_3 \geq 5$

    $y_1 + y_2 \quad\quad \geq 4$

    $2y_1 + y_2 + 3y_3 \geq 15$

    with $y_1 \geq 0, y_2 \geq 0, y_3 \geq 0.$

    Change this to a maximization problem by letting $z = -w$. The problem can now be stated as equivalently as follows.

    Maximize $z = -4y_1 - 3y_2 - 2y_3$

    subject to: $y_1 + y_2 + y_3 \geq 5$

    $y_1 + y_2 \quad\quad \geq 4$

    $2y_1 + y_2 + 3y_3 \geq 15$

    with $y_1 \geq 0, y_2 \geq 0, y_3 \geq 0.$

7. Minimize $w = y_1 + 2y_2 + y_3 + 5y_4$

    subject to: $y_1 + y_2 + y_3 + y_4 \geq 50$

    $3y_1 + y_2 + 2y_3 + y_4 \geq 100$

    with $y_1 \geq 0, y_2 \geq 0, y_3 \geq 0,$

    $y_4 \geq 0.$

    Change this to a maximization problem by letting $z = -w$. The problem can now be stated equivalently as follows.

    Maximize $z = -y_1 - 2y_2 - y_3 - 5y_4$

    subject to: $y_1 + y_2 + y_3 + y_4 \geq 50$

    $3y_1 + y_2 + 2y_3 + y_4 \geq 100$

    with $y_1 \geq 0, y_2 \geq 0, y_3 \geq 0,$

    $y_4 \geq 0.$

9. Find $x_1 \geq 0$ and $x_2 \geq 0$ such that

    $x_1 + 2x_2 \geq 24$

    $x_1 + x_2 \leq 40$

    and $z = 12x_1 + 10x_2$ is maximized. Subtracting the surplus variable x_3 and adding the slack variable x_4 leads to the equations

$$x_1 + 2x_2 - x_3 \quad\quad = 24$$
$$x_1 + x_2 \quad\quad + x_4 = 40.$$

The initial simplex tableau is as follows.

$$
\begin{array}{ccccc}
x_1 & x_2 & x_3 & x_4 & z \\
\end{array}
$$

$$
\left[
\begin{array}{ccccc|c}
① & 2 & -1 & 0 & 0 & 24 \\
1 & 1 & 0 & 1 & 0 & 40 \\
\hline
-12 & -10 & 0 & 0 & 1 & 0 \\
\end{array}
\right]
$$

The initial basic solution is not feasible, since $x_3 = -24$ is negative, so row transformations must be used. We pivot on the 1 in the first row, first column, since it is the positive entry that is farthest to the left in the first row (the row containing the -1) and since, in the first column, $24/1 = 24$ is a smaller quotient than $40/1 = 40$. After row transformations, we obtain the following tableau.

$$
\begin{array}{ccccc}
x_1 & x_2 & x_3 & x_4 & z \\
\end{array}
$$

$$
\begin{array}{l}
\\
-R_1 + R_2 \to R_2 \\
\\
12R_1 + R_3 \to R_3 \\
\end{array}
\left[
\begin{array}{ccccc|c}
1 & 2 & -1 & 0 & 0 & 24 \\
0 & -1 & ① & 1 & 0 & 16 \\
\hline
0 & 14 & -12 & 0 & 1 & 288 \\
\end{array}
\right]
$$

The basic solution is now feasible, but the problem is not yet finished since there is a negative indicator. Continue now in the usual way. The 1 in the third column is the next pivot. After row transformations, we get the following tableau.

$$
\begin{array}{ccccc}
x_1 & x_2 & x_3 & x_4 & z \\
\end{array}
$$

$$
\begin{array}{l}
R_1 + R_2 \to R_1 \\
\\
\\
12R_2 + R_3 \to R_3 \\
\end{array}
\left[
\begin{array}{ccccc|c}
1 & 1 & 0 & 1 & 0 & 40 \\
0 & -1 & 1 & 1 & 0 & 16 \\
\hline
0 & 2 & 0 & 12 & 1 & 480 \\
\end{array}
\right]
$$

This is a final tableau, since the entries in the last row are all non-negative. The maximum value is 480 and occurs when $x_1 = 40$, $x_2 = 0$, $x_3 = 16$, and $x_4 = 0$.

11. Find $x_1 \geq 0$, $x_2 \geq 0$, and $x_3 \geq 0$ such that

$$x_1 + x_2 + x_3 \leq 150$$
$$x_1 + x_2 + x_3 \geq 100$$

and $z = 2x_1 + 5x_2 + 3x_3$ is maximized.

The initial tableau is as follows.

$$
\begin{array}{cccccc}
x_1 & x_2 & x_3 & x_4 & x_5 & z \\
\end{array}
$$

$$
\left[
\begin{array}{cccccc|c}
1 & 1 & 1 & 1 & 0 & 0 & 150 \\
1 & ① & 1 & 0 & -1 & 0 & 100 \\
\hline
-2 & -5 & -3 & 0 & 0 & 1 & 0 \\
\end{array}
\right]
$$

Note that x_4 is a slack variable, while x_5 is a surplus variable. The initial basic solution is not feasible, since $x_5 = -100$ is negative. Pivot on the 1 in the second row, second column. (Note that any column with a non-negative entry in the second row could be used as the pivot column.)

$$
\begin{array}{cccccc}
x_1 & x_2 & x_3 & x_4 & x_5 & z \\
\end{array}
$$

$$
\begin{array}{l}
-R_2 + R_1 \to R_1 \\
\\
\\
5R_2 + R_3 \to R_3 \\
\end{array}
\left[
\begin{array}{cccccc|c}
0 & 0 & 0 & 1 & ① & 0 & 50 \\
1 & 1 & 1 & 0 & -1 & 0 & 100 \\
\hline
3 & 0 & 2 & 0 & -5 & 1 & 500 \\
\end{array}
\right]
$$

The basic solution is now feasible, but there is a negative indicator. Pivot on the 1 in the first row, fifth column.

$$
\begin{array}{c}
\\
R_1 + R_2 \rightarrow R_2 \\
\\
5R_1 + R_3 \rightarrow R_3
\end{array}
\quad
\begin{array}{cccccc|c}
x_1 & x_2 & x_3 & x_4 & x_5 & z & \\
0 & 0 & 0 & 1 & 1 & 0 & 50 \\
1 & 1 & 1 & 1 & 0 & 0 & 150 \\
\hline
3 & 0 & 2 & 5 & 0 & 1 & 750
\end{array}
$$

This is a final tableau. The maximum value is 750 when $x_1 = 0$, $x_2 = 150$, $x_3 = 0$, $x_4 = 0$, and $x_5 = 50$.

13. Find $x_1 \geq 0$ and $x_2 \geq 0$ such that

$$x_1 + x_2 \leq 100$$
$$x_1 + x_2 \geq 50$$
$$2x_1 + x_2 \leq 110$$

and $z = -2x_1 + 3x_2$ is maximized. The initial tableau is as follows.

$$
\begin{array}{cccccc|c}
x_1 & x_2 & x_3 & x_4 & x_5 & z & \\
1 & 1 & 1 & 0 & 0 & 0 & 100 \\
1 & ① & 0 & -1 & 0 & 0 & 50 \\
2 & 1 & 0 & 0 & 1 & 0 & 110 \\
\hline
2 & -3 & 0 & 0 & 0 & 1 & 0
\end{array}
$$

The initial basic solution is not feasible since $x_4 = -50$.
Pivot on the 1 in the second row, second column.

$$
\begin{array}{c}
-R_2 + R_1 \rightarrow R_1 \\
\\
-R_2 + R_3 \rightarrow R_3 \\
3R_2 + R_4 \rightarrow R_4
\end{array}
\quad
\begin{array}{cccccc|c}
x_1 & x_2 & x_3 & x_4 & x_5 & z & \\
0 & 0 & 1 & ① & 0 & 0 & 50 \\
1 & 1 & 0 & -1 & 0 & 0 & 50 \\
1 & 0 & 0 & 1 & 1 & 0 & 60 \\
\hline
5 & 0 & 0 & -3 & 0 & 1 & 150
\end{array}
$$

The basic solution is now feasible, but there is a negative indicator. Pivot on the 1 in the first row, fourth column.
(It would also be acceptable to pivot on the 1 in the second row, first column.)

$$
\begin{array}{c}
\\
-R_1 + R_2 \rightarrow R_2 \\
-R_1 + R_3 \rightarrow R_3 \\
\\
3R_1 + R_4 \rightarrow R_4
\end{array}
\quad
\begin{array}{cccccc|c}
x_1 & x_2 & x_3 & x_4 & x_5 & z & \\
0 & 0 & 1 & 1 & 0 & 0 & 50 \\
1 & 1 & 1 & 0 & 0 & 0 & 100 \\
1 & 0 & -1 & 0 & 1 & 0 & 10 \\
\hline
5 & 0 & 3 & 0 & 0 & 1 & 300
\end{array}
$$

This is a final tableau.
The maximum value is 300 when $x_1 = 0$, $x_2 = 100$, $x_3 = 0$, $x_4 = 50$, and $x_5 = 10$.

15. Find $y_1 \geq 0$, $y_2 \geq 0$ such that

$$10y_1 + 5y_2 \geq 100$$
$$20y_1 + 10y_2 \geq 150$$

and $w = 4y_1 + 5y_2$ is minimized. Let $z = -w$, and then $z = -4y_1 - 5y_2$ is to be maximized. The initial tableau is as follows.

$$
\begin{array}{ccccc|c}
y_1 & y_2 & y_3 & y_4 & z & \\
10 & 5 & -1 & 0 & 0 & 100 \\
⑳ & 10 & 0 & -1 & 0 & 150 \\
\hline
4 & 5 & 0 & 0 & 1 & 0
\end{array}
$$

The initial basic solution is not feasible since $y_3 = -100$ and $y_4 = -150$. Pivot on the 20 in the second row, first column.

$$
\begin{array}{c}
2R_1 - R_2 \rightarrow R_1 \\
\\
5R_3 - R_2 \rightarrow R_3
\end{array}
\quad
\begin{array}{ccccc|c}
y_1 & y_2 & y_3 & y_4 & z & \\
0 & 0 & -2 & ① & 0 & 50 \\
20 & 10 & 0 & -1 & 0 & 150 \\
\hline
0 & 15 & 0 & 1 & 5 & -150
\end{array}
$$

The basic solution is still not feasible. Pivot on the 1 in the first row, fourth column.

$$
\begin{array}{c}
\\
R_1 + R_2 \rightarrow R_2 \\
-R_1 + R_3 \rightarrow R_3
\end{array}
\quad
\begin{array}{ccccc|c}
y_1 & y_2 & y_3 & y_4 & z & \\
0 & 0 & -2 & 1 & 0 & 50 \\
20 & 10 & -2 & 0 & 0 & 200 \\
\hline
0 & 15 & 2 & 0 & 5 & -200
\end{array}
$$

The basic solution is now feasible. Additionally, all of the indicators are nonnegative, so we are finished pivoting. For ease in reading the basic solution, create a 1 in the columns corresponding to y_1 and z.

$$
\begin{array}{c}
\\
\frac{1}{20}R_2 \rightarrow R_2 \\
\frac{1}{5}R_3 \rightarrow R_3
\end{array}
\begin{array}{cccccc}
y_1 & y_2 & y_3 & y_4 & z & \\
\left[\begin{array}{ccccc|c}
0 & 0 & -2 & 1 & 0 & 50 \\
1 & .5 & -.1 & 0 & 0 & 10 \\
\hline
0 & 3 & .4 & 0 & 1 & -40
\end{array}\right]
\end{array}
$$

This is a final tableau. The maximum value of $z = -w$ is -40. Therefore, the minimum value of w is 40, and it occurs when $y_1 = 10$, $y_2 = 0$, $y_3 = 0$, and $y_4 = 50$.

17. Minimize $w = 2y_1 + y_2 + 3y_3$
    subject to: $y_1 + y_2 + y_3 \geq 100$
    $\qquad\qquad 2y_1 + y_2 \qquad \geq 50$
    with $\qquad y_1 \geq 0,\ y_2 \geq 0,\ y_3 \geq 0$.
    Let $z = -w$, and then $w = -2y_1 - y_2 - 3y_3$ is to be maximized.
    The initial tableau is as follows.

$$
\begin{array}{cccccc}
y_1 & y_2 & y_3 & y_4 & y_5 & z \\
\left[\begin{array}{cccccc|c}
1 & 1 & 1 & -1 & 0 & 0 & 100 \\
② & 1 & 0 & 0 & -1 & 0 & 50 \\
\hline
2 & 1 & 3 & 0 & 0 & 1 & 0
\end{array}\right]
\end{array}
$$

The initial basic solution is not feasible. Pivot on the 2 in the second row, first column.

$$
\begin{array}{c}
2R_1 - R_2 \rightarrow R_1 \\
\\
-R_2 + R_3 \rightarrow R_3
\end{array}
\begin{array}{cccccc}
y_1 & y_2 & y_3 & y_4 & y_5 & z \\
\left[\begin{array}{cccccc|c}
0 & 1 & 2 & -2 & 1 & 0 & 150 \\
2 & ① & 0 & 0 & -1 & 0 & 50 \\
\hline
0 & 0 & 3 & 0 & 1 & 1 & -50
\end{array}\right]
\end{array}
$$

The basic solution is still not feasible. Pivot on the 1 in the

second row, second column.

$$
-R_2 + R_1 \rightarrow R_1
\begin{array}{cccccc}
y_1 & y_2 & y_3 & y_4 & y_5 & z \\
\left[\begin{array}{cccccc|c}
-2 & 0 & ② & -2 & 2 & 0 & 100 \\
2 & 1 & 0 & 0 & -1 & 0 & 50 \\
\hline
0 & 0 & 3 & 0 & 1 & 1 & -50
\end{array}\right]
\end{array}
$$

The basic solution is still not feasible. Pivot on the 2 in the first row, third column.

$$
-3R_1 + 2R_3 \rightarrow R_3
\begin{array}{cccccc}
y_1 & y_2 & y_3 & y_4 & y_5 & z \\
\left[\begin{array}{cccccc|c}
-2 & 0 & 2 & -2 & ② & 0 & 100 \\
2 & 1 & 0 & 0 & -1 & 0 & 50 \\
\hline
6 & 0 & 0 & 6 & -4 & 2 & -400
\end{array}\right]
\end{array}
$$

The basic solution is now feasible, but there is a negative indicator. Pivot on the 2 in the first row, fifth column.

$$
\begin{array}{c}
\\
R_1 + 2R_2 \rightarrow R_2 \\
2R_1 + R_3 \rightarrow R_3
\end{array}
\begin{array}{cccccc}
y_1 & y_2 & y_3 & y_4 & y_5 & z \\
\left[\begin{array}{cccccc|c}
-2 & 0 & 2 & -2 & 2 & 0 & 100 \\
2 & 2 & 2 & -2 & 0 & 0 & 200 \\
\hline
2 & 0 & 4 & 2 & 0 & 2 & -200
\end{array}\right]
\end{array}
$$

All of the indicators are nonnegative, so we are finished pivoting. For ease in reading the basic solution, create a 1 in the columns corresponding to y_2, y_5, and z.

$$
\begin{array}{c}
\frac{1}{2}R_1 \rightarrow R_1 \\
\frac{1}{2}R_2 \rightarrow R_2 \\
\frac{1}{2}R_3 \rightarrow R_3
\end{array}
\begin{array}{cccccc}
y_1 & y_2 & y_3 & y_4 & y_5 & z \\
\left[\begin{array}{cccccc|c}
-1 & 0 & 1 & -1 & 1 & 0 & 50 \\
1 & 1 & 1 & -1 & 0 & 0 & 100 \\
\hline
1 & 0 & 2 & 1 & 0 & 1 & -100
\end{array}\right]
\end{array}
$$

This is a final tableau. The maximum value of $z = -w$ is -100. Therefore, the minimum value of w is 100, and it occurs when $y_1 = 0$, $y_2 = 100$, $y_3 = 0$, $y_4 = 0$, and $y_5 = 50$.

19. Maximize $z = 3x_1 + 2x_2$

subject to: $x_1 + x_2 = 50$

$4x_1 + 2x_2 \geq 120$

$5x_1 + 2x_2 \leq 200$

with $x_1 \geq 0, x_2 \geq 0$.

The artificial variable a_1 is used to rewrite $x_1 + x_2 = 50$ as $x_1 + x_2 + a_1 = 50$; note that a_1 must equal 0 for this equation to be a true statement. Also, the surplus variable x_3 and the slack variable x_4 are needed. The initial tableau is a follows.

$$
\begin{array}{cccccc|c}
x_1 & x_2 & a_1 & x_3 & x_4 & z & \\
\hline
1 & 1 & 1 & 0 & 0 & 0 & 50 \\
④ & 2 & 0 & -1 & 0 & 0 & 120 \\
5 & 2 & 0 & 0 & 1 & 0 & 200 \\
\hline
-3 & -2 & 0 & 0 & 0 & 1 & 0
\end{array}
$$

The initial basic solution is not feasible. Pivot on the 4 in the second row, first column.

$$
\begin{array}{c}
4R_1 - R_2 \rightarrow R_1 \\
\\
-5R_2 + 4R_3 \rightarrow R_3 \\
3R_2 + 4R_4 \rightarrow R_4
\end{array}
\begin{array}{cccccc|c}
x_1 & x_2 & a_1 & x_3 & x_4 & z & \\
\hline
0 & 2 & 4 & 1 & 0 & 0 & 80 \\
4 & 2 & 0 & -1 & 0 & 0 & 120 \\
0 & -2 & 0 & ⑤ & 2 & 0 & 200 \\
\hline
2 & -2 & 0 & -3 & 0 & 4 & 360
\end{array}
$$

The basic solution is now feasible, but there are negative indicators. Pivot on the 5 in the third row, fourth column (which is the column with the most negative indicator and the row with the smallest nonnegative quotient).

$$
\begin{array}{c}
5R_1 - R_3 \rightarrow R_1 \\
5R_2 + R_3 \rightarrow R_2 \\
\\
3R_3 + 5R_4 \rightarrow R_4
\end{array}
\begin{array}{cccccc|c}
x_1 & x_2 & a_1 & x_3 & x_4 & z & \\
\hline
0 & ⑫ & 20 & 0 & -2 & 0 & 200 \\
20 & 8 & 0 & 0 & 2 & 0 & 800 \\
0 & -2 & 0 & 5 & 2 & 0 & 200 \\
\hline
0 & -16 & 0 & 0 & 6 & 20 & 2400
\end{array}
$$

Pivot on the 12 in the first row, second column.

$$
\begin{array}{c}
\\
-2R_1 + 3R_2 \rightarrow R_2 \\
R_1 + 6R_3 \rightarrow R_3 \\
4R_1 + 3R_4 \rightarrow R_4
\end{array}
\begin{array}{cccccc|c}
x_1 & x_2 & a_1 & x_3 & x_4 & z & \\
\hline
0 & 12 & 20 & 0 & -2 & 0 & 200 \\
60 & 0 & -40 & 0 & 10 & 0 & 2000 \\
0 & 0 & 20 & 30 & 10 & 0 & 1400 \\
\hline
0 & 0 & 80 & 0 & 10 & 60 & 8000
\end{array}
$$

We now have $a_1 = 0$, so drop the a_1 column.

$$
\begin{array}{ccccc|c}
x_1 & x_2 & x_3 & x_4 & z & \\
\hline
0 & 12 & 0 & -2 & 0 & 200 \\
60 & 0 & 0 & 10 & 0 & 2000 \\
0 & 0 & 30 & 10 & 0 & 1400 \\
\hline
0 & 0 & 0 & 10 & 60 & 8000
\end{array}
$$

We are finished pivoting. Create a 1 in the columns corresponding to x_1, x_2, x_3, and z.

$$
\begin{array}{c}
\frac{1}{12}R_1 \rightarrow R_1 \\
\frac{1}{60}R_2 \rightarrow R_2 \\
\frac{1}{30}R_3 \rightarrow R_3 \\
\\
\frac{1}{60}R_4 \rightarrow R_4
\end{array}
\begin{array}{ccccc|c}
x_1 & x_2 & x_3 & x_4 & z & \\
\hline
0 & 1 & 0 & -\frac{1}{6} & 0 & 16\frac{2}{3} \\
1 & 0 & 0 & \frac{1}{6} & 0 & 33\frac{1}{3} \\
0 & 0 & 1 & \frac{1}{3} & 0 & 46\frac{2}{3} \\
\hline
0 & 0 & 0 & \frac{1}{6} & 1 & 133\frac{1}{3}
\end{array}
$$

The maximum value is $133\frac{1}{3}$ when $x_1 = 33\frac{1}{3}$, $x_2 = 16\frac{2}{3}$, $x_3 = 46\frac{2}{3}$, and $x_4 = 0$.

21. Minimize $w = 15y_1 + 12y_2$

subject to: $y_1 + 2y_2 \leq 12$

$3y_1 + y_2 \geq 18$

$y_1 + y_2 = 10$

with $y_1 \geq 0, y_2 \geq 0$.

Let $z = -w$, and then $z = -15y_1 - 12y_2$ is to be maximized. Introduce the slack variable y_3, the surplus variable y_4, and the artificial variable a_1. The initial tableau is as follows.

$$\begin{array}{cccccc} y_1 & y_2 & y_3 & y_4 & a_1 & z \\ \left[\begin{array}{cccccc|c} 1 & 2 & 1 & 0 & 0 & 0 & 12 \\ ③ & 1 & 0 & -1 & 0 & 0 & 18 \\ 1 & 1 & 0 & 0 & 1 & 0 & 10 \\ \hline 15 & 12 & 0 & 0 & 0 & 1 & 0 \end{array}\right] \end{array}$$

The initial basic solution is not feasible. Pivot on the 3 in the second row, first column.

$$\begin{array}{c} \\ 3R_1 - R_2 \to R_1 \\ \\ 3R_3 - R_2 \to R_3 \\ \\ -5R_2 + R_4 \to R_4 \end{array} \begin{array}{cccccc} y_1 & y_2 & y_3 & y_4 & a_1 & z \\ \left[\begin{array}{cccccc|c} 0 & 5 & 3 & 1 & 0 & 0 & 18 \\ 3 & 1 & 0 & -1 & 0 & 0 & 18 \\ 0 & 2 & 0 & ① & 3 & 0 & 12 \\ \hline 0 & 7 & 0 & 5 & 0 & 1 & -90 \end{array}\right] \end{array}$$

The basic solution is now feasible, but the artificial variable still remains, so we are not finished. Try pivoting on the 1 in the third row, fourth column.

$$\begin{array}{c} -R_3 + R_1 \to R_1 \\ R_3 + R_2 \to R_2 \\ \\ -5R_3 + R_4 \to R_4 \end{array} \begin{array}{cccccc} y_1 & y_2 & y_3 & y_4 & a_1 & z \\ \left[\begin{array}{cccccc|c} 0 & 3 & 3 & 0 & -3 & 0 & 6 \\ 3 & 3 & 0 & 0 & 3 & 0 & 30 \\ 0 & 2 & 0 & 1 & 3 & 0 & 12 \\ \hline 0 & -3 & 0 & 0 & -15 & 1 & -150 \end{array}\right] \end{array}$$

We now have $a_1 = 0$, so drop the a_1 column.

$$\begin{array}{ccccc} y_1 & y_2 & y_3 & y_4 & z \\ \left[\begin{array}{ccccc|c} 0 & ③ & 3 & 0 & 0 & 6 \\ 3 & 3 & 0 & 0 & 0 & 30 \\ 0 & 2 & 0 & 1 & 0 & 12 \\ \hline 0 & -3 & 0 & 0 & 1 & -150 \end{array}\right] \end{array}$$

Pivot on the 3 in the first row, second column.

$$\begin{array}{c} \\ -R_1 + R_2 \to R_2 \\ -2R_1 + 3R_3 \to R_3 \\ R_1 + R_4 \to R_4 \end{array} \begin{array}{ccccc} y_1 & y_2 & y_3 & y_4 & z \\ \left[\begin{array}{ccccc|c} 0 & 3 & 3 & 0 & 0 & 6 \\ 3 & 0 & -3 & 0 & 0 & 24 \\ 0 & 0 & -6 & 3 & 0 & 24 \\ \hline 0 & 0 & 3 & 0 & 1 & -144 \end{array}\right] \end{array}$$

We are finished pivoting. Create a 1 in the columns corresponding to y_1, y_2, and y_4.

$$\begin{array}{c} \frac{1}{3}R_1 \to R_1 \\ \frac{1}{3}R_2 \to R_2 \\ \frac{1}{3}R_3 \to R_3 \\ \\ \end{array} \begin{array}{ccccc} y_1 & y_2 & y_3 & y_4 & z \\ \left[\begin{array}{ccccc|c} 0 & 1 & 1 & 0 & 0 & 2 \\ 1 & 0 & -1 & 0 & 0 & 8 \\ 0 & 0 & -2 & 1 & 0 & 8 \\ \hline 0 & 0 & 3 & 0 & 1 & -144 \end{array}\right] \end{array}$$

This is a final tableau. The maximum value of $z = -w$ is -144. Therefore, the minimum value of w is 144, and it occurs when $y_1 = 8$, $y_2 = 2$, $y_3 = 0$, and $y_4 = 8$.

25. Let $y_1 =$ the number of barrels from S_1 to D_1;

$y_2 =$ the number of barrels from S_2 to D_1;

$y_3 =$ the number of barrels from S_1 to D_2; and

$y_4 =$ the number of barrels from S_2 to D_2.

The problem is to minimize

$$w = 30y_1 + 25y_2 + 20y_3 + 22y_4$$

subject to:

$$y_1 + y_2 \ge 3000$$
$$y_3 + y_4 \ge 5000$$
$$y_1 + y_3 \le 5000$$
$$y_2 + y_4 \le 5000$$

with $y_1 \ge 0$, $y_2 \ge 0$, $y_3 \ge 0$, $y_4 \ge 0$.

Let $z = -w$, and maximize z. The initial tableau is as follows.

$$
\begin{array}{ccccccccc|c}
y_1 & y_2 & y_3 & y_4 & y_5 & y_6 & y_7 & y_8 & z & \\
\hline
① & 1 & 0 & 0 & -1 & 0 & 0 & 0 & 0 & 3000 \\
0 & 0 & 1 & 1 & 0 & -1 & 0 & 0 & 0 & 5000 \\
1 & 0 & 1 & 0 & 0 & 0 & 1 & 0 & 0 & 5000 \\
0 & 1 & 0 & 1 & 0 & 0 & 0 & 1 & 0 & 5000 \\
\hline
30 & 25 & 20 & 22 & 0 & 0 & 0 & 0 & 1 & 0
\end{array}
$$

The initial basic solution is not feasible. Pivot on the 1 in the first row, first column.

$$
\begin{array}{r}
\\
\\
-R_1 + R_3 \to R_3 \\
\\
\\
-20R_1 + R_5 \to R_5
\end{array}
\begin{array}{ccccccccc|c}
y_1 & y_2 & y_3 & y_4 & y_5 & y_6 & y_7 & y_8 & z & \\
\hline
1 & 1 & 0 & 0 & -1 & 0 & 0 & 0 & 0 & 3000 \\
0 & 0 & 1 & 1 & 0 & -1 & 0 & 0 & 0 & 5000 \\
0 & -1 & ① & 0 & 1 & 0 & 1 & 0 & 0 & 2000 \\
0 & 1 & 0 & 1 & 0 & 0 & 0 & 1 & 0 & 5000 \\
\hline
0 & -5 & 20 & 22 & 30 & 0 & 0 & 0 & 1 & -90,000
\end{array}
$$

The basic solution is still not feasible. Pivot on the 1 in the third row, third column.

$$
\begin{array}{r}
\\
-R_3 + R_2 \to R_2 \\
\\
\\
-20R_3 + R_5 \to R_5
\end{array}
\begin{array}{ccccccccc|c}
y_1 & y_2 & y_3 & y_4 & y_5 & y_6 & y_7 & y_8 & z & \\
\hline
1 & 1 & 0 & 0 & -1 & 0 & 0 & 0 & 0 & 3000 \\
0 & ① & 0 & 1 & -1 & -1 & -1 & 0 & 0 & 3000 \\
0 & -1 & 1 & 0 & 1 & 0 & 1 & 0 & 0 & 2000 \\
0 & 1 & 0 & 1 & 0 & 0 & 0 & 1 & 0 & 5000 \\
\hline
0 & 15 & 0 & 22 & 10 & 0 & -20 & 0 & 1 & -130,000
\end{array}
$$

The basic solution is still not feasible. Pivot on the 1 in the second row, second column.

$$
\begin{array}{r}
-R_2 + R_1 \to R_1 \\
\\
R_2 + R_3 \to R_3 \\
-R_2 + R_4 \to R_4 \\
-15R_2 + R_5 \to R_5
\end{array}
\begin{array}{ccccccccc|c}
y_1 & y_2 & y_3 & y_4 & y_5 & y_6 & y_7 & y_8 & z & \\
\hline
1 & 0 & 0 & -1 & 0 & ① & 1 & 0 & 0 & 0 \\
0 & 1 & 0 & 1 & -1 & -1 & -1 & 0 & 0 & 3000 \\
0 & 0 & 1 & 1 & 0 & -1 & 0 & 0 & 0 & 5000 \\
0 & 0 & 0 & 0 & 1 & 1 & 1 & 1 & 0 & 2000 \\
\hline
0 & 0 & 0 & 7 & 25 & 15 & -5 & 0 & 1 & -175,000
\end{array}
$$

The basic solution is now feasible, but there is a negative indicator. Pivot on the 1 in the first row, seventh column.

$$
\begin{array}{c}
\begin{array}{cccccccccc}
\quad y_1 & y_2 & y_3 & y_4 & y_5 & y_6 & y_7 & y_8 & z & \\
\end{array} \\
\begin{array}{r}
\\
R_1 + R_2 \to R_2 \\
\\
-R_1 + R_4 \to R_4 \\
\\
5R_1 + R_5 \to R_5
\end{array}
\left[
\begin{array}{ccccccccc|c}
1 & 0 & 0 & -1 & 0 & 1 & 1 & 0 & 0 & 0 \\
1 & 1 & 0 & 0 & -1 & 0 & 0 & 0 & 0 & 3000 \\
0 & 0 & 1 & 1 & 0 & -1 & 0 & 0 & 0 & 5000 \\
-1 & 0 & 0 & 1 & 1 & 0 & 0 & 1 & 0 & 2000 \\
\hline
5 & 0 & 0 & 2 & 25 & 20 & 0 & 0 & 1 & -175{,}000
\end{array}
\right]
\end{array}
$$

This is a final tableau. The minimum value of w is 175,000 when $y_1 = 0$, $y_2 = 3000$, $y_3 = 5000$, $y_4 = 0$, $y_5 = 0$, $y_6 = 0$, $y_7 = 0$, and $y_8 = 2000$. To minimize the shipping cost, ship 5000 barrels of oil from supplier S_1 to distributor D_2, and ship 3000 barrels of oil from supplier S_2 to distributor D_1. The minimum cost is $175,000.

27. Let x_1 = the number of kilograms of sauce and
and x_2 = the number of kilograms of whole tomatoes.

The problem is to

$$\text{minimize } w = 3.25y_1 + 4y_2$$
$$\text{subject to:} \quad y_1 + y_2 \le 3{,}000{,}000$$
$$y_1 \qquad\quad \ge \quad 80{,}000$$
$$y_2 \ge \quad 800{,}000$$
$$\text{with} \qquad y_1 \ge 0, \ y_2 \ge 0.$$

Let $z = -w$, and maximize z.

The initial tableau is as follows.

$$
\begin{array}{cccccc}
y_1 & y_2 & y_3 & y_4 & y_5 & z \\
\end{array}
$$
$$
\left[
\begin{array}{ccccc|c}
1 & 1 & 1 & 0 & 0 & 0 & 3{,}000{,}000 \\
① & 0 & 0 & -1 & 0 & 0 & 80{,}000 \\
0 & 1 & 0 & 0 & -1 & 0 & 800{,}000 \\
\hline
3.25 & 4 & 0 & 0 & 0 & 1 & 0
\end{array}
\right]
$$

The initial basic solution is not feasible. Pivot on the 1 in the second row, first column.

$$
\begin{array}{cccccc}
\quad y_1 & y_2 & y_3 & y_4 & y_5 & z \\
\end{array}
$$
$$
\begin{array}{r}
-R_2 + R_1 \to R_1 \\
\\
\\
-3.25R_2 + R_4 \to R_4
\end{array}
\left[
\begin{array}{cccccc|c}
0 & 1 & 1 & 1 & 0 & 0 & 2{,}920{,}000 \\
1 & 0 & 0 & -1 & 0 & 0 & 80{,}000 \\
0 & ① & 0 & 0 & -1 & 0 & 800{,}000 \\
\hline
0 & 4 & 0 & 3.25 & 0 & 1 & -260{,}000
\end{array}
\right]
$$

The basic solution is still not feasible. Pivot on the 1 in the third row, second column.

$$
\begin{array}{c}
 \\
-R_3 + R_1 \to R_1 \\
 \\
 \\
4R_3 + R_4 \to R_4
\end{array}
\begin{array}{cccccc}
y_1 & y_2 & y_3 & y_4 & y_5 & z \\
\end{array}
\left[
\begin{array}{cccccc|r}
0 & 0 & 1 & 1 & 1 & 0 & 2{,}120{,}000 \\
1 & 0 & 0 & -1 & 0 & 0 & 80{,}000 \\
0 & 1 & 0 & 0 & -1 & 0 & 800{,}000 \\
\hline
0 & 0 & 0 & 3.25 & 4 & 1 & -3{,}460{,}000
\end{array}
\right]
$$

The basic solution is now feasible. Additionally, there are no negative indicators, so this is a final tableau. The minimum value of w is 3,460,000 when $y_1 = 80{,}000$, $y_2 = 800{,}000$, $y_3 = 2{,}120{,}000$, $y_4 = 0$, and $y_5 = 0$.

Brand X should use 800,000 kg of tomatoes for whole tomatoes and 80,000 kg of tomatoes for sauce for a minimum cost of $3,460,000.

29. Let y_1 = the number of small tubes
and y_2 = the number of large tubes.

The problem is to minimize $.15y_1 + .12y_2$

$$
\begin{aligned}
\text{subject to: } \quad & y_1 \ge 800 \\
& y_2 \ge 500 \\
& y_1 + y_2 \ge 1500 \\
& y_1 \ge 2y_2 \\
\text{with} \quad & y_1 \ge 0, \; y_2 \ge 0.
\end{aligned}
$$

The last constraint can be rewritten as $y_1 - 2y_2 \ge 0$.

Let $z = -w$, and maximize z. The initial tableau is as follows.

$$
\begin{array}{ccccccc}
y_1 & y_2 & y_3 & y_4 & y_5 & y_6 & z \\
\end{array}
\left[
\begin{array}{ccccccc|r}
\textcircled{1} & 0 & -1 & 0 & 0 & 0 & 0 & 800 \\
0 & 1 & 0 & -1 & 0 & 0 & 0 & 500 \\
1 & 1 & 0 & 0 & -1 & 0 & 0 & 1500 \\
1 & -2 & 0 & 0 & 0 & -1 & 0 & 0 \\
\hline
.15 & .12 & 0 & 0 & 0 & 0 & 1 & 0
\end{array}
\right]
$$

Continue by pivoting on the circled entry in the above tableau.

$$
\begin{array}{c}
 \\
 \\
-R_1 + R_3 \to R_3 \\
-R_1 + R_4 \to R_4 \\
-.15R_1 + R_5 \to R_5
\end{array}
\begin{array}{ccccccc}
y_1 & y_2 & y_3 & y_4 & y_5 & y_6 & z \\
\end{array}
\left[
\begin{array}{ccccccc|r}
1 & 0 & -1 & 0 & 0 & 0 & 0 & 800 \\
0 & \textcircled{1} & 0 & -1 & 0 & 0 & 0 & 500 \\
0 & 1 & 1 & 0 & -1 & 0 & 0 & 700 \\
0 & -2 & 1 & 0 & 0 & -1 & 0 & -800 \\
\hline
0 & .12 & .15 & 0 & 0 & 0 & 1 & -120
\end{array}
\right]
$$

$$
\begin{array}{c}
\begin{array}{ccccccc}
y_1 & y_2 & y_3 & y_4 & y_5 & y_6 & z
\end{array} \\
\begin{array}{ccc}
 & & \\
 & & \\
-R_2 + R_3 \to R_3 \\
2R_2 + R_4 \to R_4 \\
 & & \\
-.12R_2 + R_5 \to R_5
\end{array}
\left[
\begin{array}{ccccccc|c}
1 & 0 & -1 & 0 & 0 & 0 & 0 & 800 \\
0 & 1 & 0 & -1 & 0 & 0 & 0 & 500 \\
0 & 0 & 1 & 1 & -1 & 0 & 0 & 200 \\
0 & 0 & 1 & -2 & 0 & -1 & 0 & 200 \\
\hline
0 & 0 & .15 & .12 & 0 & 0 & 1 & -180
\end{array}
\right]
\end{array}
$$

$$
\begin{array}{c}
\begin{array}{ccccccc}
y_1 & y_2 & y_3 & y_4 & y_5 & y_6 & z
\end{array} \\
\begin{array}{ccc}
R_1 + R_4 \to R_1 \\
 & & \\
-R_4 + R_3 \to R_3 \\
 & & \\
 & & \\
-.15R_4 + R_5 \to R_5
\end{array}
\left[
\begin{array}{ccccccc|c}
1 & 0 & 0 & -2 & 0 & -1 & 0 & 1000 \\
0 & 1 & 0 & -1 & 0 & 0 & 0 & 500 \\
0 & 0 & 0 & 3 & -1 & 1 & 0 & 0 \\
0 & 0 & 1 & -2 & 0 & -1 & 0 & 200 \\
\hline
0 & 0 & 0 & .42 & 0 & .15 & 1 & -210
\end{array}
\right]
\end{array}
$$

This is a final tableau. The minimum value of w is 210 when $y_1 = 1000$, $y_2 = 500$, $y_3 = 200$, $y_4 = 0$, $y_5 = 0$, and $y_6 = 0$.

1000 small and 500 large test tubes should be ordered for a minimum cost of $210.

31. Let x_1 = the amount to invest in securities;

   x_2 = the amount to invest in bonds; and

   x_3 = the amount to invest in mutual funds.

The problem is to maximize $.07x_1 + .06x_2 + .10x_3$

   subject to: $x_1 + x_2 + x_3 \leq 100,000$

   $x_1 \qquad\qquad \geq 40,000$

   $x_2 + x_3 \geq 50,000$

   with $x_1 \geq 0,\ x_2 \geq 0,\ x_3 \geq 0$.

The initial tableau is as follows.

$$
\begin{array}{c}
\begin{array}{ccccccc}
x_1 & x_2 & x_3 & x_4 & x_5 & x_6 & z
\end{array} \\
\left[
\begin{array}{ccccccc|c}
1 & 1 & 1 & 1 & 0 & 0 & 0 & 100,000 \\
① & 0 & 0 & 0 & -1 & 0 & 0 & 40,000 \\
0 & 1 & 1 & 0 & 0 & -1 & 0 & 50,000 \\
\hline
-.07 & -.06 & -.1 & 0 & 0 & 0 & 1 & 0
\end{array}
\right]
\end{array}
$$

Continue by pivoting on the circled entry in each tableau.

$$\begin{array}{r} \\ -R_2 + R_1 \to R_1 \\ \\ \\ .07R_2 + R_4 \to R_4 \end{array} \begin{array}{c} \begin{array}{ccccccc|c} x_1 & x_2 & x_3 & x_4 & x_5 & x_6 & z & \end{array} \\ \left[\begin{array}{ccccccc|c} 0 & 1 & 1 & 1 & 1 & 0 & 0 & 60,000 \\ 1 & 0 & 0 & 0 & -1 & 0 & 0 & 40,000 \\ 0 & \textcircled{1} & 1 & 0 & 0 & -1 & 0 & 50,000 \\ \hline 0 & -.06 & -.1 & 0 & -.07 & 0 & 1 & 2800 \end{array}\right] \end{array}$$

$$\begin{array}{r} \\ -R_3 + R_1 \to R_1 \\ \\ \\ .06R_3 + R_4 \to R_4 \end{array} \begin{array}{c} \begin{array}{ccccccc|c} x_1 & x_2 & x_3 & x_4 & x_5 & x_6 & z & \end{array} \\ \left[\begin{array}{ccccccc|c} 0 & 0 & 0 & 1 & \textcircled{1} & 0 & 0 & 10,000 \\ 1 & 0 & 0 & 0 & -1 & 0 & 0 & 40,000 \\ 0 & 1 & 1 & 0 & 0 & -1 & 0 & 50,000 \\ \hline 0 & 0 & -.04 & 0 & -.07 & -.06 & 1 & 5800 \end{array}\right] \end{array}$$

$$\begin{array}{r} \\ \\ R_1 + R_2 \to R_2 \\ \\ .07R_1 + R_4 \to R_4 \end{array} \begin{array}{c} \begin{array}{ccccccc|c} x_1 & x_2 & x_3 & x_4 & x_5 & x_6 & z & \end{array} \\ \left[\begin{array}{ccccccc|c} 0 & 0 & 0 & 1 & 1 & 1 & 0 & 10,000 \\ 1 & 0 & 0 & 1 & 0 & 1 & 0 & 50,000 \\ 0 & 1 & \textcircled{1} & 0 & 0 & -1 & 0 & 50,000 \\ \hline 0 & 0 & -.04 & .07 & 0 & .01 & 1 & 6500 \end{array}\right] \end{array}$$

$$\begin{array}{r} \\ \\ \\ \\ .04R_3 + R_4 \to R_4 \end{array} \begin{array}{c} \begin{array}{ccccccc|c} x_1 & x_2 & x_3 & x_4 & x_5 & x_6 & z & \end{array} \\ \left[\begin{array}{ccccccc|c} 0 & 0 & 0 & 1 & 1 & \textcircled{1} & 0 & 10,000 \\ 1 & 0 & 0 & 1 & 0 & 1 & 0 & 50,000 \\ 0 & 1 & 1 & 0 & 0 & -1 & 0 & 50,000 \\ \hline 0 & .04 & 0 & .07 & 0 & -.03 & 1 & 8500 \end{array}\right] \end{array}$$

$$\begin{array}{r} \\ \\ -R_1 + R_2 \to R_2 \\ R_1 + R_3 \to R_3 \\ .03R_1 + R_4 \to R_4 \end{array} \begin{array}{c} \begin{array}{ccccccc|c} x_1 & x_2 & x_3 & x_4 & x_5 & x_6 & z & \end{array} \\ \left[\begin{array}{ccccccc|c} 0 & 0 & 0 & 1 & 1 & 1 & 0 & 10,000 \\ 1 & 0 & 0 & 0 & -1 & 0 & 0 & 40,000 \\ 0 & 1 & 1 & 1 & 1 & 0 & 0 & 60,000 \\ \hline 0 & .04 & 0 & .1 & .03 & 0 & 1 & 8800 \end{array}\right] \end{array}$$

This is a final tableau. The maximum value is 8800 when $x_1 = 40,000$, $x_2 = 0$, $x_3 = 60,000$, $x_4 = 0$, $x_5 = 0$, and $x_6 = 10,000$.

\$40,000 should be invested in government securities and \$60,000 in mutual funds for a maximum annual interest of \$8800.

33. Let y_1 = the number of #1 pills
and y_2 = the number of #2 pills to be purchased.

Organize the given information in a table.

	Vitamin A	Vitamin B_1	Vitamin C	Cost
#1	8	1	2	$.10
#2	2	1	7	$.20
Total Needed	16	5	20	

The problem is to minimize $w = .1y_1 + .2y_2$

$$\text{subject to:} \qquad 8y_1 + 2y_2 \geq 16$$
$$y_1 + y_2 \geq 5$$
$$2y_1 + 7y_2 \geq 20$$

with $y_1 \geq 0$, $y_2 \geq 0$.

Let $z = -w$, and maximize z. The initial tableau is as follows.

$$
\begin{array}{cccccc}
y_1 & y_2 & y_3 & y_4 & y_5 & z \\
\end{array}
$$
$$
\left[
\begin{array}{cccccc|c}
⑧ & 2 & -1 & 0 & 0 & 0 & 16 \\
1 & 1 & 0 & -1 & 0 & 0 & 5 \\
2 & 7 & 0 & 0 & -1 & 0 & 20 \\
\hline
.1 & .2 & 0 & 0 & 0 & 1 & 0
\end{array}
\right]
$$

Continue by pivoting on the circled entry in each tableau.

$$
\begin{array}{cccccc}
y_1 & y_2 & y_3 & y_4 & y_5 & z \\
\end{array}
$$
$$
\begin{array}{l}
\\
8R_2 - R_1 \rightarrow R_2 \\
4R_3 - R_1 \rightarrow R_3 \\
\\
80R_4 - R_1 \rightarrow R_4
\end{array}
\left[
\begin{array}{cccccc|c}
8 & 2 & -1 & 0 & 0 & 0 & 16 \\
0 & 6 & 1 & -8 & 0 & 0 & 24 \\
0 & ㉖ & 1 & 0 & -4 & 0 & 64 \\
\hline
0 & 14 & 1 & 0 & 0 & 80 & -16
\end{array}
\right]
$$

$$
\begin{array}{cccccc}
y_1 & y_2 & y_3 & y_4 & y_5 & z \\
\end{array}
$$
$$
\begin{array}{l}
13R_1 - R_3 \rightarrow R_1 \\
13R_2 - 3R_3 \rightarrow R_2 \\
\\
-7R_3 + 13R_1 \rightarrow R_4
\end{array}
\left[
\begin{array}{cccccc|c}
104 & 0 & -14 & 0 & 4 & 0 & 144 \\
0 & 0 & ⑩ & -104 & 12 & 0 & 120 \\
0 & 26 & 1 & 0 & -4 & 0 & 64 \\
\hline
0 & 0 & 6 & 0 & 28 & 1040 & -656
\end{array}
\right]
$$

$$
\begin{array}{cccccc}
y_1 & y_2 & y_3 & y_4 & y_5 & z \\
\end{array}
$$
$$
\begin{array}{l}
5R_1 + 7R_2 \rightarrow R_1 \\
\\
10R_3 - R_2 \rightarrow R_3 \\
\\
-3R_3 + 5R_4 \rightarrow R_4
\end{array}
\left[
\begin{array}{cccccc|c}
520 & 0 & 0 & -728 & 144 & 0 & 1560 \\
0 & 0 & 10 & -104 & 12 & 0 & 120 \\
0 & 260 & 0 & 104 & 52 & 0 & 520 \\
\hline
0 & 0 & 0 & 312 & 104 & 5200 & 3640
\end{array}
\right]
$$

Create a 1 in the columns corresponding to y_1, y_2, y_3, and z.

$$
\begin{array}{c}
\\
\frac{1}{520}R_1 \to R_1 \\
\frac{1}{10}R_2 \to R_2 \\
\frac{1}{260}R_3 \to R_3 \\
\frac{1}{5200}R_4 \to R_4
\end{array}
\begin{array}{ccccccc}
y_1 & y_2 & y_3 & y_4 & y_5 & z & \\
\left[\begin{array}{cccccc|c}
1 & 0 & 0 & -1.4 & .28 & 0 & 3 \\
0 & 0 & 1 & -10.4 & 1.2 & 0 & 12 \\
0 & 1 & 0 & .4 & .2 & 0 & 2 \\
\hline
0 & 0 & 0 & .06 & .02 & 1 & -.7
\end{array}\right]
\end{array}
$$

This is a final tableau. The minimum value is .7 when y_1 = 3, y_2 = 2, y_3 = 12, y_4 = 0, and y_5 = 0.

Mark should buy 3 of pill #1 and 2 of pill #2 for a minimum cost of 70¢.

35. Let y_1 = the number of units of ingredient I;
       y_2 = the number of units of ingredient II; and
       y_3 = the number of units of ingredient III.

The problem is to minimize $w = 4y_1 + 7y_2 + 5y_3$

$$
\begin{aligned}
\text{subject to:}\quad 4y_1 + \ y_2 + 10y_3 &\geq 10 \\
3y_1 + 2y_2 + \ \ y_3 &\geq 12 \\
4y_2 + \ 5y_3 &\geq 20
\end{aligned}
$$

with $y_1 \geq 0$, $y_2 \geq 0$, $y_3 \geq 0$.

Let $z = -w$, and maximize z. The initial tableau is as follows.

$$
\begin{array}{ccccccc}
y_1 & y_2 & y_3 & y_4 & y_5 & y_6 & z \\
\left[\begin{array}{ccccccc|c}
④ & 1 & 10 & -1 & 0 & 0 & 0 & 10 \\
3 & 2 & 1 & 0 & -1 & 0 & 0 & 12 \\
0 & 4 & 5 & 0 & 0 & -1 & 0 & 20 \\
\hline
4 & 7 & 5 & 0 & 0 & 0 & 1 & 0
\end{array}\right]
\end{array}
$$

Continue by pivoting on the circled entry in each tableau.

$$
\begin{array}{c}
\\
\\
-3R_1 + 4R_2 \to R_2 \\
\\
\\
-R_1 + R_4 \to R_4
\end{array}
\begin{array}{ccccccc}
y_1 & y_2 & y_3 & y_4 & y_5 & y_6 & z \\
\left[\begin{array}{ccccccc|c}
4 & 1 & 10 & -1 & 0 & 0 & 0 & 10 \\
0 & ⑤ & -26 & 3 & -4 & 0 & 0 & 18 \\
0 & 4 & 5 & 0 & 0 & -1 & 0 & 20 \\
\hline
0 & 6 & -5 & 1 & 0 & 0 & 1 & -10
\end{array}\right]
\end{array}
$$

$$
\begin{array}{c}
\begin{array}{ccccccc}
y_1 & y_2 & y_3 & y_4 & y_5 & y_6 & z
\end{array}\\
\begin{array}{l}
5R_1 - R_2 \to R_1\\
\\
-4R_2 + 5R_3 \to R_3\\
-6R_2 + 5R_4 \to R_4
\end{array}
\left[
\begin{array}{ccccccc|c}
20 & 0 & 76 & -8 & 4 & 0 & 0 & 32\\
0 & 5 & -26 & 3 & -4 & 0 & 0 & 18\\
0 & 0 & 129 & -12 & \boxed{16} & -5 & 0 & 28\\
0 & 0 & 131 & -13 & 24 & 0 & 5 & -158
\end{array}
\right]
\end{array}
$$

$$
\begin{array}{c}
\begin{array}{ccccccc}
y_1 & y_2 & y_3 & y_4 & y_5 & y_6 & z
\end{array}\\
\begin{array}{l}
4R_1 - R_3 \to R_1\\
4R_2 + R_3 \to R_2\\
\\
-3R_3 + 2R_4 \to R_4
\end{array}
\left[
\begin{array}{ccccccc|c}
80 & 0 & 175 & -20 & 0 & 5 & 0 & 100\\
0 & 20 & 25 & 0 & 0 & -5 & 0 & 100\\
0 & 0 & \boxed{129} & -12 & 16 & -5 & 0 & 28\\
0 & 0 & -125 & 10 & 0 & 15 & 10 & -400
\end{array}
\right]
\end{array}
$$

$$
\begin{array}{c}
\begin{array}{ccccccc}
y_1 & y_2 & y_3 & y_4 & y_5 & y_6 & z
\end{array}\\
\begin{array}{l}
129R_1 - 175R_3 \to R_1\\
129R_2 - 25R_3 \to R_2\\
\\
125R_3 + 129R_4 \to R_4
\end{array}
\left[
\begin{array}{ccccccc|c}
10{,}320 & 0 & 0 & -480 & -2800 & 1520 & 0 & 8000\\
0 & 2580 & 0 & \boxed{300} & -400 & -520 & 0 & 12{,}200\\
0 & 0 & 129 & -12 & 16 & -5 & 0 & 28\\
0 & 0 & 0 & -210 & 2000 & 259{,}310 & 1290 & -48{,}100
\end{array}
\right]
\end{array}
$$

$$
\begin{array}{c}
\begin{array}{ccccccc}
y_1 & y_2 & y_3 & y_4 & y_5 & y_6 & z
\end{array}\\
\begin{array}{l}
5R_1 + 8R_2 \to R_1\\
\\
R_2 + 25R_3 \to R_3\\
7R_2 + 10R_4 \to R_4
\end{array}
\left[
\begin{array}{ccccccc|c}
51{,}600 & 20{,}640 & 0 & 0 & -17{,}200 & 3440 & 0 & 137{,}600\\
0 & 2580 & 0 & 300 & -400 & -520 & 0 & 12{,}200\\
0 & 2580 & 3225 & 0 & 0 & -645 & 0 & 12{,}900\\
0 & 18{,}060 & 0 & 0 & 17{,}200 & 2{,}589{,}460 & 12{,}290 & -395{,}600
\end{array}
\right]
\end{array}
$$

Create a 1 in the columns corresponding to y_1, y_3, y_4, and z.

$$
\begin{array}{c}
\begin{array}{ccccccc}
y_1 & y_2 & y_3 & y_4 & y_5 & y_6 & z
\end{array}\\
\begin{array}{l}
\dfrac{1}{51{,}600}R_1 \to R_1\\[4pt]
\dfrac{1}{300}R_2 \to R_2\\[4pt]
\dfrac{1}{3225}R_3 \to R_3\\[4pt]
\dfrac{1}{12{,}900}R_4 \to R_4
\end{array}
\left[
\begin{array}{ccccccc|c}
1 & .4 & 0 & 0 & -.33 & .067 & 0 & 2\frac{2}{3}\\
0 & 8.6 & 0 & 1 & -1.33 & -1.733 & 0 & 40\frac{2}{3}\\
0 & .8 & 1 & 0 & 0 & -.2 & 0 & 4\\
0 & 1.4 & 0 & 0 & 1.33 & 201 & 1 & -30\frac{2}{3}
\end{array}
\right]
\end{array}
$$

This is a final tableau. The minimum value is $30\frac{2}{3}$ when $y_1 = 2\frac{2}{3}$, $y_2 = 0$, $y_3 = 4$, $y_4 = 40\frac{2}{3}$, $y_5 = 0$, and $y_6 = 0$.

The biologist can meet his needs at a minimum cost of $30.67 by using $2\frac{2}{3}$ units of ingredient I and 4 units of ingredient III. (Ingredient II should not be used at all.)

37. Let y_1 = the number of ounces of ingredient I;

y_2 = the number of ounces of ingredient II; and

y_3 = the number of ounces of ingredient III.

The problem is to

minimize $w = .30y_1 + .09y_2 + .27y_3$

subject to: $y_1 + y_2 + y_3 \geq 10$

$y_1 + y_2 + y_3 \leq 15$

$y_1 \geq \frac{1}{4}y_2$

$y_3 \geq y_1$

with $y_1 \geq 0, \ y_2 \geq 0, \ y_3 \geq 0.$

Use a computer to find that the minimum is $w = 1.55$ when $y_1 = 1\frac{2}{3}$, $y_2 = 6\frac{2}{3}$, and $y_3 = 1\frac{2}{3}$.

The additive should consist of $1\frac{2}{3}$ oz of ingredient I, $6\frac{2}{3}$ oz of ingredient II, and $1\frac{2}{3}$ oz of ingredient III, for a minimum cost of \$1.55.   The amount of additive that should be used per gallon of gasoline is $1\frac{2}{3} + 6\frac{2}{3} + 1\frac{2}{3} =$ 10 oz.

Section 4.4

1. To find the transpose of a matrix, the rows of the original matrix are written as the columns of the trans-pose.

The transpose of $\begin{bmatrix} 1 & 2 & 3 \\ 3 & 2 & 1 \\ 1 & 10 & 0 \end{bmatrix}$

is $\begin{bmatrix} 1 & 3 & 1 \\ 2 & 2 & 10 \\ 3 & 1 & 0 \end{bmatrix}$.

3. The transpose of $\begin{bmatrix} -1 & 4 & 6 & 12 \\ 13 & 25 & 0 & 4 \\ -2 & -1 & 11 & 3 \end{bmatrix}$

is $\begin{bmatrix} -1 & 13 & -2 \\ 4 & 25 & -1 \\ 6 & 0 & 11 \\ 12 & 4 & 3 \end{bmatrix}$.

5. Maximize $z = 4x_1 + 3x_2 + 2x_3$

subject to: $x_1 + x_2 + x_3 \leq 5$

$x_1 + x_2 \qquad \leq 4$

$2x_1 + x_2 + 3x_3 \leq 15$

with $x_1 \geq 0, \ x_2 \geq 0, \ x_3 \geq 0.$

Begin by writing the augmented matrix for the given problem.

$$\left[\begin{array}{ccc|c} 1 & 1 & 1 & 5 \\ 1 & 1 & 0 & 4 \\ 2 & 1 & 3 & 15 \\ \hline 4 & 3 & 2 & 0 \end{array} \right]$$

Find the transpose of this matrix.

$$\left[\begin{array}{ccc|c} 1 & 1 & 2 & 4 \\ 1 & 1 & 1 & 3 \\ 1 & 0 & 3 & 2 \\ \hline 5 & 4 & 15 & 0 \end{array} \right]$$

The dual problem is stated from this second matrix as follows (using y instead of x).

Minimize $w = 5y_1 + 4y_2 + 15y_3$

subject to: $y_1 + y_2 + 2y_3 \geq 4$

$y_1 + y_2 + y_3 \geq 3$

$y_1 \qquad + 3y_3 \geq 2$

with $y_1 \geq 0, \ y_2 \geq 0, \ y_3 \geq 0.$

7. Minimize $w = y_1 + 2y_2 + y_3 + 5y_4$

subject to: $y_1 + y_2 + y_3 + y_4 \geq 50$

$3y_1 + y_2 + 2y_3 + y_4 \geq 100$

with $y_1 \geq 0, \ y_2 \geq 0, \ y_3 \geq 0, \ y_4 \geq 0.$

Write the augmented matrix for the given problem.

$$\begin{bmatrix} 1 & 1 & 1 & 1 & 50 \\ 3 & 1 & 2 & 1 & 100 \\ \hline 1 & 2 & 1 & 5 & 0 \end{bmatrix}$$

Find the transpose of this matrix.

$$\begin{bmatrix} 1 & 3 & 1 \\ 1 & 1 & 2 \\ 1 & 2 & 1 \\ 1 & 1 & 5 \\ \hline 50 & 100 & 0 \end{bmatrix}$$

The dual problem is stated from this second matrix as follows (using x instead of y).

Maximize $z = 50x_1 + 100x_2$

subject to: $x_1 + 3x_2 \leq 1$

$x_1 + x_2 \leq 2$

$x_1 + 2x_2 \leq 1$

$x_1 + x_2 \leq 5$

with $x_1 \geq 0$, $x_2 \geq 0$.

9. Find $y_1 \geq 0$ and $y_2 \geq 0$ such that

$$2y_1 + 3y_2 \geq 6$$

$$2y_1 + y_2 \geq 7$$

and $w = 5y_1 + 2y_2$ is minimized.

Write the augmented matrix for this problem.

$$\begin{bmatrix} 2 & 3 & 6 \\ 2 & 1 & 7 \\ \hline 5 & 2 & 0 \end{bmatrix}$$

Find the transpose of this matrix.

$$\begin{bmatrix} 2 & 2 & 5 \\ 3 & 1 & 2 \\ \hline 6 & 7 & 0 \end{bmatrix}$$

Use this matrix to write the dual problem.

Find $x_1 \geq 0$ and $x_2 \geq 0$ such that

$$2x_1 + 2x_2 \leq 5$$

$$3x_1 + x_2 \leq 2$$

and $z = 6x_1 + 7x_2$ is maximized. Introduce slack variables x_3 and x_4. The initial tableau is as follows.

$$\begin{array}{ccccc} x_1 & x_2 & x_3 & x_4 & z \\ \end{array}$$
$$\begin{bmatrix} 2 & 2 & 1 & 0 & 0 & 5 \\ 3 & ① & 0 & 1 & 0 & 2 \\ \hline -6 & -7 & 0 & 0 & 1 & 0 \end{bmatrix}$$

Pivot on the 1 in the second row, second column, since that column has the most negative indicator and that row has the smallest nonnegative quotient.

$$\begin{array}{ccccc} x_1 & x_2 & x_3 & x_4 & z \\ \end{array}$$
$$\begin{array}{l} -2R_2 + R_1 \to R_1 \\ \\ 7R_2 + R_3 \to R_3 \end{array} \begin{bmatrix} -4 & 0 & 1 & -2 & 0 & 1 \\ 3 & 1 & 0 & 1 & 0 & 2 \\ \hline 15 & 0 & 0 & 7 & 1 & 14 \end{bmatrix}$$

The minimum value of w is the same as the maximum value of z. The minimum value of w is 14 when $y_1 = 0$ and $y_2 = 7$.
(Note that the values of y_1 and y_2 are given by the entries in the bottom row of the columns corresponding to the slack variables in the final tableau.)

11. Find $y_1 \geq 0$ and $y_2 \geq 0$ such that

$$10y_1 + 5y_2 \geq 100$$

$$20y_1 + 10y_2 \geq 150$$

and $w = 4y_1 + 5y_2$ is minimized.

The dual problem is as follows.
Find $x_1 \geq 0$ and $x_2 \geq 0$ such that

$$10x_1 + 20x_2 \leq 4$$
$$5x_1 + 10x_2 \leq 5$$

and $z = 100x_1 + 150x_2$ is maximized.
The initial simplex tableau is as follows.

$$
\begin{array}{ccccc|c}
x_1 & x_2 & x_3 & x_4 & z & \\
10 & ⓩ0 & 1 & 0 & 0 & 4 \\
5 & 10 & 0 & 1 & 0 & 5 \\
\hline
-100 & -150 & 0 & 0 & 1 & 0
\end{array}
$$

Pivot on the 20 in the first row, second column.

$$
\begin{array}{ccccc|c}
x_1 & x_2 & x_3 & x_4 & z & \\
⑩0 & 20 & 1 & 0 & 0 & 4 \\
0 & 0 & -1 & 2 & 0 & 6 \\
-50 & 0 & 15 & 0 & 2 & 60
\end{array}
$$

$2R_2 - R_1 \rightarrow R_2$
$15R_1 + 2R_3 \rightarrow R_3$

Pivot on the 10 in the first row, first column.

$$
\begin{array}{ccccc|c}
x_1 & x_2 & x_3 & x_4 & z & \\
10 & 20 & 1 & 0 & 0 & 4 \\
0 & 0 & -1 & 2 & 0 & 6 \\
0 & 100 & 20 & 0 & 2 & 80
\end{array}
$$

$5R_1 + R_3 \rightarrow R_3$

Create a 1 in the columns corresponding to x_1, x_4, and z.

$$
\begin{array}{ccccc|c}
x_1 & x_2 & x_3 & x_4 & z & \\
1 & 2 & \frac{1}{10} & 0 & 0 & \frac{2}{5} \\
0 & 0 & -\frac{1}{2} & 1 & 0 & 3 \\
0 & 50 & 10 & 0 & 1 & 40
\end{array}
$$

$\frac{1}{10}R_1 \rightarrow R_1$
$\frac{1}{2}R_2 \rightarrow R_2$
$\frac{1}{2}R_3 \rightarrow R_3$

The minimum value of w is 40 when $y_1 = 10$ and $y_2 = 0$. (These values of y_1 and y_2 are read from the last row of the columns corresponding to x_3 and x_4 in the final tableau.)

13. Minimize $w = 2y_1 + y_2 + 3y_3$
    subject to: $y_1 + y_2 + y_3 \geq 100$
    $\qquad\qquad 2y_1 + y_2 \qquad \geq 50$
    with $y_1 \geq 0$, $y_2 \geq 0$, $y_3 \geq 0$.

The dual problem is as follows.
Maximize $z = 100x_1 + 50x_2$
subject to: $x_1 + 2x_2 \leq 2$
$\qquad\qquad x_1 + x_2 \leq 1$
$\qquad\qquad x_1 \qquad \leq 3$
with $x_1 \geq 0$, $x_2 \geq 0$, $x_3 \geq 0$.

The initial simplex tableau is as follows.

$$
\begin{array}{cccccc|c}
x_1 & x_2 & x_3 & x_4 & x_5 & z & \\
1 & 2 & 1 & 0 & 0 & 0 & 2 \\
①1 & 1 & 0 & 1 & 0 & 0 & 1 \\
1 & 0 & 0 & 0 & 1 & 0 & 3 \\
\hline
-100 & -50 & 0 & 0 & 0 & 1 & 0
\end{array}
$$

Pivot on the 1 in the second row, first column.

$$
\begin{array}{cccccc|c}
x_1 & x_2 & x_3 & x_4 & x_5 & z & \\
0 & 1 & 1 & -1 & 0 & 0 & 1 \\
1 & 1 & 0 & 1 & 0 & 0 & 1 \\
0 & -1 & 0 & -1 & 1 & 0 & 2 \\
0 & 50 & 0 & 100 & 0 & 1 & 100
\end{array}
$$

$-R_2 + R_1 \rightarrow R_1$
$-R_2 + R_3 \rightarrow R_3$
$100R_2 + R_4 \rightarrow R_4$

The minimum value of w is 100 when $y_1 = 0$, $y_2 = 100$, and $y_3 = 0$.

15. Minimize $z = x_1 + 2x_2$
    subject to: $-2x_1 + x_2 \geq 1$
    $\qquad\qquad x_1 - 2x_2 \geq 1$
    with $x_1 \geq 0$, $x_2 \geq 0$.

A quick sketch of the constraints $-2x_1 + x_2 \geq 1$ and $x_1 - 2x_2 \geq 1$ will verify that the two corresponding half planes do not overlap in the first quadrant of the x_1x_2-plane. Therefore, this problem (P) has no feasible solutions. The dual of the given problem is as follows.

Maximize $w = y_1 + y_2$
subject to: $-2y_1 + y_2 \leq 1$
$\qquad\qquad y_1 - 2y_2 \leq 2$
with $\qquad y_1 \geq 0, \ y_2 \geq 0.$

A quick sketch here will verify that there is a feasible region in the y_1y_2-plane, and it is unbounded. Therefore, there is no maximum value of w in this problem (D). (P) has no feasible solutions and the objective function of (D) is unbounded is choice (a).

17. Maximize $x_1 + 1.5x_2 = z$
subject to: $x_1 + 2x_2 \leq 200$
$\qquad\qquad 4x_1 + 3x_2 \leq 600$
$\qquad\qquad\qquad\quad x_2 \leq 90$
with $\qquad x_1 \geq 0, \ x_2 \geq 0.$

The final simplex tableau of this problem is as follows.

x_1	x_2	x_3	x_4	x_5	z	
0	1	.8	-.2	0	0	40
1	0	-.6	.4	0	0	120
0	0	-.8	.2	1	0	50
0	0	.6	.1	0	1	180

(a) The corresponding dual problem is as follows.

Minimize $w = 200y_1 + 600y_2 + 90y_3$
subject to:
$\qquad y_1 + 4y_2 \qquad \geq 1$
$\qquad 2y_1 + 3y_2 + y_3 \geq 1.5$
with $y_1 \geq 0, \ y_2 \geq 0, \ y_3 \geq 0.$

(b) From the given final tableau, the optimal solution to the dual problem is $y_1 = .6$, $y_2 = .1$, $y_3 = 0$, and $w = 180$.

(c) The shadow value for felt is .6; an increase in supply of 10 units of felt will increase profit to $180 + .6(10) = \$186$.

(d) The shadow values are .1 for stuffing and 0 for trim. If stuffing and trim are each decreased by 10 units, the profit will be $180 - .1(10) - 0(10) = \$179$.

19. Let y_1 = the number of packages of Sun Hill and
$\qquad y_2$ = the number of packages of Bear Valley.

The problem is to
minimize $w = 3y_1 + 2y_2$
subject to: $10y_1 + 2y_2 \geq 20$
$\qquad\qquad 4y_1 + 4y_2 \geq 24$
$\qquad\qquad 2y_1 + 8y_2 \geq 24$
with $\qquad y_1 \geq 0, \ y_2 \geq 0.$

The constraints can be simplified and the problem restated as follows.

Minimize $w = 3y_1 + 2y_2$
subject to: $5y_1 + y_2 \geq 10$
$\qquad\qquad y_1 + y_2 \geq 6$
$\qquad\qquad y_1 + 4y_2 \geq 12$
with $\qquad y_1 \geq 0, \ y_2 \geq 0.$

The dual problem is as follows.

Maximize $z = 10x_1 + 6x_2 + 12x_3$

subject to: $5x_1 + x_2 + x_3 \leq 3$

$$x_1 + x_2 + 4x_3 \leq 2$$

with $x_1 \geq 0, x_2 \geq 0.$

The initial simplex tableau is as follows.

$$\begin{array}{cccccc} x_1 & x_2 & x_3 & x_4 & x_5 & z \end{array}$$

$$\left[\begin{array}{cccccc|c} 5 & 1 & 1 & 1 & 0 & 0 & 3 \\ 1 & 1 & ④ & 0 & 1 & 0 & 2 \\ \hline -10 & -6 & -12 & 0 & 0 & 1 & 0 \end{array}\right]$$

Continue by pivoting on the circled entry in each tableau.

$$\begin{array}{cccccc} x_1 & x_2 & x_3 & x_4 & x_5 & z \end{array}$$

$$\begin{array}{c} 4R_1 - R_2 \to R_1 \\ \\ 3R_2 + R_3 \to R_3 \end{array} \left[\begin{array}{cccccc|c} ⑲ & 3 & 0 & 4 & -1 & 0 & 10 \\ 1 & 1 & 4 & 0 & 1 & 0 & 2 \\ \hline -7 & -3 & 0 & 0 & 3 & 1 & 6 \end{array}\right]$$

$$\begin{array}{cccccc} x_1 & x_2 & x_3 & x_4 & x_5 & z \end{array}$$

$$\begin{array}{c} \\ 19R_2 - R_1 \to R_2 \\ \\ 7R_1 + 19R_3 \to R_3 \end{array} \left[\begin{array}{cccccc|c} 19 & 3 & 0 & 4 & -1 & 0 & 10 \\ 0 & ⑯ & 76 & -4 & 20 & 0 & 28 \\ \hline 0 & -36 & 0 & 28 & 50 & 19 & 184 \end{array}\right]$$

$$\begin{array}{cccccc} x_1 & x_2 & x_3 & x_4 & x_5 & z \end{array}$$

$$\begin{array}{c} 16R_1 - 3R_2 \to R_1 \\ \\ 9R_2 - 4R_3 \to R_3 \end{array} \left[\begin{array}{cccccc|c} 304 & 0 & -228 & 76 & -76 & 0 & 76 \\ 0 & 16 & 76 & -4 & 20 & 0 & 28 \\ \hline 0 & 0 & 684 & 76 & 380 & 76 & 988 \end{array}\right]$$

We are finished pivoting. Create a 1 in the columns corresponding to x_1, x_3, and z.

$$\begin{array}{cccccc} x_1 & x_2 & x_3 & x_4 & x_5 & z \end{array}$$

$$\begin{array}{c} \frac{1}{304}R_1 \to R_1 \\ \\ \frac{1}{16}R_2 \to R_2 \\ \\ \frac{1}{76}R_3 \to R_3 \end{array} \left[\begin{array}{cccccc|c} 1 & 0 & -\frac{3}{4} & \frac{1}{4} & -\frac{1}{4} & 0 & \frac{1}{4} \\ 0 & 1 & \frac{19}{4} & -\frac{1}{4} & \frac{5}{4} & 0 & \frac{7}{4} \\ \hline 0 & 0 & 9 & 1 & 5 & 1 & 13 \end{array}\right]$$

The minimum value of w is 13 when $y_1 = 1$ and $y_2 = 5$.

You should buy 1 package of Sun Hill and 5 packages of Bear Valley, for a minimum cost of \$13.

21. Organize the information in a table.

	Units of Nutrient A (per bag)	Units of Nutrient B (per bag)	Cost (per bag)
Feed 1	1	2	\$3
Feed 2	3	1	\$2
Minimum	7	4	

Let y_1 = the number of bags of feed 1 and y_2 = the number of bags of feed 2.

(a) We want the cost to equal \$7 for 7 units of A and 4 units of B exactly. Therefore, use a system of equations rather than a system of inequalities.

$$3y_1 + 2y_2 = 7$$
$$y_1 + 3y_2 = 7$$
$$2y_1 + y_2 = 4$$

Use Gauss–Jordan elimination to solve this system of equations.

$$\begin{bmatrix} 3 & 2 & 7 \\ 1 & 3 & 7 \\ 2 & 1 & 4 \end{bmatrix}$$

$$\begin{array}{c} 3R_2 - R_1 \to R_2 \\ -2R_1 + 3R_3 \to R_3 \end{array} \begin{bmatrix} 3 & 2 & 7 \\ 0 & 7 & 14 \\ 0 & -1 & -2 \end{bmatrix}$$

$$\begin{array}{c} 7R_1 - 2R_2 \to R_1 \\ R_2 + 7R_3 \to R_3 \end{array} \begin{bmatrix} 21 & 0 & 21 \\ 0 & 7 & 14 \\ 0 & 0 & 0 \end{bmatrix}$$

$$\begin{array}{c} \frac{1}{21}R_1 \to R_1 \\ \frac{1}{7}R_2 \to R_2 \end{array} \begin{bmatrix} 1 & 0 & 1 \\ 0 & 1 & 2 \\ 0 & 0 & 0 \end{bmatrix}$$

Thus, $y_1 = 1$ and $y_2 = 2$, so use 1 bag of feed 1 and 2 bags of feed 2. The cost will be $3(1) + 2(2) = \$7$ as desired. The number of units of A is $1(1) + 3(2) = 7$ and the number of units of B is $2(1) + 1(2) = 4$.

(b)

	Units of Nutrient A (per bag)	Units of Nutrient B (per bag)	Cost (per bag)
Feed 1	1	2	\$3
Feed 2	3	1	\$2
Minimum	5	4	

The problem is to

minimize $w = 3y_1 + 2y_2$

subject to: $y_1 + 3y_2 \geq 5$

$\qquad\qquad 2y_1 + y_2 \geq 4$

with $\qquad y_1 \geq 0, y_2 \geq 0.$

The dual problem is as follows.

Maximize $z = 5x_1 + 4x_2$

subject to: $x_1 + 2x_2 \leq 3$

$\qquad\qquad 3x_1 + x_2 \leq 2$

with $\qquad x_1 \geq 0, x_2 \geq 0.$

The initial simplex tableau is as follows.

$$\begin{array}{ccccc} x_1 & x_2 & x_3 & x_4 & z \end{array}$$
$$\left[\begin{array}{ccccc|c} 1 & 2 & 1 & 0 & 0 & 3 \\ ③ & 1 & 0 & 1 & 0 & 2 \\ \hline -5 & -4 & 0 & 0 & 1 & 0 \end{array}\right]$$

Pivot as indicated.

$$\begin{array}{ccccccc} & x_1 & x_2 & x_3 & x_4 & z \end{array}$$
$$\begin{array}{c} 3R_1 - R_2 \rightarrow R_1 \\ \\ 5R_2 + 3R_3 \rightarrow R_3 \end{array} \left[\begin{array}{ccccc|c} 0 & ⑤ & 3 & -1 & 0 & 7 \\ 3 & 1 & 0 & 1 & 0 & 2 \\ 0 & -7 & 0 & 5 & 3 & 10 \end{array}\right]$$

$$\begin{array}{ccccc} x_1 & x_2 & x_3 & x_4 & z \end{array}$$
$$\begin{array}{c} \\ 5R_2 - R_1 \rightarrow R_2 \\ \\ 7R_1 + 5R_3 \rightarrow R_3 \end{array} \left[\begin{array}{ccccc|c} 0 & 5 & 3 & -1 & 0 & 7 \\ 15 & 0 & -3 & 6 & 0 & 3 \\ \hline 0 & 0 & 21 & 18 & 15 & 99 \end{array}\right]$$

$$\begin{array}{ccccc} x_1 & x_2 & x_3 & x_4 & z \end{array}$$
$$\begin{array}{c} \frac{1}{5}R_1 \rightarrow R_1 \\ \frac{1}{15}R_2 \rightarrow R_2 \\ \frac{1}{15}R_3 \rightarrow R_3 \end{array} \left[\begin{array}{ccccc|c} 0 & 1 & \frac{3}{5} & -\frac{1}{5} & 0 & \frac{7}{5} \\ 1 & 0 & -\frac{1}{5} & \frac{2}{5} & 0 & \frac{1}{5} \\ \hline 0 & 0 & \frac{7}{5} & \frac{6}{5} & 1 & \frac{33}{5} \end{array}\right]$$

Reading from the final column of the final tableau, $x_2 = \$1.40$ is the cost of nutrient B and $x_1 = \$.20$ is the cost of nutrient A. With 5 units of A and 4 units of B, this gives a minimum cost of $5(\$.20) + 4(\$1.40) = \$6.60$ as given in the lower right corner. 1.4 (or $7/5$) bags of feed 1 and 1.2 (or $6/5$) bags of feed 2 should be used.

23. Let y_1 = the number of ounces of ingredient I;

$\qquad y_2$ = the number of ounces of ingredient II; and

$\qquad y_3$ = the number of ounces of ingredient III.

As seen in Exercise 37 in Section 4.3, the problem is to

minimize $w = .30y_1 + .09y_2 + .27y_3$

subject to:

$\qquad\qquad y_1 + y_2 + y_3 \geq 10$

$\qquad\qquad y_1 + y_2 + y_3 \leq 15$

$\qquad\qquad y_1 \geq \frac{1}{4}y_2$

$\qquad\qquad y_3 \geq y_1$

with $\quad y_1 \geq 0, y_2 \geq 0, y_3 \geq 0.$

Two of the constraints need to be rearranged.   The problem can be restated as follows.

Minimize $w = .30y_1 + .09y_2 + .27y_3$ subject to:

$$y_1 + y_2 + y_3 \geq 10$$
$$-y_1 - y_2 - y_3 \geq -15$$
$$4y_1 - y_2 \qquad \geq 0$$
$$-y_1 \qquad + y_3 \geq 0$$

with $y_1 \geq 0,\ y_2 \geq 0,\ y_3 \geq 0$.

The dual problem is as follows.

Maximize $z = 10x_1 - 15x_2$ subject to:

$$x_1 - x_2 + 4x_3 - x_4 \leq .30$$
$$x_1 - x_2 - x_3 \qquad \leq .09$$
$$x_1 - x_2 \qquad - x_4 \leq .27$$

with $x_1 \geq 0,\ x_2 \geq 0,\ x_3 \geq 0,\ x_4 \geq 0$. The initial simplex tableau is as follows.

x_1	x_2	x_3	x_4	x_5	x_6	x_7	z	
1	-1	4	-1	1	0	0	0	.30
1	-1	-1	0	0	1	0	0	.09
1	-1	0	-1	0	0	1	0	.27
-10	15	0	0	0	0	0	1	0

Use a computer to perform the simplex algorithm and find that the minimum is $w = 1.55$ when $y_1 = \frac{5}{3}$, $y_2 = \frac{20}{3}$, and $y_3 = \frac{5}{3}$. The additive should consist of $1\frac{2}{3}$ oz of ingredient I, $6\frac{2}{3}$ oz ingredient II, and $1\frac{2}{3}$ oz of ingredient III for a minimum cost of $1.55. The amount of

additive that should be used per gallon of gasoline is

$$\frac{5}{3} + \frac{20}{3} + \frac{5}{3} = 10 \text{ oz.}$$

Chapter 4   Review Exercises

3.   Maximize $z = 5x_1 + 3x_2$ subject to:
$$2x_1 + 5x_2 \leq 50$$
$$x_1 + 3x_2 \leq 25$$
$$4x_1 + x_2 \leq 18$$
$$x_1 + x_2 \leq 12$$
with $x_1 \geq 0,\ x_2 \geq 0$.

(a) Adding the slack variables x_3, x_4, x_5, and x_6, we obtain the following equations.

$$2x_1 + 5x_2 + x_3 \qquad\qquad\qquad = 50$$
$$x_1 + 3x_2 \qquad + x_4 \qquad\qquad = 25$$
$$4x_1 + x_2 \qquad\qquad + x_5 \qquad = 18$$
$$x_1 + x_2 \qquad\qquad\qquad + x_6 = 12$$

(b) The initial tableau is

x_1	x_2	x_3	x_4	x_5	x_6	z	
2	5	1	0	0	0	0	50
1	3	0	1	0	0	0	25
4	1	0	0	1	0	0	18
1	1	0	0	0	1	0	12
-5	-3	0	0	0	0	1	0

5.   Maximize $z = 5x_1 + 8x_2 + 6x_3$ subject to:
$$x_1 + x_2 + x_3 \leq 90$$
$$2x_1 + 5x_2 + x_3 \leq 120$$
$$x_1 + 3x_2 \qquad \geq 80$$
with $x_1 \geq 0,\ x_2 \geq 0,\ x_3 \geq 0$.

(a) Adding the slack variables x_4 and x_5 and subtracting the surplus variable x_6, we obtain the following equations.

$$
\begin{aligned}
x_1 + x_2 + x_3 + x_4 &= 90 \\
2x_1 + 5x_2 + x_3 \quad\ + x_5 &= 120 \\
x_1 + 3x_2 \qquad\qquad\quad - x_6 &= 80
\end{aligned}
$$

(b) The initial tableau is

	x_1	x_2	x_3	x_4	x_5	x_6	z	
	1	1	1	1	0	0	0	90
	2	5	1	0	1	0	0	120
	1	3	0	0	0	-1	0	80
	-5	-8	-6	0	0	0	1	0

7.

	x_1	x_2	x_3	x_4	x_5	z	
	1	2	3	1	0	0	28
	②	4	1	0	1	0	32
	-5	-2	-3	0	0	1	0

The most negative entry in the last row is -5 and the smaller of the two quotients is $32/2 = 16$. Hence, the 2 in the second row, first column, is the first pivot. Performing row transformations leads to the following tableau.

	x_1	x_2	x_3	x_4	x_5	z	
$2R_1-R_2 \to R_1$	0	0	⑤	2	-1	0	24
	2	4	1	0	1	0	32
$5R_2+2R_3 \to R_3$	0	16	-1	0	5	2	160

Pivot on the 5 in the first row, third column.

	x_1	x_2	x_3	x_4	x_5	z	
$5R_2-R_1 \to R_2$	0	0	5	2	-1	0	24
	10	20	0	-2	6	0	136
$5R_3+R_1 \to R_3$	0	80	0	2	24	10	824

Create a 1 in the columns corresponding to x_1, x_3, and z.

	x_1	x_2	x_3	x_4	x_5	z	
$\frac{1}{5}R_1 \to R_1$	0	0	1	.4	-.2	0	4.8
$\frac{1}{10}R_2 \to R_2$	1	2	0	-.2	.6	0	13.6
$\frac{1}{10}R_3 \to R_3$	0	8	0	.2	2.4	1	82.4

The maximum value is 82.4 when $x_1 = 13.6$, $x_2 = 0$, $x_3 = 4.8$, $x_4 = 0$, and $x_5 = 0$.

9.

	x_1	x_2	x_3	x_4	x_5	x_6	z	
	1	2	2	1	0	0	0	50
	③	1	0	0	1	0	0	20
	1	0	2	0	0	-1	0	15
	-5	-3	-2	0	0	0	1	0

The initial basic solution is not feasible since $x_6 = -15$. In the third row, where the negative coefficient appears, the nonnegative entry that appears farthest to the left is the 1 in the first column; in the first column, the smallest nonnegative quotient is $\frac{20}{3}$. Pivot on the 3 in the second row, first column.

$$\begin{array}{c} \\ 3R_1-R_2{\rightarrow}R_1 \\ \\ \\ 3R_3-R_2{\rightarrow}R_3 \\ \\ 5R_2+3R_4{\rightarrow}R_4 \end{array} \begin{array}{ccccccc} x_1 & x_2 & x_3 & x_4 & x_5 & x_6 & z \\ \left[\begin{array}{ccccccc|c} 0 & 5 & 6 & 3 & 0 & 0 & 0 & 130 \\ 3 & 1 & 0 & 0 & 1 & 0 & 0 & 20 \\ 0 & -1 & ⑥ & 0 & 0 & -3 & 0 & 25 \\ \hline 0 & -4 & -6 & 0 & 5 & 0 & 3 & 100 \end{array}\right] \end{array}$$

Continue by pivoting on each circled entry.

$$\begin{array}{c} \\ R_1-R_3{\rightarrow}R_1 \\ \\ \\ \\ R_3+R_4{\rightarrow}R_4 \end{array} \begin{array}{ccccccc} x_1 & x_2 & x_3 & x_4 & x_5 & x_6 & z \\ \left[\begin{array}{ccccccc|c} 0 & ⑥ & 0 & 3 & 0 & 3 & 0 & 105 \\ 3 & 1 & 0 & 0 & 1 & 0 & 0 & 20 \\ 0 & -1 & 6 & 0 & 0 & -3 & 0 & 25 \\ \hline 0 & -5 & 0 & 0 & 5 & -3 & 3 & 125 \end{array}\right] \end{array}$$

The basic solution is now feasible, but there are negative indicators. Continue pivoting.

$$\begin{array}{c} \\ \\ 6R_2-R_1{\rightarrow}R_2 \\ 6R_3+R_1{\rightarrow}R_3 \\ \\ 5R_1+6R_4{\rightarrow}R_4 \end{array} \begin{array}{ccccccc} x_1 & x_2 & x_3 & x_4 & x_5 & x_6 & z \\ \left[\begin{array}{ccccccc|c} 0 & 6 & 0 & 3 & 0 & ③ & 0 & 105 \\ 18 & 0 & 0 & -3 & 6 & -3 & 0 & 15 \\ 0 & 0 & 36 & 3 & 0 & -15 & 0 & 255 \\ \hline 0 & 0 & 0 & 15 & 30 & -3 & 18 & 1275 \end{array}\right] \end{array}$$

$$\begin{array}{c} \\ \\ R_1+R_2{\rightarrow}R_2 \\ 5R_1+R_3{\rightarrow}R_3 \\ \\ R_1+R_4{\rightarrow}R_4 \end{array} \begin{array}{ccccccc} x_1 & x_2 & x_3 & x_4 & x_5 & x_6 & z \\ \left[\begin{array}{ccccccc|c} 0 & 6 & 0 & 3 & 0 & 3 & 0 & 105 \\ 18 & 6 & 0 & 0 & 6 & 0 & 0 & 120 \\ 0 & 30 & 36 & 18 & 0 & 0 & 0 & 780 \\ \hline 0 & 6 & 0 & 18 & 30 & 0 & 18 & 1380 \end{array}\right] \end{array}$$

Create a 1 in the columns corresponding to x_1, x_3, x_6 and z.

$$\begin{array}{c} \\ \frac{1}{3}R_1{\rightarrow}R_1 \\ \frac{1}{18}R_2{\rightarrow}R_2 \\ \frac{1}{36}R_3{\rightarrow}R_3 \\ \frac{1}{18}R_4{\rightarrow}R_4 \end{array} \begin{array}{ccccccc} x_1 & x_2 & x_3 & x_4 & x_5 & x_6 & z \\ \left[\begin{array}{ccccccc|c} 0 & 2 & 0 & 1 & 0 & 1 & 0 & 35 \\ 1 & .33 & 0 & 0 & .33 & 0 & 0 & 6.67 \\ 0 & .83 & 1 & .5 & 0 & 0 & 0 & 21.67 \\ \hline 0 & .33 & 0 & 1 & 1.67 & 0 & 1 & 76.67 \end{array}\right] \end{array}$$

The maximum value is about 76.67 when $x_1 \approx 6.67$, $x_2 = 0$, $x_3 \approx 21.67$, $x_4 = 0$, $x_5 = 0$, and $x_6 = 35$.

11. Minimize $w = 10y_1 + 15y_2$
subject to: $y_1 + y_2 \geq 17$
$\qquad\qquad 5y_1 + 8y_2 \geq 42$
with $y_1 \geq 0$, $y_2 \geq 0$.

Let $z = -w$, and maximize z. The problem can be restated as follows. Maximize $z = -10y_1 - 15y_2$ subject to: $y_1 + y_2 \geq 17$
$\qquad\qquad 5y_1 + 8y_2 \geq 42$
with $y_1 \geq 0$, $y_2 \geq 0$.

13. Minimize $w = 7y_1 + 2y_2 + 3y_3$
subject to: $y_1 + y_2 + 2y_3 \geq 48$
$\qquad\quad y_1 + y_2 \qquad\quad \geq 12$
$\qquad\qquad\qquad\quad y_3 \geq 10$
$\qquad\quad 3y_1 \qquad + \quad y_3 \geq 30$
with $y_1 \geq 0$, $y_2 \geq 0$, $y_3 \geq 0$.

Let $z = -w$, and maximize z. The problem can be restated as follows.
Maximize $z = -7y_1 - 2y_2 - 3y_3$
subject to: $y_1 + y_2 + 2y_3 \geq 48$
$\qquad\quad y_1 + y_2 \qquad\quad \geq 12$
$\qquad\qquad\qquad\quad y_3 \geq 10$
$\qquad\quad 3y_1 \qquad + \quad y_3 \geq 30$
with $y_1 \geq 0$, $y_2 \geq 0$, $y_3 \geq 0$.

15.

$$\begin{array}{ccccccc} y_1 & y_2 & y_3 & y_4 & y_5 & y_6 & z \\ \left[\begin{array}{ccccccc|c} 0 & 0 & 3 & 0 & 1 & 1 & 0 & 2 \\ 1 & 0 & -2 & 0 & 2 & 0 & 0 & 8 \\ 0 & 1 & 7 & 0 & 0 & 0 & 0 & 12 \\ 0 & 0 & 1 & 1 & -4 & 0 & 0 & 1 \\ \hline 0 & 0 & 5 & 0 & 8 & 0 & 1 & -62 \end{array}\right] \end{array}$$

From this final tableau, read that the maximum value of $z = -w$ is -62 when $y_1 = 8$, $y_2 = 12$, $y_3 = 0$, $y_4 = 1$, $y_5 = 0$, and $y_6 = 2$. Therefore the minimum value of w is 62 when $y_1 = 8$, $y_2 = 12$, $y_3 = 0$, $y_4 = 1$, $y_5 = 0$, and $y_6 = 2$.

19.

$$\left[\begin{array}{cccccc|c} 4 & 2 & 3 & 1 & 0 & 0 & 9 \\ 5 & 4 & 1 & 0 & 1 & 0 & 10 \\ \hline -6 & -7 & -5 & 0 & 0 & 1 & 0 \end{array}\right]$$

(a) The 1 in column 4 and the 1 in column 5 indicate that the constraints involve $\leq$. The problem being solved with this tableau is to

maximize $z = 6x_1 + 7x_2 + 5x_3$

subject to: $4x_1 + 2x_2 + 3x_3 \leq 9$

$5x_1 + 4x_2 + x_3 \leq 10$

with $x_1 \geq 0$, $x_2 \geq 0$, $x_3 \geq 0$.

(b) If the 1 in row 1, column 4 were -1 rather than 1, then the first constraint would have a surplus variable rather than a slack variable, which means the first constraint would be $4x_1 + 2x_2 + 3x_3 \geq 9$ instead of $4x_1 + 2x_2 + 3x_3 \leq 9$.

(c)

x_1	x_2	x_3	x_4	x_5	z	
3	0	5	2	-1	0	8
11	10	0	-1	3	0	21
47	0	0	13	11	10	227

From this tableau, the solution is

$x_1 = 0$, $x_2 = \dfrac{21}{10} = 2.1$, $x_3 = \dfrac{8}{5} = 1.6$,

and $z = \dfrac{227}{10} = 22.7$.

(d) The dual of the original problem is as folows.

Minimize $w = 9y_1 + 10y_2$

subject to: $4y_1 + 5y_2 \geq 6$

$2y_1 + 4y_2 \geq 7$

$3y_1 + y_2 \geq 5$

with $y_1 \geq 0$, $y_2 \geq 0$.

(e) From the tableau in part (c), the solution of the dual in part (d) is $y_1 = \dfrac{13}{10} = 1.3$, $y_2 = \dfrac{11}{10} = 1.1$, and

$z = \dfrac{227}{10} = 22.7$.

21. (a) Let x_1 = the amount invested in the oil leases;

x_2 = the amount invested in bonds;

and x_3 = the amount invested in stock.

(b) We want to maximize

$z = .15x_1 + .09x_2 + .05x_3$.

(c) The constraints are as follows.

$x_1 + x_2 + x_3 \leq 50,000$

$x_1 + x_2 \leq 15,000$

$x_1 + x_3 \leq 25,000$

23. (a) Let x_1 = the number of 5-gallon bags;

x_2 = the number of 10-gallon bags; and

x_3 = the number of 20-gallon bags.

(b) We want to maximize

$z = x_1 + .9x_2 + .95x_3$.

(c) The constraints are as follows.

$x_1 + 1.1x_2 + 1.5x_3 \leq 8$

$x_1 + 1.2x_2 + 1.3x_3 \leq 8$

$2x_1 + 3x_2 + 4x_3 \leq 8$

25. Based on the information given in Exercise 21, the initial tableau is as follows.

$$
\begin{array}{ccccccc}
x_1 & x_2 & x_3 & x_4 & x_5 & x_6 & z \\
\end{array}
$$

$$
\begin{bmatrix}
1 & 1 & 1 & 1 & 0 & 0 & 0 & 50{,}000 \\
① & 1 & 0 & 0 & 1 & 0 & 0 & 15{,}000 \\
1 & 0 & 1 & 0 & 0 & 1 & 0 & 25{,}000 \\
-.15 & -.09 & -.05 & 0 & 0 & 0 & 1 & 0
\end{bmatrix}
$$

Continue by pivoting on each circled entry.

$$
\begin{array}{ccccccc}
x_1 & x_2 & x_3 & x_4 & x_5 & x_6 & z \\
\end{array}
$$

$$
\begin{array}{l}
R_1 - R_2 \to R_1 \\
\\
\\
R_3 - R_2 \to R_2 \\
\\
.15R_2 + R_4 \to R_4
\end{array}
\begin{bmatrix}
0 & 0 & 1 & 1 & -1 & 0 & 0 & 35{,}000 \\
1 & 1 & 0 & 0 & 1 & 0 & 0 & 15{,}000 \\
0 & -1 & ① & 0 & -1 & 1 & 0 & 10{,}000 \\
0 & .06 & -.05 & 0 & .15 & 0 & 1 & 2250
\end{bmatrix}
$$

$$
\begin{array}{ccccccc}
x_1 & x_2 & x_3 & x_4 & x_5 & x_6 & z \\
\end{array}
$$

$$
\begin{array}{l}
R_1 - R_3 \to R_1 \\
\\
\\
\\
.05R_3 + R_4 \to R_4
\end{array}
\begin{bmatrix}
0 & 1 & 0 & 1 & 0 & -1 & 0 & 25{,}000 \\
1 & 1 & 0 & 0 & 1 & 0 & 0 & 15{,}000 \\
0 & -1 & 1 & 0 & -1 & 1 & 0 & 10{,}000 \\
0 & .01 & 0 & 0 & .1 & .05 & 1 & 2750
\end{bmatrix}
$$

The maximum value is z = 2750 when

$$
\begin{aligned}
x_1 &= 15{,}000, \\
x_2 &= 0, \\
x_3 &= 10{,}000, \\
x_4 &= 25{,}000, \\
x_5 &= 0, \text{ and} \\
x_6 &= 0.
\end{aligned}
$$

He should invest \$15,000 in oil leases and \$10,000 in stock for a maximum return of \$2750.

27. Based on the information in Exercise 23, the initial tableau is as follows.

$$
\begin{array}{ccccccc}
x_1 & x_2 & x_3 & x_4 & x_5 & x_6 & z \\
\end{array}
$$

$$
\begin{bmatrix}
1 & 1.1 & 1.5 & 1 & 0 & 0 & 0 & 8 \\
1 & 1.2 & 1.3 & 0 & 1 & 0 & 0 & 8 \\
② & 3 & 4 & 0 & 0 & 1 & 0 & 8 \\
-1 & -.9 & -.95 & 0 & 0 & 0 & 1 & 0
\end{bmatrix}
$$

Pivot on the circled entry.

$$
\begin{array}{c}
\\
2R_1 - R_3 \rightarrow R_1 \\
2R_2 - R_3 \rightarrow R_2 \\
\\
\\
2R_4 + R_3 \rightarrow R_4
\end{array}
\begin{array}{cccccccc}
x_1 & x_2 & x_3 & x_4 & x_5 & x_6 & z & \\
\left[\begin{array}{ccccccc|c}
0 & -.8 & -1 & 2 & 0 & -1 & 0 & 8 \\
0 & -.6 & -1.4 & 0 & 2 & -1 & 0 & 8 \\
2 & 3 & 4 & 0 & 0 & 1 & 0 & 8 \\
\hline
0 & 1.2 & 2.1 & 0 & 0 & 1 & 2 & 8
\end{array}\right]
\end{array}
$$

The maximum value is $z = \dfrac{8}{2} = 4$

when $x_1 = \dfrac{8}{2} = 4$, $x_2 = 0$, $x_3 = 0$,

$x_4 = \dfrac{8}{2} = 4$, $x_5 = \dfrac{8}{2} = 4$, and $x_6 = 0$.

For a maximum profit of \$4 per unit, 4 units of 5-gallon bags (and none of the others) should be made.

Extended Application 1

1. **(a)** We obtain

    (i) $.4x_1 + .23x_2 + .805x_3 + .998x_5$
    $+ .04x_6 + .5x_{10} + .625x_{11} \geq 14$

    (ii) $.054x_1 + .069x_2 + .025x_4 + .078x_6$
    $+ .28x_7 + .97x_8 \geq 6$

    (iii) $.707x_9 + .1x_{10} \geq 16$

    (iv) $.35x_{10} + .315x_{11} \geq .35$

    (v) $x_1 + x_2 + x_3 + x_4 + x_5 + x_6 + x_7$
    $+ x_8 + x_9 + x_{10} + x_{11} \leq 99.6$

    Solving on a computer gives a minimum cost of \$12.55 when $x_2 = 58.6957$, $x_8 = 2.01031$, $x_9 = 22.4895$, $x_{10} = 1$, $x_{12} = 1$, and $x_{13} = 1$.

    (b) Replace 14 with 17 in inequality (i) and 16 with 16.5 in inequality (iii).

The minimum cost is \$15.05 when $x_2 = 71.7391$, $x_8 = 3.14433$, $x_9 = 23.1966$, $x_{10} = 1$, $x_{12} = 1$, and $x_{13} = 1$.

2. Use a computer to solve the problem in the text.
    The minimum cost is \$14.31 when $x_2 = 67.394$, $x_8 = 3.459$, $x_9 = 22.483$, $x_{10} = 1$, $x_{12} = .25$, and $x_{13} = .15$.

Extended Application 2

1. $w_1 = \dfrac{570}{600} = .95$

    $w_2 = \dfrac{500}{600} = .83$

    $w_3 = \dfrac{450}{600} = .75$

    $w_4 = \dfrac{600}{600} = 1.00$

    $w_5 = \dfrac{520}{600} = .87$

    $w_6 = \dfrac{565}{600} = .94$

2. Use the simplex method with a computer to find the merit increase for each employee. The correct answers are $x_1 = 100$, $x_2 = 0$, $x_3 = 0$, $x_4 = 90$, $x_5 = 0$, and $x_6 = 210$.

CHAPTER 4   TEST

1. For the maximization problem below:

    (a) Determine the number of slack variables needed.

    (b) Determine the number of surplus variables needed.

    (c) Convert each constraint into a linear equation.

    Maximize $z = 50x_1 + 80x_2$

    subject to: $x_1 + 2x_2 \leq 32$

    $\qquad\qquad 3x_1 + 4x_2 \leq 84$

    $\qquad\qquad\qquad x_2 \leq 12$

    with $\qquad x_1 \geq 0, \; x_2 \geq 0.$

2. For the maximization problem below:

    (a) Set up the initial simplex tableau.

    (b) Determine the initial basic solution.

    (c) Find the first pivot element and justify your choice.

    Maximize $z = 10x_1 + 5x_2$

    subject to: $6x_1 + 2x_2 \leq 36$

    $\qquad\qquad 2x_1 + 4x_2 \leq 32$

    with $\qquad x_1 \geq 0, \; x_2 \geq 0.$

3. Solve the problem with given initial tableau.

x_1	x_2	x_3	x_4	x_5	x_6	z	
1	1	1	1	0	0	0	1000
40	20	30	0	1	0	0	3200
1	2	1	0	0	1	0	160
-100	-300	-200	0	0	0	1	0

4. Use the simplex method to solve the following problem.

Mammoth Micros markets computers with single-sided and double-sided disk drives. They obtain these drives from Large Disks, Inc. and Double Drives Are Us. Large Disk charges $250 for a single-sided and $350 for a double-sided disk. Double Drives charges $290 and $320. Each month Large Disks can supply at most 1000 drives in all. Double Drives can supply at most 2000. Mammoth needs at least 1200 single and 1600 double drives. How many of each type should they buy from each company to minimize their total costs? What is the minimum cost?

5. Maximize $z = 6x_1 - 2x_2$

   subject to: $x_1 + x_2 \leq 10$

   $\qquad\qquad 3x_1 + 2x_2 \geq 24$

   with $\qquad x_1 \geq 0, \; x_2 \geq 0.$

6. For the following minimization problem,

   (a) state the dual problem, and

   (b) solve the problem using the simplex method.

   Minimize $w = 5y_1 + 7y_2$

   subject to: $y_1 \qquad\quad \geq 4$

   $\qquad\qquad y_1 + y_2 \geq 8$

   $\qquad\qquad y_1 + 2y_2 \geq 10$

   with $\qquad y_1 \geq 0, \; y_2 \geq 0.$

CHAPTER 4   TEST ANSWERS

1. **(a)** 3 slack variables are needed.

    (b) 0 surplus variables are needed.

    (c)
$$x_1 + 2x_2 + x_3 \qquad\qquad = 32$$
$$3x_1 + 4x_2 \qquad + x_4 \qquad = 84$$
$$x_2 \qquad\qquad + x_5 = 12$$

2. **(a)**

$$\begin{array}{ccccc|c}
x_1 & x_2 & x_3 & x_4 & z & \\
6 & 2 & 1 & 0 & 0 & 36 \\
2 & 4 & 0 & 1 & 0 & 32 \\
\hline
-10 & -5 & 0 & 1 & 1 & 0
\end{array}$$

    (b) $x_3 = 36$, $x_4 = 32$, $z = 0$

    (c) Column one has the most negative indicator. The smallest nonnegative quotient occurs in row 1, since $\frac{36}{6}$ is smaller than $\frac{32}{2}$. Pivot on the 6.

3. The final tableau is

$$\begin{array}{ccccccc|c}
x_1 & x_2 & x_3 & x_4 & x_5 & x_6 & z & \\
1 & 0 & 1 & 2 & 0 & -1 & 0 & 40 \\
10 & 0 & 0 & -40 & 1 & 10 & 0 & 800 \\
0 & 1 & 0 & -1 & 0 & 1 & 0 & 60 \\
\hline
100 & 0 & 0 & 100 & 0 & 100 & 1 & 26,000
\end{array}$$

    The maximum value is 26,000 when $x_1 = 0$, $x_2 = 60$, and $x_3 = 40$.

4. They should buy 1000 single-sided from Large Disks, and they should buy 200 single-sided and 1600 double-sided from Doubles Are Us for a minimum cost of $820,000.

5. The maximum value is 60 when $x_1 = 10$ and $x_2 = 0$.

6. **(a)** Maximize $z = 4x_1 + 8x_2 + 10x_3$
        subject to: $x_1 + x_2 + x_3 \le 5$
                    $x_2 + 2x_3 \le 7$
        with $x_1 \ge 0$, $x_2 \ge 0$, $x_3 \ge 0$.

    (b) The minimum value is 44 when $y_1 = 6$ and $y_2 = 2$.

CHAPTER 5 SETS AND PROBABILITY

Section 5.1

1. $3 \in \{2, 5, 7, 9, 10\}$

    False. The number 3 is not an element of the set.

3. $9 \notin \{2, 1, 5, 8\}$

    True. 9 is not an element of the set.

5. $\{2, 5, 8, 9\} = \{2, 5, 9, 8\}$

    True. The sets contain exactly the same elements, so they are equal.

7. $\{$all whole numbers greater than 7 and less than $10\} = \{8, 9\}$

    True. 8 and 9 are the only such numbers.

9. $\{x | x$ is an odd integer, $6 \leq x \leq 18\}$
    $= \{7, 9, 11, 15, 17\}$

    False. The number 13 should be included.

11. $\emptyset$ and $\{\ \}$ are notations that represent the empty set, and they correspond to choices (b) and (c).
    (The other two choices do not represent the empty set since $\{\emptyset\}$ contains an element and 0 is not a set.)

13. $A \subseteq U$

    Every element of A is also an element of U.

15. $A \not\subseteq E$

    A contains elements that do not belong to E, namely 2, 4, 8, 10, and 12.

17. $\emptyset \subseteq A$

    The empty set is a subset of every set.

19. $D \subseteq B$

    Every element of D is also an element of B.

21. There are exactly $2^6 = 64$ subsets of A. A set with n distinct elements has 2^n subsets, and A has $n = 6$ elements.

23. There are exactly $2^3 = 8$ subsets of C. A set with n distinct elements has 2^n subsets, and C has $n = 3$ elements.

25. $\{4, 5, 6\}$

    Since the set has 3 elements, there are $2^3 = 8$ subsets.

27. $\{x | x$ is a counting number between 6 and $12\}$

    The set contains five elements: 7, 8, 9, 10, and 11. Therefore, there are $2^5 = 32$ subsets.

31. $\{8, 11, 15\} \cap \{8, 11, 19, 20\} = \{8, 11\}$

$\{8, 11\}$ is the set of all elements belonging to both of the first two sets, so it is the intersection of those sets.

33. $\{6, 12, 14, 16\} \cap \{6, 14, 19\} = \{6, 14\}$

$\{6, 14\}$ is the set of all elements belonging to both of the first two sets, so it is the intersection of those sets.

35. $\{3, 5, 9, 10\} \cup \emptyset = \{3, 5, 9, 10\}$

The empty set contains no elements, so the union of any set with the empty set will result in an answer set that is identical to the original set. (On the other hand, $\{3, 5, 9, 10\} \cap \emptyset = \emptyset$.)

37. $\{1, 2, 4\} \cup \{1, 2\} = \{1, 2, 4\}$

The answer set $\{1, 2, 4\}$ consists of all elements belonging to the first set, to the second set, or to both sets, and therefore it is the union of the first two sets. (On the other hand,
$$\{1, 2, 4\} \cap \{1, 2\} = \{1, 2\}.)$$

39. $X \cap Y = \{2, 3, 4, 5\} \cap \{3, 5, 7, 9\}$
$$= \{3, 5\}$$

41. X' consists of those elements of U which are not in X.
$$X' = \{7, 9\}$$

43. $X' \cap Y'$

$X' = \{7, 9\}$ and $Y' = \{2, 4\}$ so $X' \cap Y' = \emptyset$.

45. $X \cup (Y \cap Z)$

$X = \{2, 3, 4, 5\}$ and $Y \cap Z = \{5, 7, 9\}$.

Hence, the union of these two sets is $X \cup (Y \cap Z) = \{2, 3, 4, 5, 7, 9\}$
$$= U.$$

47. M' consists of all students in U who art not in M, so M' consists of all students in this school not taking this course.

49. $N \cap P$ is the set of all students in this school taking both accounting and zoology.

51. $A = \{2, 4, 6, 8, 10, 12\}$,
$B = \{2, 4, 8, 10\}$, $C = \{4, 8, 12\}$,
$D = \{2, 10\}$, $E = \{6\}$,
$U = \{2, 4, 6, 8, 10, 12, 14\}$

A pair of sets is disjoint if the two sets have no elements in common. The pairs of these sets that are disjoint are B and E, C and E, D and E, and C and D.

53. Since B is the set of all stocks in the list with a price-to-earnings ratio of a least 13, B' is the set of all stocks on the list whose price-to-earnings ratio is less than 13.

However, none of the stocks in the list has a price-to-earnings ratio that is less than 13, so B' = ∅.

55. A ∩ B is the set of all stocks with a dividend greater than $3 and a price-to-earnings ratio of at least 13. A ∩ B = {IBM, Mobil}, and (A ∩ B)' is the set of all stocks in the list that are not elements of A ∩ B. Therefore, (A ∩ B)' = {ATT, GE, Hershey, Nike}.

57. F = {networks with more than 5000 subscribers}
    = {HBO, Showtime, Cinemax}

59. H = {networks that show movies}
    = {HBO, Showtime, Cinemax, Disney Channel, Movie Channel}

61. H ∩ G = {networks that show movies and non-cartoon comedy}
    = {Showtime, Cinemax}

63. U = {s, d, c, g, i, m, h}
    and O = {i, m, h, g}, so
    O' = {s, d, c}.

65. N ∩ O = {s, d, c, g} ∩ {i, m, h, g}
    = {g}

67. N ∩ O' = {s, d, c, g} ∩ {s, d, c}
    = {s, d, c}

Section 5.2

1. B ∩ A' is the set of all elements in B *and* not in A.
   See the Venn diagram in the back of the textbook.

3. A' ∪ B is the set of all elements which do not belong to *or* which do belong to B, or both.
   See the Venn diagram in the back of the textbook.

5. B' ∪ (A' ∩ B')
   First find A' ∩ B', the set of elements not in A *and* not in B.

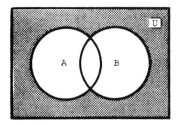

For the union, we want those elements in B' *or* (A' ∩ B'), or both.

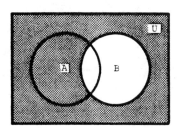

7. U′ is the empty set Ø.
    See the Venn diagram in the back
    of the textbook.

9. Three sets divide the universal set
    into 8 regions. (Examples of this
    situation will be seen in Exercises
    11–17.)

11. (A ∩ C′) ∪ B

    First find A ∩ C′, the region in A
    and not in C.

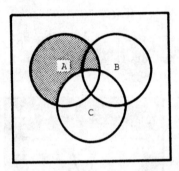

    For the union, we want the region in
    (A ∩ C′) *or* in B, or both.

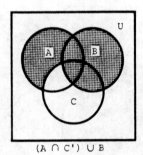

    (A ∩ C′) ∪ B

13. A′ ∩ (B ∩ C)

    First find A′, the region not in A.

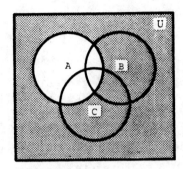

    Then find B ∩ C, the region where B
    and C overlap.

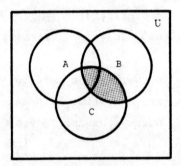

    Now intersect these regions.

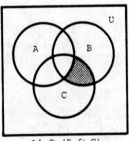

    A′ ∩ (B ∩ C)

15. $(A \cap B') \cup C$

First find $A \cap B'$, the region in A *and* not in B.

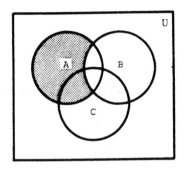

For the union, we want the region in $(A \cap B')$ *or* in C, or both.

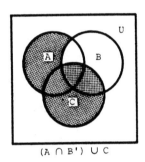

$(A \cap B') \cup C$

17. $A' \cap (B' \cup C)$

First find A'.

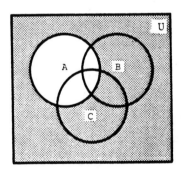

Then find $B' \cup C$, the region not in B *or* in C, or both.

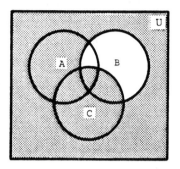

Now intersect these regions.

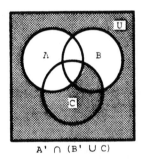

$A' \cap (B' \cup C)$

19. $n(A \cup B) = n(A) + n(B) - (A \cap B)$
$= 5 + 8 - 4$
$= 9$

21. $n(A \cup B) = n(A) + n(B) - n(A \cap B)$
$20 = n(A) + 7 - 3$
$20 = n(A) + 4$
$16 = n(A)$

23. $n(U) = 38$
$n(A) = 16$
$n(A \cap B) = 12$
$n(B') = 20$

First put 12 in A ∩ B. Since
n(A) = 16, and 12 are in A ∩ B,
there must be 4 elements in A that
are not in A ∩ B. n(B') = 20, so
there are 20 not in B. We already
have 4 not in B (but in A), so there
must be another 16 outside B *and*
outside A. So far we have accounted
for 32, and n(U) = 38, so 6 must be
in B but not in any region yet
identified. Thus n(A' ∩ B) = 6.
See the Venn diagram in the back of
the textbook.

25. n(A ∪ B) = 17
     n(A ∩ B) = 3
     n(A) = 8
     n(A' ∪ B') = 21

Start with n(A ∩ B) = 3. Since
n(A) = 8, there must be 5 more in A
not in B. n(A ∪ B) = 17; we already
have 8, so 9 more must be in B not
yet counted. A' ∪ B' consists of
all the region not in A ∩ B, where
we have 3. So far 5 + 9 = 14 are in
this region, so another 21 - 14 = 7
must be outside both A and B.
See the Venn diagram in the back of
the textbook.

27. n(A) = 28
     n(B) = 34
     n(C) = 25
     n(A ∩ B) = 14
     n(B ∩ C) = 15
     n(A ∩ C) = 11
     n(A ∩ B ∩ C) = 9
     n(U) = 59

We start with n(A ∩ B ∩ C) = 9.
If n(A ∩ B) = 14, an additional 5
are in A ∩ B but not in A ∩ B ∩ C.
Similarly, N(B ∩ C) = 15, so
15 - 9 = 6 are in B ∩ C but not in
A ∩ B ∩ C. Also, n(A ∩ C) = 11, so
11 - 9 = 2 are in A ∩ C but not in
A ∩ B ∩ C. Now we turn our attention to n(A) = 28. So far we have
2 + 9 + 5 = 16 in A; there must be
another 28 - 16 = 12 in A not yet
counted. Similarly, n(B) = 34; we
have 5 + 9 + 6 = 20 so far, and
34 - 20 = 14 more must be put in B.
For C, n(C) = 25; we have
2 + 9 + 6 = 17 counted so far. Then
there must be 8 more in C not yet
counted. The count now stands at
56, and n(U) = 59, so 3 must be
outside the three sets.
See the Venn diagram in the back of
the textbook.

29. n(A ∩ B) = 6
     n(A ∩ B ∩ C) = 4
     n(A ∩ C) = 7
     n(B ∩ C) = 4
     n(A ∩ C') = 11
     n(B ∩ C') = 8
     n(C) = 15
     n(A' ∩ B' ∩ C') = 5

Start with n(A ∩ B) = 6 and
n(A ∩ B ∩ C) = 4 to get 6 - 4 = 2
in that portion of A ∩ B outside of
C. From n(B ∩ C) = 4, there are
4 - 4 = 0 elements in that portion

of B ∩ C outside of A. Use
n(A ∩ C) = 7 to get 7 − 4 = 3 elements in that portion of A ∩ C outside of B. Since n(A ∩ C') = 11,
there are 11 − 2 = 9 elements in
that part of A outside of B and C.
Use n(B ∩ C') = 8 to get 8 − 2 = 6
elements in that part of B outside
of A and C. Since n(C) = 15, there
are 15 − 3 − 4 − 0 = 8 elements in C
outside of A and B. Finally, 5 must
be outside all three sets, since
n(A' ∩ B' ∩ C') = 5.
See the Venn diagram in the back of
the textbook.

31. (A ∪ B)' = A' ∩ B'

For (A ∪ B)', first find A ∪ B.

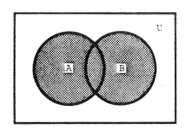

Now find (A ∪ B)', the region outside A ∪ B.

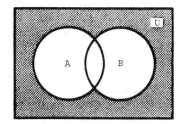

For A' ∩ B', first find A' and B'
individually.

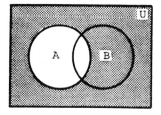 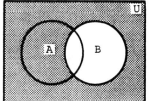

Then A' ∩ B' is the region where A'
and B' overlap, which is the entire
region outside A ∪ B (the same
result as in the second diagram).
Therefore, (A ∪ B)' = A' ∩ B'.

33. A ∩ (B ∪ C) = (A ∩ B) ∪ (A ∩ C)

First find A and B ∪ C individually.

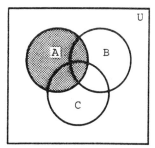 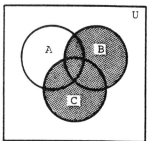

Then A ∩ (B ∪ C) is the region where
the above two diagrams overlap.

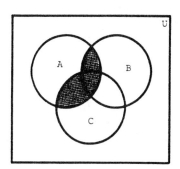

Next find A ∩ B and A ∩ C
individually.

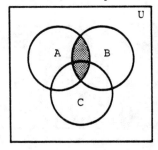 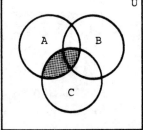

Then (A ∩ B) ∪ (A ∩ C) is the union
of the above two diagrams.

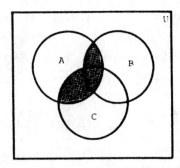

The Venn diagram for A ∩ (B ∪ C) is
identical to the Venn diagram for
(A ∩ B) ∪ (A ∩ C), so conclude that
A ∩ (B ∪ C) = (A ∩ B) ∪ (A ∩ C).

35. Let M be the set of those who use a
microwave oven. E be the set of
those who use an electric range, and
G be the set of those who use a gas
range. We are given the following
information.

n(U) = 140
n(M) = 58
N(E) = 63
N(G) = 58
n(M ∩ E) = 19
n(M ∩ G) = 17
n(G ∩ E) = 4
n(M ∩ G ∩ E) = 1
n(M′ ∩ G′ ∩ E′) = 2

Since n(M ∩ G ∩ E) = 1, there is
1 element in the region where the
three sets overlap.
Since n(M ∩ E) = 19, there are
19 − 1 = 18 elements in M ∩ E but
not in M ∩ E ∩ G.
Since n(M ∩ G) = 17, there are
17 − 1 = 16 elements in M ∩ G but
not in M ∩ E ∩ G.
Since n(G ∩ E) = 4, there are
4 − 1 = 3 elements in G ∩ E but
not in M ∩ E ∩ G.
Now consider n(M) = 58. So far we
have 16 + 1 + 18 = 35 in M; there
must be another 58 − 35 = 23 in M
not yet counted.
Similarly, n(E) = 63; we have
18 + 1 + 13 = 32 counted so far.
There must be 63 − 32 = 31 more in
E not yet counted.
Also, n(G) = 58; we have
16 + 1 + 13 = 30 counted so far.
There must be 58 − 30 = 28 more
in G not yet counted.
Lastly, n(M′ ∩ G′ ∩ E′) = 2 indicates that there are 2 elements outside of all three sets.

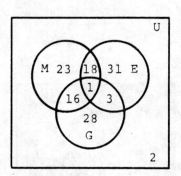

Note that the numbers in the Venn diagram add up to 142 even though $n(U) = 140$. Jeff has made some error and he should definitely be reassigned.

37. **(a)** $n(Y \cap R) = 40$

since 40 is the number in the table where the Y row and R column meet.

(b) $n(M \cap D) = 30$

since 30 is the number in the table where the M row and the D column meet.

(c) $n(D \cap Y) = 15$

$n(M) = 80$ since that is the total in the M row.

$m(M \cap (D \cap Y)) = 0$ since no person can simultaneously have an age in the range 21–25 *and* have an age in the range 26–35. By the union rule for sets,

$n(M \cup (D \cap Y))$
$= n(M) + n(D \cap Y) - n(M \cap (D \cap Y))$
$= 80 + 15 - 0$
$= 95.$

(d) $Y' \cap (D \cup N)$ consists of all people in the D column or in the N column who are at the same time not in the Y row. Therefore,

$n(Y' \cap (D \cup N)) = 30 + 50 + 20 + 10$
$= 110.$

(e) $n(N) = 45$
$n(O) = 70$
$n(O') = 220 - 70 = 150$
$n(O' \cap N) = 15 + 20 = 35$

By the union rule,

$n(O' \cup N) = n(O') + n(N) - n(O' \cap N)$
$= 150 + 45 - 35$
$= 160.$

(f) $M' \cap (R' \cap N')$ consists of all people who are not in the R column and not in the N column and who are at the same time not in the M row. Therefore,

$n(M' \cap (R' \cap N')) = 15 + 50 = 65.$

(g) $M \cup (D \cap Y)$ consists of all people age 21–25 who drink diet cola *or* anyone age 26–35.

39. Let T be the set of tall pea plants, G be the set of plants with green peas, and S be the set of plants with smooth peas. We are given the following information.

$n(T) = 22$
$n(G) = 25$
$n(S) = 39$
$n(T \cap G) = 9$
$n(T \cap S) = 17$
$n(G \cap S) = 20$
$n(T \cap G \cap S) = 6$
$n(T' \cap G' \cap S') = 4$

Start with the last two restricted regions, $T \cap G \cap S$ and $T' \cap G' \cap S'$. With $n(G \cap S) = 20$, there are $20 - 6 = 14$ yet to be labeled; $n(T \cap S) = 17$ puts $17 - 6 = 11$ in $T \cap S \cap G'$; also, $n(T \cap G) = 9$ puts $9 - 6 = 3$ in $T \cap G \cap S'$. Now fill in $T \cap (G' \cap S') = 22 - 20 = 2$,

$G \cap (T' \cap S') = 25 - 23 = 2$, and

$S \cap (T' \cap G') = 39 - 31 = 8$.

$N(U) = 2 + 3 + 2 + 11 + 6 + 14 + 8 + 4$

$= 50$

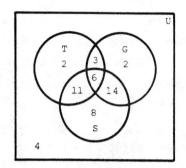

(a) $n(U) = 50$

(b) $n(T \cap S' \cap G') = 2$

(c) $n(T' \cap S \cap G) = 14$

41. First fill in the Venn diagram, starting with the region common to all three sets.

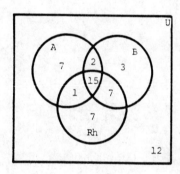

(a) The total of these numbers in the diagram is 54.

(b) $7 + 3 + 7 = 17$ had only one antigen.

(c) $1 + 2 + 7 = 10$ had exactly two antigens.

(d) A person with O–positive blood has only the Rh antigen, so this number is 7.

(e) A person with AB–positive blood has all three antigens, so this number is 15.

(f) A person with B–negative blood has only the B antigen, so this number is 3.

(g) A person with O–negative blood has none of the antigens. There are 12 such people.

(h) A person with A–positive blood has the A and Rh antigens, but not the B–antigen. The number is 1.

43. Let W be the set of women, C the set of those who speak Cantonese, and F the set of those who set off fire-crackers. We are given the following information.

$$n(W) = 120$$
$$n(C) = 150$$
$$n(F) = 170$$
$$n(W' \cap C) = 108$$
$$n(W' \cap F') = 100$$
$$n(W \cap C' \cap F) = 18$$
$$n(W' \cap C' \cap F') = 78$$
$$n(W \cap C \cap F) = 30$$

Note that

$n(W' \cap C \cap F')$

$= n(W' \cap F') - n(W' \cap C' \cap F')$

$= 100 - 78 = 22$.

Furthermore,

n(W′ ∩ C ∩ F)

  = n(W′ ∩ C) − n(W′ ∩ C ∩ F′)

  = 108 − 22 = 86.

We now have

n(W ∩ C ∩ F′)

 = n(C) − n(W′ ∩ C ∩ F)

   − n(W ∩ C ∩ F) − n(W′ ∩ C ∩ F′)

 = 150 − 86 − 30 − 22 = 12.

With all of the overlaps of W, C, and F determined, we can now compute n(W ∩ C′ ∩ F′) = 60 and n(W′ ∩ C′ ∩ F) = 36.

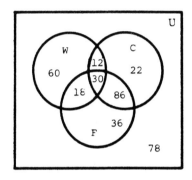

(a) Adding up the disjoint components, we find the total attendance to be

60 + 12 + 18 + 30 + 22 + 86 + 36 + 78 = 342.

(b) n(C′) = 342 − n(C)

        = 342 − 150 = 192

(c) n(W ∩ F′) = 60 + 12 = 72

(d) n(W′ ∩ C ∩ F) = 86

Section 5.3

3. The sample space is the set of the twelve months, {January, February, March, ..., December}.

5. The possible number of points earned could be any whole number from 0 to 80. The sample space is the set
{0, 1, 2, 3, ..., 80}.

7. The possible decisions are to go ahead with a new oil shale plant or to cancel it. The sample space is the set {go ahead, cancel}.

9. Let h = heads and t = tails for the coin; the die can display 6 different numbers. There are 12 possible outcomes in the sample space, which is the set
{(h, 1), (t, 1), (h, 2), (t, 2),
 (h, 3), (t, 3), (h, 4), (t, 4),
 (h, 5), (t, 5), (h, 6), (t, 6)}.

13. Use the first letter of each name. The sample space is the set
{AB, AC, AD, AE, BC,
  BD, BE, CD, CE, DE}.

(a) One of the committee members must be Chinn. This event is {AC, BC, CD, CE}.

(b) Alam, Bartolini, and Chinn may be on any committee; Dickson and Ellsberg may not be on the same committee. This event is {AB, AC, AD, AE, BC, BD, BE, CD, CE}.

(c) Both Alam and Chinn are on the committee. This event is $\{AC\}$.

15. Each outcome consists of two of the numbers 1, 2, 3, 4, and 5, without regard for order. For example, let (2, 5) represent the outcome that the slips of paper marked with 2 and 5 are drawn. There are 10 pairs in this sample space, which is

$\{(1, 2),\ (1, 3),\ (1, 4),\ (1, 5),$
$\ (2, 3),\ (2, 4),\ (2, 5),\ (3, 4),$
$\ (3, 5),\ (4, 5)\}.$

(a) Both numbers in the outcome pair are even. This event is $\{(2, 4)\}$, which is called a simple event since it consists of only one outcome.

(b) One number in the pair is even and the other number is odd. This event is

$\{(1, 2),\ (1, 4),\ (2, 3),$
$\ (2, 5),\ (3, 4),\ (4, 5)\}.$

(c) Each slip of paper has a different number written on it, so it is not possible to draw two slips marked with the same number. This event is $\emptyset$, which is called an impossible event since it contains no outcomes.

17. There are 6 possibilities for the first number and 6 for the second number. For example, let (2, 5) represent the outcome of observing 2 on the first toss and 5 on the

second toss. There are 36 ordered pairs in this sample space, which is

$\{(1, 1),\ (1, 2),\ (1, 3),\ (1, 4),$
$\ (1, 5),\ (1, 6),\ (2, 1),\ (2, 2),$
$\ (2, 3),\ (2, 4),\ (2, 5),\ (2, 6),$
$\ (3, 1),\ (3, 2),\ (3, 3),\ (3, 4),$
$\ (3, 5),\ (3, 6),\ (4, 1),\ (4, 2),$
$\ (4, 3),\ (4, 4),\ (4, 5),\ (4, 6),$
$\ (5, 1),\ (5, 2),\ (5, 3),\ (5, 4),$
$\ (5, 5),\ (5, 6),\ (6, 1),\ (6, 2),$
$\ (6, 3),\ (6, 4),\ (6, 5),\ (6, 6)\}.$

(a) There are six outcomes that have 3 as the first number of the ordered pair. This event is

$\{(3, 1),\ (3, 2),\ (3, 3),$
$\ (3, 4),\ (3, 5),\ (3, 6)\}.$

(b) There are five outcomes in which the sum of the two numbers is 8. This event is

$\{(2, 6),\ (3, 5),\ (4, 4),$
$\ (5, 3),\ (6, 2)\}.$

(c) The sum never exceeds 12, so this event is impossible. The event is the empty set, which we write as $\emptyset$.

19. $S = \{1, 2, 3, 4, 5, 6\}$, so $n(S) = 6$.
$E = \{2\}$, so $n(E) = 1$.

If all the outcomes in a sample space S are equally likely, then the probability of an event E is

$$P(E) = \frac{n(E)}{n(S)}.$$

In this problem,

$$P(E) = \frac{n(E)}{n(S)} = \frac{1}{6}.$$

21. $S = \{1, 2, 3, 4, 5, 6\}$, so $n(S) = 6$.

$E = \{1, 2, 3, 4\}$, so $n(E) = 4$.

$$P(E) = \frac{4}{6} = \frac{2}{3}.$$

23. $S = \{1, 2, 3, 4, 5, 6\}$, so $n(S) = 6$.

$E = \{3, 4\}$, so $n(E) = 2$.

$$P(E) = \frac{2}{6} = \frac{1}{3}.$$

25. $n(S) = 52$

Let E be the event "a 9 is drawn." There are four 9's in the deck, so $n(E) = 4$.

$$P(9) = P(E) = \frac{n(E)}{n(S)} = \frac{4}{52} = \frac{1}{13}$$

27. Let F be the event "a black 9 is drawn." There are two black 9's in the deck, so $n(F) = 2$. As before, $n(S) = 52$.

$$P(\text{black } 9) = P(F) = \frac{n(F)}{n(S)} = \frac{2}{52} = \frac{1}{26}$$

29. Let G be the event "a 9 of hearts is drawn." There is only one 9 of hearts in a deck of 52 cards, so $n(G) = 1$. Again, $n(S) = 52$.

$$P(9 \text{ of hearts}) = P(G) = \frac{n(G)}{n(S)} = \frac{1}{52}$$

31. Let H be the event "a 2 or a queen is drawn." There are four 2's and four queens in the deck, so $n(H) = 8$. Also, $n(S) = 52$.

$$P(2 \text{ or queen}) = P(H) = \frac{n(H)}{n(S)} = \frac{8}{52} = \frac{2}{13}$$

33. Let J be the event "a red face card is drawn." There are three face cards (jack, queen, and king) in each of the two red suits (hearts and diamonds), so $n(J) = 6$. Also, $n(S) = 52$.

$$P(\text{red face card}) = P(J) = \frac{n(J)}{n(S)}$$
$$= \frac{6}{52} = \frac{3}{26}$$

35. Since there are $2 + 3 + 5 + 8 = 18$ marbles in the jar and the experiment consists of drawing one of them at random, $n(S) = 18$. 2 of the marbles are white, so

$$P(\text{white}) = \frac{2}{18} = \frac{1}{9}.$$

37. 5 of the marbles are yellow and $n(S) = 18$, so

$$P(\text{yellow}) = \frac{5}{18}.$$

39. $2 + 3 + 5 = 10$ of the marbles are not black and $n(S) = 18$, so

$$P(\text{not black}) = \frac{10}{18} = \frac{5}{9}.$$

41. E: worker is female

F: worker has worked less than 5 yr

G: worker contributes to a voluntary retirement plan

(a) E′ occurs when E does not, so E is the event "worker is male."

(b) E ∩ F occurs when both E and F occur, so E ∩ F is the event "worker is female and has worked less than 5 yr."

(c) $E \cup G'$ is the event "worker is female or does not contribute to a voluntary retirement plan, or both."

43. First, calculate that

$$9.0 + 1.2 + 2.7 + 1.0 + 1.1 = 15.0$$

is the total amount of funding (in billions of dollars).

(a) P(federal government)

$$= \frac{9.0}{15.0} = \frac{9}{15} = \frac{3}{5}$$

(b) P(industry) $= \frac{1.0}{15.0} = \frac{1}{15}$

(c) P(institutional)

$$= \frac{2.7}{15.0} = \frac{27}{150} = \frac{9}{50}$$

45. E: person smokes

F: person has a family of heart disease

G: person is overweight

(a) $E \cup F$ occurs when E or F or both occur, so $E \cup F$ is the event "person smokes or has a family history of heart disease, or both."

(b) $E' \cap F$ occurs when E does not occur and F does occur, so $E' \cap F$ is the event "person does not smoke and has a family history of heart disease."

(c) $F' \cup G'$ is the event "person does not have a family history of heart disease or is not overweight, or both."

Section 5.4

3. No, it is possible to wear both glasses and sandals. The events are not mutually exclusive.

5. Yes, teenagers cannot be over 30. The events are mutually exclusive.

7. No, a postal worker can be male. The events are not mutually exclusive.

9. When the two dice are rolled, there are 36 equally likely outcomes. Let 5–3 represent the outcome "the first die shows a 5 and the second die shows a 3," and so on.

(a) Rolling a sum of 8 occurs when the outcome is 2–6, 3–5, 4–4, 5–3, or 6–2. Therefore, since there are five such outcomes, the probability of this event is

$$P(\text{sum is 8}) = \frac{5}{36}.$$

(b) A sum of 9 occurs when the outcome is 3–6, 4–5, 5–4, or 6–3, so

$$P(\text{sum is 9}) = \frac{4}{36} = \frac{1}{9}.$$

(c) A sum of 10 occurs when the outcome is 4–6, 5–5, or 6–4, so

$$P(\text{sum is 10}) = \frac{3}{36} = \frac{1}{12}.$$

(d) A sum of 13 does not occur in any of the 36 outcomes, so

$$P(\text{sum is } 16) = \frac{0}{36} = 0.$$

11.   **(a)** P(sum is not more than 5)

$$= P(2) + P(3) + P(4) + P(5)$$

$$= \frac{1}{36} + \frac{2}{36} + \frac{3}{36} + \frac{4}{36}$$

$$= \frac{10}{36} = \frac{5}{18}$$

(b) P(sum is not less than 8)

$$= P(8) + P(9) + P(10)$$
$$+ P(11) + (12)$$

$$= \frac{5}{36} + \frac{4}{36} + \frac{3}{36} + \frac{2}{36} + \frac{1}{36}$$

$$= \frac{15}{36} = \frac{5}{12}$$

(c) P(sum is between 3 and 7)

$$= P(4) + P(5) + P(6)$$

$$= \frac{3}{36} + \frac{4}{36} + \frac{5}{36}$$

$$= \frac{12}{36} = \frac{1}{3}$$

13.   **(a)** There are a total of 12 aces, 2's and 3's in a deck of 52, so

$$P(A \text{ or } 2 \text{ or } 3) = \frac{12}{52} = \frac{3}{13}.$$

(b) There are 13 diamonds plus three 7's in other suits, so

$$P(D \text{ or } 7) = \frac{16}{52} = \frac{4}{13}.$$

Alternatively, using the union rule for probability,

$$P(D) + P(7) - P(7 \text{ of diamonds})$$
$$= \frac{13}{52} + \frac{4}{52} - \frac{1}{52} = \frac{16}{52} = \frac{4}{13}.$$

(c) There are 26 black cards plus 2 red aces, so

$$P = \frac{28}{52} = \frac{7}{13}.$$

(d) There are 13 hearts plus 3 additional jacks in other suits, so

$$P = \frac{16}{52} = \frac{4}{13}.$$

15.   **(a)** There are 3 uncles plus 2 cousins out of 10, so

$$P = \frac{5}{10} = \frac{1}{2}.$$

(b) There are 3 uncles, 2 brothers, and 2 cousins, for a total of 7 out of 10, so

$$P = \frac{7}{10}.$$

(c) There are 2 aunts, 2 cousins, and 1 mother, for a total of 5 out of 10, so

$$P = \frac{5}{10} = \frac{1}{2}.$$

17.   **(a)** The sample space for this experiment is listed in part (b) below. The only outcomes in which both numbers are even are (2, 4) and (4, 2), so

$$P = \frac{2}{20} = \frac{1}{10}.$$

(b) The sample space is

$$\{(\underline{1, 2}), (1, 3), (\underline{1, 4}), (\underline{1, 5}),$$
$$(\underline{2, 1}), (\underline{2, 3}), (\underline{2, 4}), (\underline{2, 5}),$$
$$(3, 1), (\underline{3, 2}), (\underline{3, 4}), (\underline{3, 5}),$$
$$(\underline{4, 1}), (\underline{4, 2}), (\underline{4, 3}), (\underline{4, 5}),$$
$$(\underline{5, 1}), (\underline{5, 2}), (\underline{5, 3}), (\underline{5, 4})\}.$$

The 18 underlined pairs are the outcomes in which one number is even or greater than 3, so

$$P = \frac{18}{20} = \frac{9}{10}.$$

(c) The sum is 5 in the outcomes (1, 4), (2, 3), (3, 2), and (4, 1). The second draw is 2 in the outcomes (1, 2), (3, 2), (4, 2), and (5, 2). There are 7 distinct outcomes out of 20, so

$$P = \frac{7}{20}.$$

19. $P(Z) = .42$, $P(Y) = .38$, $P(Z \cup Y) = .61$

Begin by using the union rule for probability.

$$P(Z \cup Y) = P(Z) + P(Y) - P(Z \cap Y)$$
$$.61 = .42 + .38 - P(Z \cap Y)$$
$$.61 = .80 - P(Z \cap Y)$$
$$-.19 = -P(Z \cap Y)$$
$$.19 = P(Z \cap Y)$$

This gives the first value to be labelled in the Venn diagram. Then the part of Z outside Y must contain .42 - .19 = .23, and the part of Y outside Z must contain .38 - .19 = .19. Observe that .23 + .19 + .19 = .61, which agrees with the given information that P(Z ∪ Y) = .61. The part of U outside both Y and Z must contain 1 - .61 = .39.

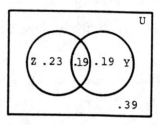

This Venn diagram may now be used to find the following probabilities.

(a) Z′ ∩ Y′ is the event represented by the part of the Venn diagram that is outside Z and outside Y.

$$P(Z' \cap Y') = .39$$

(b) Z′ ∪ Y′ is everything outside Z or outside Y or both, which is all of U except Z ∩ Y.

$$P(Z' \cup Y') = .39 + .19 + .23$$
$$= .81$$

(c) Z′ ∪ Y is everything outside Z or inside Y or both.

$$P(Z' \cup Y) = .19 + .19 + .39$$
$$= .77$$

(d) Z ∩ Y′ is everything inside Z and outside Y.

$$P(Z \cap Y') = .23$$

21. Let E be the event "a 5 is rolled."

$$P(E) = \frac{1}{6} \text{ and } P(E') = \frac{5}{6}.$$

The odds in favor of rolling a 5 are

$$\frac{P(E)}{P(E')} = \frac{1/6}{5/6} = \frac{1}{5},$$

which is written "1 to 5."

23. Let E be the event "1, 2, 3, or 4 is rolled." Here P(E) = 4/6 = 2/3 and P(E') = 1/3. The odds in favor of E are

$$\frac{P(E)}{P(E')} = \frac{2/3}{1/3} = \frac{2}{1},$$

which is written "2 to 1."

25. **(a)** Yellow: there are 3 ways to win and 12 ways to lose. The odds in favor of drawing yellow are 3 to 12, or 1 to 4.

(b) Blue: there are 8 ways to win and 7 ways to lose; the odds in favor of drawing blue are 8 to 7.

(c) White: there are 4 ways to win and 11 ways to lose; the odds in favor of drawing white are 4 to 11.

27. We are given that P(profit) = .74, so P(no profit) = .26. Hence, the odds against making a profit are .26 to .74 or 13 to 37.

29. The odds of winning are 3 to 2; that means there are 3 ways to win and 2 ways to lose, out of a total of 2 + 3 = 5 ways altogether. Hence, the probability of losing is $\frac{2}{5}$.

31. It is possible to establish an exact probability for this event, so this is not an empirical probability.

33. It is not possible to establish an exact probability for this event, so this is an empirical probability.

35. It is not possible to establish an exact probability for this event, so this is an empirical probability.

37. The gambler's claim is a mathematical fact, so this is not an empirical proability.

41. **(a)** The event "in the 70s" corresponds to two rows in the given probability distribution table.

P(in the 70s)
= P(70 - 74) + P(75 - 79)
= .28 + .22
= .50

(b) P(not in the 60s)
= 1 - P(in the 60s)
= 1 - (.08 + .15)
= 1 - .23 = .77

(c) P(not in the 60s or 70s)
= P(below 60) + P(80 - 84)
+ P(85 - 89) + P(90 - 94)
+ P(95 - 99) + P(100 or more)
= .01 + .08 + .06 + .04
+ .02 + .06
= .27

43. **(a)** P(less than \$350)
= 1 - P(\$350 or more)
= 1 - (.08 + .03)
= 1 - .11 = .89

(b) P($75 or more)

   = P($75 – $99.99)

   + P($100 – $199.99)

   + P($200 – $349.99)

   + P($350 – $499.99)

   + P($500 or more)

   = .11 + .09 + .07 + .08 + .03

   = .38

(c) P($200 or more)

   = P($200 – $340.99)

   + P($350 – $499.99)

   + P($500 or more)

   = .07 + .08 + .03

   = .18

45. The probability assignment is possible because the probability of each outcome is a number between 0 and 1, and the sum of the probabilities of all the outcomes is 1.

47. The probability assignment is not possible. All of the probabilities are between 0 and 1, but the sum of the probabilities is $\frac{13}{12}$ which is larger than 1.

49. The probability assignment is not possible. One of the probabilities is negative instead of being between 0 and 1, and the sum of the probabilities is not 1.

51. (a) P($500 or more)

   = 1 – P(less than $500)

   = 1 – (.31 + .18)

   = 1 – .49 = .51

(b) P(less than $1000)

   = .31 + .18 + .18

   = .67

(c) P($500 to $2999)

   = .18 + .13 + .08

   = .39

(d) P($3000 or more)

   = .05 + .06 + .01

   = .12

53. P(C) = .049, P(M ∩ C) = .042, P(M ∪ C) = .534

Place the given information in a Venn diagram by starting with .042 in the intersection of the regions for M and C.

Since P(C) = .049,

   .049 – .042 = .007

goes inside region C, but outside the intersection of C and M. Thus,

   P(C ∩ M′) = .007.

Since P(M ∪ C) = .534,

   .534 – .042 – .007 = .485

goes inside region M, but outside the intersection of C and M. Thus, P(M ∩ C′) = .485. The labeled regions have probability

   .485 + .042 + .007 = .534.

Since the entire region of the Venn diagram must have probability 1, the region outside M and C, or $M' \cap C'$, has probability

$$1 - .534 = .466.$$

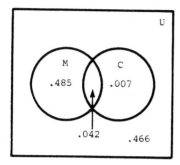

(a) P(C') = 1 - P(C)

= 1 - .049

= .951

(b)   P(M) = .485 + .042

= .527

(c) P(M') = 1 - P(M)

= 1 - .527

= .473

(d) $P(M' \cap C') = .466$

(e) $P(C \cap M') = .007$

(f) $P(C \cup M')$

= $P(C) + P(M') - P(C \cap M')$

= .049 + .473 - .007

= .515

55. **(a)** Now red is no longer dominant, and RW or WR results in pink, so P(red) = P(RR) = 1/4.

(b) P(pink) = P(RW) + P(WR)

= $\frac{1}{4} + \frac{1}{4} = \frac{1}{2}$

(c) P(white) = P(WW) = 1/4

57. Since 55 of the workers were women, 130 - 55 = 75 were men. Since 3 of the women earned more than $40,000, 55 - 3 = 52 of them earned less than $40,000. Since 62 of the men earned less than $40,000, 75 - 62 = 13 earned more than $40,000.

These data for the 130 workers can be summarized in the following table.

	Men	Women
Under $40,000	62	52
Over $40,000	13	3

(a) P(a woman earning less than $40,000)

= $\frac{52}{130}$ = .4

(b) P(a man earning more than $40,000)

= $\frac{13}{130}$ = .1

(c) P(a man or is earning more than $40,000)

= $\frac{62 + 13 + 3}{130}$

= $\frac{78}{130}$ = .6

(d) P(a woman or is earning less $40,000)

than

= $\frac{52 + 3 + 62}{130}$

= $\frac{117}{130}$ = .9

59. Let A be the set of refugees who came to escape abject poverty and B be the set of refugees who came to escape political oppression. Then $P(A) = .80$, $P(B) = .90$, and $P(A \cap B) = .70$.

$$P(A \cup B) = P(A) + P(B) - P(A \cap B)$$
$$= .80 + .90 - .70 = 1$$
$$P(A' \cap B') = 1 - P(A \cap B)$$
$$= 1 - 1 = 0$$

The probability that a refugee in the camp was not poor nor seeking political asylum is 0.

61. This exercise should be solved by computer methods. The solution will vary according to the computer program that is used. The theoretical probabilities may be found in the following way.

When a coin is tossed 5 times, the sample space is

$S = \{$hhhhh, hhhht, hhhth, hhthh,
hthhh, thhhh, hhhtt, hhtht,
hthht, thhht, hhtth, hthth,
thhth, ththh, tthhh, htthh,
hhttt, hthtt, httht, htttt,
ttthh, ttthh, tthht, ththt,
thhtt, thtth, tttth, ttthht,
tthtt, thttt, htttt, ttttt$\}$.

(a) P(4 heads) $= \dfrac{5}{32} = .15625$

(b) P(hhthh) $= \dfrac{1}{32} = .03125$

63. This exercise should be solved by computer methods. The solution will vary according to the computer program that is used. The answer is .632.

Section 5.5

1. Let A be the event "the number is 2" and B be the event "the number is odd."

The problem seeks the conditional probability $P(A|B)$. Use the definition

$$P(A|B) = \frac{P(A \cap B)}{P(B)}.$$

Here, $P(A \cap B) = 0$ and $P(B) = 1/2$. Thus,

$$P(A|B) = \frac{0}{1/2} = 0.$$

3. Let A be the event "the number is even" and B be the event "the number is 6." Then

$$P(A|B) = \frac{P(A \cap B)}{P(B)} = \frac{1/6}{1/6} = 1.$$

5. Let A be the event "sum of 6" and B be the event "double." 6 of the 36 ordered pairs have a sum of 6, so $P(B) = 6/36 = 1/6$. There is only one outcome, 3-3, in $A \cap B$, so $P(A \cap B) = 1/36$. Thus

$$P(A|B) = \frac{1/36}{1/6} = \frac{6}{36} = \frac{1}{6}.$$

7. Use a reduced sample space. After the first card drawn is a heart, there remain 51 cards, of which 12 are hearts. Thus,

P(heart on 2nd|heart on 1st)

$$= \frac{12}{51} = \frac{4}{17}.$$

9. Use a reduced sample space. After the first card drawn is a jack, there remain 51 cards, of which 11 are face cards. Thus,

P(face card on 2nd|jack on 1st)

$$= \frac{11}{51}.$$

11. P(a jack and a 10)

= P(jack followed by 10)

+ P(10 followed by jack)

$$= \frac{4}{52} \cdot \frac{4}{51} + \frac{4}{52} \cdot \frac{4}{51}$$

$$= \frac{16}{2652} + \frac{16}{2652}$$

$$= \frac{32}{2652} \approx .012$$

13. P(two black cards)

= P(black on 1st)

• P(black on 2nd|black on 1st)

$$= \frac{26}{52} \cdot \frac{25}{51}$$

$$= \frac{650}{2652} \approx .245$$

19. First draw the tree diagram.

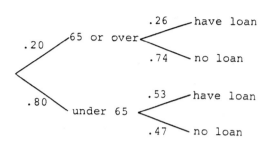

(a) P(person is 65 or over and has a loan)

= P(65 or over)

• P(has loan|65 or over)

= (.20)(.26) = .052

(b) P(person has a loan)

= P(65 or over and has loan)

+ P(under 65 and has loan)

= (.20)(.26) + (.80)(.53)

= .052 + .424 = .476

21. Draw the tree diagram.

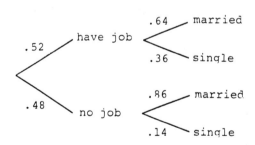

(a) P(married)

= P(job and married)

+ P(no job and married)

= (.52)(.64) + (.48)(.86)

= .3328 + .4128 = .7456

(b) P(job and single)

= (.52)(.36) = .1872

23.   Draw the tree diagram.

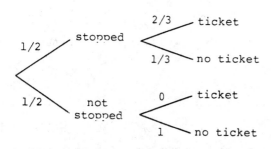

(a) P(no ticket)

= P(stopped and no ticket)

+ P(not stopped and no
ticket)

$= \left(\frac{1}{2}\right)\left(\frac{1}{3}\right) + \left(\frac{1}{2}\right)(1)$

$= \frac{1}{6} + \frac{1}{2} = \frac{2}{3}$

(b) If his activities on the 3 consecutive weekends are independent,
then

P(no ticket in 3 weekends)

= P(no ticket) • P(no ticket)

• P(no ticket)

$= \frac{2}{3} \cdot \frac{2}{3} \cdot \frac{2}{3}$

$= \frac{8}{27} \approx .296$

25.   P(component fails) = .03

(a) Let n represent the number of
these components to be connected in
parallel; that is, the component has
n - 1 backup components.

P(at least one component works)

= 1 - P(no component works)

= 1 - P(all n components fail)

$= 1 - (.03)^n$

If this probability is to be at
least .999999, then

$1 - (.03)^n \geq .999999$

$-(.03)^n \geq -.000001$

$(.03)^n \leq .000001,$

and the smallest whole number value
of n for which this inequality holds
true is n = 4.

4 - 1 = 3 backup components must be
used.

27.   Let A be the event "student studies"
and B be the event "student gets a
good grade." We are told that
P(A) = .6, P(B) = .7, and
P(A ∩ B) = .52.

P(A) • P(B) = (.6)(.7) = .42

Since P(A) • P(B) is not equal to
P(A ∩ B), A and B are not independent. Rather, they are dependent
events.

29. The probability that a customer cashing a check will fail to make a deposit is

$$P(D' \mid C) = \frac{n(D' \cap C)}{n(C)} = \frac{30}{80} = \frac{3}{8}.$$

31. The probability that a customer making a deposit will not cash a check is

$$P(C' \mid D) = \frac{n(C' \cap D)}{n(D)} = \frac{20}{70} = \frac{2}{7}.$$

33. **(a)** Since the separate flights are independent, the probability of 3 flights in a row is

$$(.98)(.98)(.98) = .941192 \approx .94.$$

35. Since 60% of production comes from assembly line B, P(A) = .40 (the remaining 40%). Also P(pass inspection|A) = .95, so P(not pass|A) = .05. Therefore,

P(A ∩ not pass)

   = P(A) • P(not pass|A)

   = (.40)(.05) = .02.

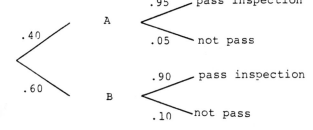

37. The sample space is {RW, WR, RR, WW}. The event "red" is {RW, WR, RR} and the event "mixed" is {RW, WR}.

$$P(\text{mixed}|\text{red}) = \frac{n(\text{mixed and red})}{n(\text{red})}$$

$$= \frac{2}{3}.$$

39. P(3 girls|3rd is a girl)

$$= \frac{P(3 \text{ girls and 3rd is girl})}{P(3\text{rd is girl})}$$

$$= \frac{1/8}{1/2} = \frac{1}{4}$$

41. P(3 girls|at least 2 girls)

$$= \frac{P(3 \text{ girls and at least 2 girls})}{P(\text{at least 2 girls})}$$

$$= \frac{P(3 \text{ girls})}{P(\text{at least 2 girls})}$$

$$= \frac{P(3 \text{ girls})}{P(2 \text{ girls}) + P(3 \text{ girls})}$$

$$= \frac{1/8}{(3/8) + (1/8)}$$

$$= \frac{1/8}{4/8} = \frac{1}{4}$$

(Note that P(3 girls) = P(GGG)

$$= \frac{1}{2} \cdot \frac{1}{2} \cdot \frac{1}{2} = \frac{1}{8}$$

and

P(2 girls) = P(GGB)

         + P(BGG)

         + P(GBG)

$$= \frac{1}{8} + \frac{1}{8} + \frac{1}{8} = \frac{3}{8}.)$$

43. P(M) = .527, the total of the M column.

45. $P(M \cap C) = .042$, the entry in the M column and C row.

47. $P(M|C) = \dfrac{P(M \cap C)}{P(C)} = \dfrac{.042}{.049} \approx .857$

(The exact value is $\dfrac{.042}{.049} = \dfrac{42}{49} = \dfrac{6}{7}$.)

49. $P(M'|C) = \dfrac{P(M' \cap C)}{P(C)}$

$= \dfrac{.007}{.049} \approx .143$

(The exact value is $\dfrac{.007}{.049} = \dfrac{7}{49} = \dfrac{1}{7}$.)

51. From the table,

$P(C \cap D) = .0004$ and
$P(C) \cdot P(D) = (.0800)(.0050)$
$\qquad\qquad = .0004.$

Since $P(C \cap D) = P(C) \cdot P(D)$, C and D are independent events; color blindness and deafness are independent events.

53. First draw the tree diagram.

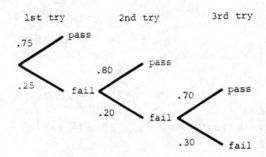

(a) P(fails both 1st and 2nd tests) = P(fails 1st) · P(fails 2nd|fails 1st)
$\qquad\qquad\qquad\qquad\qquad = .25(.20)$
$\qquad\qquad\qquad\qquad\qquad = .05$

(b) P(fails three times in a row) = $(.25)(.2)(.3) = .015$

(c) P(requires at least 2 tries) = P(does not pass on 1st try) = .25

55. There are 6 males of which 1 is a beagle and 1 a cocker spaniel, so there are 4 male poodles and hence 2 female poodles. The table is as follows.

	Beagle	Cocker Spaniel	Poodle	Totals
Male	1	1	4	6
Female	2	0	2	4
Totals	3	1	6	10

$$P(\text{beagle}|\text{male}) = \frac{n(\text{beagle and male})}{n(\text{male})} = \frac{1}{6}$$

57. $P(\text{cocker spaniel}|\text{female}) = \dfrac{n(\text{cocker spaniel and female})}{n(\text{female})} = \dfrac{0}{4} = 0$

59. $P(\text{female}|\text{beagle}) = \dfrac{n(\text{female and beagle})}{n(\text{beagle})} = \dfrac{2}{3}$

Section 5.6

1. Use Bayes' theorem with two possibilities M and M′.

$$P(M|N) = \frac{P(M) \cdot P(N|M)}{P(M) \cdot P(N|M) + P(M') \cdot P(N|M')} = \frac{(.4)(.3)}{(.4)(.3) + (.6)(.4)}$$

$$= \frac{.12}{.12 + .24} = \frac{.12}{.36} = \frac{12}{36} = \frac{1}{3}$$

3. Using Bayes' theorem,

$$P(R_1|Q) = \frac{P(R_1) \cdot P(Q|R_1)}{P(R_1) \cdot P(Q|R_1) + P(R_2) \cdot P(Q|R_2) + P(R_3) \cdot P(Q|R_3)}$$

$$= \frac{.05(.40)}{.05(.40) + .6(.30) + .35(.60)} = \frac{.02}{.41} = \frac{2}{41}.$$

5. Using Bayes' theorem,

$$P(R_3|Q) = \frac{P(R_3) \cdot P(Q|R_3)}{P(R_1) \cdot P(Q|R_1) + P(R_2) \cdot P(Q|R_2) + P(R_3) \cdot P(Q|R_3)}$$

$$= \frac{.35(.60)}{.05(.40) + .6(.30) + .35(.60)} = \frac{.21}{.41} = \frac{21}{41}.$$

7. We first draw the tree diagram and determine the probabilities as indicated below.

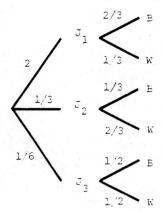

We want to determine the probability that if a white ball is drawn, it came from the second jar. This $P(J_2|W)$. Use Bayes' theorem.

$$P(J_2|W) = \frac{P(J_2) \cdot P(W|J_2)}{P(J_2) \cdot P(W|J_2) + P(J_1) \cdot P(W|J_1) + P(J_3) \cdot P(W|J_3)}$$

$$= \frac{\dfrac{1}{3} \cdot \dfrac{2}{3}}{\dfrac{1}{3} \cdot \dfrac{2}{3} + \dfrac{1}{2} \cdot \dfrac{1}{3} + \dfrac{1}{6} \cdot \dfrac{1}{2}} = \frac{\dfrac{2}{9}}{\dfrac{2}{9} + \dfrac{1}{6} + \dfrac{1}{12}} = \frac{\dfrac{2}{9}}{\dfrac{17}{36}} = \frac{8}{17}$$

9. Let G represent "good worker," B represent "bad worker," S represent "passes the test," and F represent "fails the test."
   The given information is
   $P(G) = .70$, $P(B) = P(G') = .30$, $P(S|G) = .80$ (and therefore $P(F|G) = .20$), and $P(S|B) = .40$ (and therefore $P(F|B) = .60$).
   If passing the test is made a requirement for employment, then the percent of the new hires that will turn out to be good workers is

$$P(G|S) = \frac{P(G) \cdot P(S|G)}{P(G) \cdot P(S|G) + P(B) \cdot P(S|B)} = \frac{.70(.80)}{.70(.80) + .30(.40)} = \frac{.56}{.56 + .12}$$

$$= \frac{.56}{.68} \approx .824 \quad \text{or} \quad 82.4\%.$$

11. Let Q represent "qualified" and M represent "approved by the manager." Set up the tree diagram.

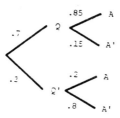

$$P(Q' \mid M) = \frac{P(Q') \cdot P(M \mid Q')}{P(Q) \cdot P(M \mid Q) + P(Q') \cdot P(M \mid Q')} = \frac{.30(.20)}{.70(.85) + .30(.20)} = \frac{.06}{.655} \approx .092$$

13. Let D represent "damaged," A represent "qualified," and B represent "unqualifed." Set up the tree diagram.

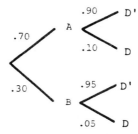

$$P(B \mid D) = \frac{P(B) \cdot P(D \mid B)}{P(B) \cdot P(D \mid B) + P(A) \cdot P(D \mid A)} = \frac{.30(.05)}{.30(.05) + .70(.10)}$$

$$= \frac{.015}{.015 + .07} = \frac{.015}{.085} \approx .176$$

15. Let A represent "slow pay" and L represent "large down payment." Set up the tree diagram.

$$P(S' \mid L) = \frac{P(S') \cdot P(L \mid S')}{P(S) \cdot P(L \mid S) + P(S') \cdot P(L \mid S')} = \frac{.98(.5)}{.02(.14) + .98(.5)} = \frac{.49}{.4928} \approx .994$$

17.   Start with the tree diagram, where the first stage refers to the companies and the second to a defective appliance.

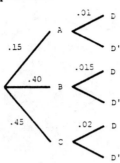

$$P(B|D) = \frac{P(B) \cdot P(D|B)}{P(A) \cdot P(D|A) + P(B(\cdot P(D|B) + P(C) \cdot P(D|C)}$$

$$= \frac{.40(.015)}{.15(.01) + .40(.015) + .45(.02)} = \frac{.0060}{.0165} \approx .364$$

19.   Let that PF mean professional football, CF mean college football, B mean baseball, and HR mean high ratings. Draw the tree diagram.

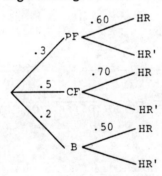

$$P(PF|HR) = \frac{P(PF) \cdot P(HR|PF)}{P(CF) \cdot P(HR|CF) + P(B) \cdot P(HR|B) + P(PF) \cdot P(HR|PF)}$$

$$= \frac{.3(.60)}{.5(.70) + .2(.50) + .3(.60)} = \frac{.18}{.35 + .10 + .18} = \frac{.18}{.63} = \frac{18}{63} = \frac{2}{7}$$

21.   Let R mean recession, M mean mediocre, B mean booming, and H mean huge profits. Draw the tree diagram.

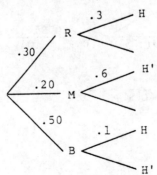

$$P(R|H) = \frac{P(R) \cdot P(H|R)}{P(R) \cdot P(H|R) + P(M) \cdot P(H|M) + P(B) \cdot P(H|B)}$$

$$= \frac{.30(.3)}{.30(.3) + .20(.6) + .50(.1)} = \frac{.09}{.26} = \frac{9}{26}$$

23. There are a total of 1260 + 700 + 560 + 280 = 2800 mortgages being studied at the bank.

    (a) $P(5\% \text{ down}|\text{default}) = \dfrac{.05\left(\frac{1260}{2800}\right)}{.05\left(\frac{1260}{2800}\right) + .03\left(\frac{700}{2800}\right) + .02\left(\frac{560}{2800}\right) + .01\left(\frac{280}{2800}\right)}$

    $$= \frac{.05(.45)}{.05(.45) + .03(.25) + .02(.2) + .01(.1)}$$

    $$= \frac{.0225}{.0225 + .0075 + .004 + .001} = \frac{.0225}{.035} \approx .643$$

    (b) A mortgage being paid to maturity is the complement of a mortgage being defaulted.

    $P(10\% \text{ down}|\text{paid to maturity}) = P(10\% \text{ down}|\text{not default})$

    $$= \frac{.97\left(\frac{700}{2800}\right)}{.95\left(\frac{1260}{2800}\right) + .97\left(\frac{700}{2800}\right) + .98\left(\frac{560}{2800}\right) + .99\left(\frac{280}{2800}\right)}$$

    $$= \frac{.97(.25)}{.95(.45) + .97(.25) + .98(.2) + .99(.1)}$$

    $$= \frac{.2425}{.4275 + .2425 + .196 + .099} = \frac{.2425}{.965} \approx .251$$

25. **(a)** Draw the tree diagram.

$$P(\text{AIDS}|\text{positive}) = \frac{.01(.95)}{.01(.95) + .99(.05)} = \frac{.0095}{.0095 + .0495} = \frac{.0095}{.059} \approx .161$$

27. P(30 - 34|man who never married)

$$= \frac{.126(.250)}{.151(.875) + .126(.433) + .126(.250) + .110(.140) + .487(.054)}$$

$$= \frac{.0315}{.259881} \approx .121$$

29. P(35 - 39|woman who never married)

$$= \frac{.103(.090)}{.142(.752) + .117(.295) + .116(.161) + .103(.090) + .522(.033)}$$

$$= \frac{.00927}{.186471} \approx .050$$

31. Draw the tree diagram.

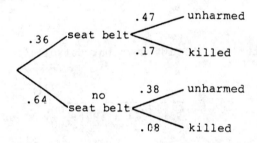

$$P(\text{no seat belt}|\text{unharmed}) = \frac{.64(.38)}{.36(.47) + .64(.38)} = \frac{.2432}{.1692 + .2432} = \frac{.2432}{.4124} \approx .590$$

33. Let V represent voting age and 65 represent 65 or over. Draw the tree diagram.

$$P(65|V) = \frac{.155(.74)}{.11(.48) + .076(.53) + .376(.68) + .283(.64) + .155(.74)} = \frac{.1147}{.64458} \approx .178$$

Chapter 5 Review Exercises

1. $9 \in \{8, 4, -3, -9, 6\}$

   False. 9 is not an element of the set.

3. $2 \notin \{0, 1, 2, 3, 4\}$

   False. 2 is an element of the set.

5. $\{3, 4, 5\} \subset \{2, 3, 4, 5, 6\}$

   True. Every element of $\{3, 4, 5\}$ is an element of $\{2, 3, 4, 5, 6\}$.

7. $\{3, 6, 9, 10\} \subset \{3, 9, 11, 13\}$

   False. 10 is an element of $\{3, 6, 9, 10\}$, but 10 is not an element of $\{3, 9, 11, 13\}$. There-fore, $\{3, 6, 9, 10\}$ is not a subset of $\{3, 9, 11, 13\}$.

9. $\{2, 8\} \not\subset \{2, 4, 6, 8\}$

   False. Since both 2 and 8 are elements of $\{2, 4, 6, 8\}$, $\{2, 8\}$ is a subset of $\{2, 4, 6, 8\}$.

11. $K = \{c, d, f, g\}$

   K has 4 elements, so it has $2^4 = 16$ subsets.

13. K' (the complement of K) is the set of all elements of U that do *not* belong to K.

   $$K' = \{a, b, e\}$$

15. $K \cap R$ (the intersection of K and R) is the set of all elements belonging to both set K and set R.

   $$K \cap R = \{c, d, g\}$$

17. $(K \cap R)' = \{a, b, e, f\}$

19. $\emptyset' = U$

21. $A \cap C$ is the set of all employees in the KO Brown Company who are in the accounting department *and* have at least 10 years in the company.

23. $A \cup D$ is the set of all employees in the KO Brown Company who are in the accounting department *or* have MBA degrees.

25. $B' \cap C'$ is the set of employees who are not in the sales department *and* have worked less than 10 years with the company.

27. $A \cup B'$ is the set of all elements which belong to A or do not belong to B, or both.

   See the Venn diagram in the back of the textbook.

29. $(A \cap B) \cup C$

   First find $A \cap B$.

Now find the union of this region with C.

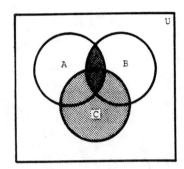

31. The sample space for rolling a die is
S = {1, 2, 3, 4, 5, 6}.

33. The sample space of the possible weights is

S = {0, .5, 1, 1.5, 2, ..., 299.5, 300}.

35. The sample space consists of all ordered pairs (a, b) where a can be 3, 5, 7, 9, or 11, and b is either R(red) or G(green). Thus,

S = {(3, R), (3, G), (5, R), (5, G),
(7, R), (7, G), (9, R), (9, G),
(11, R), (11, G)}.

37. The event that the second ball is green is

F = {(3, G), (5, G), (7, G),
(9, G), (11, G)}.

39. There are 13 hearts out of 52 cards in a deck. Thus,

$$P(\text{heart}) = \frac{13}{52} = \frac{1}{4}.$$

41. There are three face cards in each suit (jack, queen, and king) and there are four suits, so there are 3 · 4 = 12 face cards out of the 52 cards. Thus,

$$P(\text{face card}) = \frac{12}{52} = \frac{3}{13}.$$

43. There are 4 queens of which 2 are red, so

$$P(\text{red}|\text{queen}) = \frac{n(\text{red and queen})}{n(\text{queen})}$$

$$= \frac{2}{4} = \frac{1}{2}.$$

45. There are 4 kings of which all 4 are face cards. Thus,

$P(\text{face card}|\text{king})$

$$= \frac{n(\text{face card and king})}{n(\text{king})}$$

$$= \frac{4}{4} = 1.$$

51. Marilyn vos Savant's answer is that the contestant should switch doors. To understand why, recall that the puzzle begins with the contestant choosing door 1 and then the host opening door 3 to reveal a goat. When the host opens door 3 and shows the goat, that does not affect the probability of the car being behind door 1; the contestant had a $\frac{1}{3}$ probability of being correct to begin with, and he still has a $\frac{1}{3}$ probability after the host opens door 3.

The contestant knew that the host would open another door regardless of what was behind door 1, so opening either other door gives no new information about door 1. The probability of the car being behind door 1 is still $\frac{1}{3}$; with the goat behind door 3, the only other place the car could be is behind door 2, so the probability that the car is behind door 2 is now $\frac{2}{3}$. By switching to door 2, the contestant can double his chances of winning the car.

53. Let E represent the event "draw a black jack." $P(E) = 2/52 = 1/26$ and then $P(E') = 25/26$. The odds in favor of drawing a black jack are

$$\frac{P(E)}{P(E')} = \frac{1/26}{25/26} = \frac{1}{25},$$

or "1 to 25."

55. The sum is 8 for each of the 5 outcomes 2-6, 3-5, 4-4, 5-3, and 6-2. There are 36 outcomes in all in the sample space.

$$P(\text{sum is } 8) = \frac{5}{36} \approx .139$$

57. P(sum is at least 10)

= P(sum is 10) + P(sum is 11)
  + P(sum is 12)

$$= \frac{3}{36} + \frac{2}{36} + \frac{1}{36}$$

$$= \frac{6}{36} = \frac{1}{6} \approx .167$$

59. The sum can be 9 or 11.
P(sum is 9) = 4/36 and
P(sum is 11) = 2/36.

P(sum is odd number greater than 8)

$$= \frac{4}{36} + \frac{2}{36} = \frac{6}{36} = \frac{1}{6} \approx .167$$

61. Consider the reduced sample space of the 11 outcomes in which at least one die is a four. Of these, 2 have a sum of 7, 3-4 and 4-3. Therefore,

P(sum is 7|at least one die is a four)

$$= \frac{2}{11} \approx .182.$$

63. $P(E) = .51$. $P(F) = .37$,
$P(E \cap F) = .22$

(a) $P(E \cup F) = P(E) + P(F) - P(E \cap F)$
$$= .51 + .37 - .22$$
$$= .66$$

(b) Draw a Venn diagram.

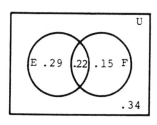

$E \cap F'$ is the portion of the diagram that is inside E and outside F.
$P(E \cap F') = .29$

(c) $E' \cup F$ is outside E or inside F, or both.

$$P(E' \cup F) = .22 + .15 + .34 = .71$$

(d) $E' \cap F'$ is outside E and outside F.

$P(E' \cap F') = .34$

65. The probability that the ball came from box B, given that it is red, is

$P(B | \text{red})$

$$= \frac{P(B)P(\text{red}|B)}{P(B)P(\text{red}|B) + P(A)P(\text{red}|A)}$$

$$= \frac{\frac{5}{8}\left(\frac{2}{5}\right)}{\frac{5}{8}\left(\frac{2}{5}\right) + \frac{3}{8}\left(\frac{5}{6}\right)}$$

$$= \frac{4}{9} \approx .444$$

67. Let C represent "competent shop" and R represent "able to repair appliance."

Draw a tree diagram and label the given information.

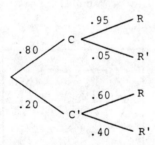

The probability that an appliance that was repaired correctly was repaired by an incompetent shop is

$$P(C' | R) = \frac{P(C' \cap R)}{P(R)}$$

$$= \frac{P(C') \cdot P(R|C')}{P(C) \cdot P(R|C) + P(C') \cdot P(R|C')}$$

$$= \frac{.20(.60)}{.80(.95) + .20(.60)}$$

$$= \frac{.12}{.76 + .12} = \frac{.12}{.88} = \frac{12}{88}$$

$$= \frac{3}{22} \approx .136.$$

69. The probability that an appliance that was repaired incorrectly was repaired by an incompetent shop is

$$P(C' | R') = \frac{P(C' \cap R')}{P(R')}$$

$$= \frac{.20(.40)}{.20(.40) + .80(.05)}$$

$$= \frac{.08}{.12} = \frac{8}{12} = \frac{2}{3}.$$

71. (a) P(no more than 3 defects)

$= P(0) + P(1) + P(2) + P(3)$

$= .31 + .25 + .18 + .12$

$= .86$

(b) P(at least 3 defects)

$= P(3) + P(4) + P(5)$

$= .12 + .08 + .06$

$= .26$

73. (a)

	N_2	T_2
N_1	$N_1 N_2$	$N_1 T_2$
T_1	$T_1 N_2$	$T_1 T_2$

Since the four combinations are equally likely, each has probability 1/4.

(b) P(two trait cells)

$$= P(T_1 T_2) = \frac{1}{4}$$

(c) P(one normal cell and one trait cell)

$= P(N_1 T_2) + P(T_1 N_2)$

$$= \frac{1}{4} + \frac{1}{4} = \frac{1}{2}$$

(d) P(not a carrier and does not have the disease)

$$= P(N_1 N_2) = \frac{1}{4}$$

75. **(a)** P(answer yes) = P(answer B) · P(answer yes|answer B)

$$+ \text{ P(answer A)} \cdot \text{P(answer yes|answer A)}$$

Divide by P(answer B).

$$\frac{\text{P(answer yes)}}{\text{P(answer B)}} = \text{P(answer yes|answer B)} + \frac{\text{P(answer A)} \cdot \text{P(yes|A)}}{\text{P(answer B)}}$$

Solve for P(answer yes|answer B).

$$\text{P(answer yes|answer B)} = \frac{\text{P(answer yes)} - \text{P(answer A)} \cdot \text{P(yes|A)}}{\text{P(answer B)}}$$

(b) Using the formula from part (a),

$$\frac{(.6) - \left(\frac{1}{2}\right)\left(\frac{1}{2}\right)}{\left(\frac{1}{2}\right)} = \frac{7}{10}.$$

77. **(a)**

Car Type	Satisfied	Not Satisfied	Totals
New	300	100	400
Used	450	150	600
Totals	750	250	1000

(b) 1000 buyers were surveyed.

(c) 300 bought a new car and were satisfied.

(d) 250 were not satisfied.

(e) 600 bought used cars.

(f) 150 who were not satisfied had bought a used car.

(g) The event is "the buyer purchased a used car given that the buyer is not satisfied."

(h) P(used car|not satisfied) = $\dfrac{\text{n(used car and not satisfied)}}{\text{n(not satisfied)}}$

$$= \frac{150}{250} = \frac{3}{5}$$

(i) P(used car and not satisfied) = $\dfrac{\text{n(used car and not satisfied)}}{\text{n(buyers)}}$

$$= \frac{150}{1000} = \frac{3}{20}$$

Extended Application

1. Using Bayes' theorem,

$$P(H_2 \mid C_1) = \frac{P(C_1 \mid H_2) \cdot P(H_2)}{P(C_1 \mid H_1) \cdot P(H_1) + P(C_1 \mid H_2) \cdot P(H_2) + P(C_1 \mid H_3) \cdot P(H_3)}$$

$$= \frac{.4(.15)}{.9(.8) + .4(.15) + .1(.05)} = \frac{.06}{.785} \approx .076.$$

2. Using Bayes' theorem,

$$P(H_1 \mid C_2) = \frac{P(C_2 \mid H_1) \cdot P(H_1)}{P(C_2 \mid H_1) \cdot P(H_1) + P(C_2 \mid H_2) \cdot P(H_2) + P(C_2 \mid H_3) \cdot P(H_3)}$$

$$= \frac{(.2)(.8)}{(.2)(.8) + (.8)(.15) + (.3)(.05)} \approx .542.$$

3. Using Bayes' theorem,

$$P(H_3 \mid C_2) = \frac{P(C_2 \mid H_3) \cdot P(H_3)}{P(C_2 \mid H_1) \cdot P(H_1) + P(C_2 \mid H_2) \cdot P(H_2) + P(C_2 \mid H_3) \cdot P(H_3)}$$

$$= \frac{.3(.05)}{.2(.8) + .8(.15) + .3(.05)} = \frac{.015}{.295} \approx .051.$$

CHAPTER 5   TEST

1. Write true or false for each statement.

   (a) $3 \in \{1, 5, 7, 9\}$ **(b)** $\{1, 3\} \not\subset \{0, 1, 2, 3, 4\}$ **(c)** $\emptyset \subset \{2\}$

   (d) A set of 6 distinct elements has exactly 64 subsets.

2. Let $U = \{1, 2, 3, 4, 5, 6, 7, 8, 9\}$, $A = \{1, 3, 4, 5\}$, $B = \{2, 4, 5\}$, and
   $C = \{1, 3, 5, 7\}$.

   Find each of the following sets.

   (a) $A \cap B'$ **(b)** $A \cap (B \cup C')$

3. Draw a Venn diagram and shade the region that represents $A \cap (B \cup C')$.

4. Draw a Venn diagram and fill in regions given that $n(U) = 25$, $n(A) = 11$,
   $n(B \cap A') = 9$, and $n(A \cap B) = 6$.

5. In the Mellonville Social Club, bridge, poker, and rummy are the 3 most
   popular card games. A recent survey of the members found that

   > 60 play poker;
   > 63 play bridge;
   > 37 play rummy;
   > 22 play poker and rummy;
   > 19 play bridge and rummy;
   > 23 play bridge only;
   > 10 play all three games;
   > 7 play none of the games.

   (a) Draw Venn diagram and fill in the numbers in each region.

   (b) How many members do not play poker?

   (c) How may members play bridge or poker?

   (d) How many members play poker but not rummy?

6. Suppose that for events A and B, P(A) = .4, P(B) = .3 and P(A ∪ B) = .68. Find each of the following probabilities.

    (a) P(A ∩ B)   **(b)** P(A′)   **(c)** P(A ∩ B′)   **(d)** P(A′ ∩ B′)

7. An urn contains 4 red, 3 blue, and 2 yellow marbles. A single marble is drawn.

    (a) Find the odds in favor of drawing a red marble.

    (b) Find the probability that a red or a blue marble is drawn.

8. Three cards are drawn without replacement from a standard deck of 52.

    (a) What is the probability that all three are spades?

    (b) What is the probability that all three spades, given that the first card drawn is a spade?

9. The probability of passing the University of Waterloo's physical fitness test is .3. If you fail the first time, your chances of passing on the second try drop to .1. Draw a tree diagram and compute the probability that a person will pass on the first or second try.

10. The Magnum Opus Publishing Company uses three printers to put out its lengthy tomes. Printer A produces 40% of their books with a 20% failure rate. Printer B produces 25% with a 10% failure rate. Printer C produces the remainder with a 40% failure rate. Given that a book is badly printed, what is the probability that it was printed by Printer C?

CHAPTER 5 TEST ANSWERS

1. **(a)** False **(b)** False **(c)** True **(d)** True

2. **(a)** $\{1, 3\}$ **(b)** $\{4, 5\}$

3. 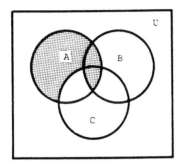 4.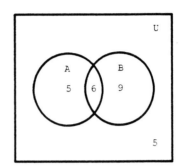

5. **(a)** 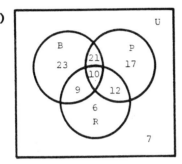 **(b)** 45 **(c)** 92 **(d)** 38

6. **(a)** .02 **(b)** .6 **(c)** .38 **(d)** .32

7. **(a)** 4 to 5 **(b)** 7/9

8. **(a)** $\left(\frac{32}{52}\right)\left(\frac{12}{51}\right)\left(\frac{11}{50}\right) \approx .013$ **(b)** $\left(\frac{12}{51}\right)\left(\frac{11}{50}\right) \approx .052$

9. 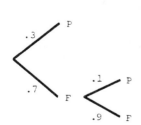 10. $\dfrac{(.35)(.4)}{(.35)(.4) + (.4)(.2) + (.25)(.1)} \approx .57$

P(pass) = .3 + .07 = .37

CHAPTER 6 COUNTING PRINCIPLES; FURTHER PROBABILITY TOPICS

Section 6.1

1. $P(4, 2) = \dfrac{4!}{(4 - 2)!} = \dfrac{4!}{2!}$

    $= \dfrac{4 \cdot 3 \cdot 2!}{2!} = 12$

3. $6! = 6 \cdot 5 \cdot 4 \cdot 3 \cdot 2 \cdot 1 = 720$

5. $0! = 1$

    This is the definition of $0!$.

7. $4! = 4 \cdot 3 \cdot 2 \cdot 1 = 24$

9. $P(13, 2) = \dfrac{13!}{(13 - 2)!} = \dfrac{13!}{11!}$

    $= \dfrac{13 \cdot 12 \cdot 11!}{11!}$

    $= 13 \cdot 12 = 156$

11. $P(25, 12) = \dfrac{25!}{(25 - 12)!}$

    $= \dfrac{25!}{13!}$

    $= 2.490952 \times 10^{15}$

13. $P(14, 5) = \dfrac{14!}{(14 - 5)!}$

    $= \dfrac{14!}{9!}$

    $= \dfrac{14 \cdot 13 \cdot 12 \cdot 11 \cdot 10 \cdot 9!}{9!}$

    $= 14 \cdot 13 \cdot 12 \cdot 11 \cdot 10$

    $= 240,240$

15. By the multiplication principle, there will be $5 \cdot 3 \cdot 2 = 30$ different home types available.

17. By the multiplication principle, there will be $3 \cdot 8 \cdot 5 = 120$ different meals possible.

19. **(a)** The number of ways 5 works can be arranged is

    $$P(5, 5) = 5! = 120.$$

    (b) If one of the 2 overtures must be chosen first, followed by arrangements of the 4 remaining pieces, then

    $$P(2, 1) \cdot P(4, 4) = 2 \cdot 24 = 48$$

    is the number of ways the program can be arranged.

21. The number of possible arrangements is

    $$P(10, 7) = \frac{10!}{3!} = 604,800.$$

23. Pick any 4 of the 380 nonmathematical courses.

    The number of possible schedules is

    $P(380, 4) = \dfrac{380!}{376!}$

    $= 380 \cdot 379 \cdot 378 \cdot 377$

    $= 20,523,714,120.$

    In scientific notation, this answer would be written as 2.05237×10^{10}.

25. The number of possible batting orders is

    $P(20, 9) = \dfrac{20!}{11!}$

    $= 60,949,324,800.$

    In scientific notation, this answer would be written as 6.09493×10^{10}.

27. **(a)** Pick one of the 5 traditional numbers followed by an arrangement of the remaining total of 7. The program can be arranged in $P(5, 1) \cdot P(7, 7) = 5 \cdot 7! = 25,200$ different ways.

(b) Pick one of the 3 original Cajun compositions to play last, preceded by an arrangement of the remaining total of 7. This program can be arranged in

$P(7, 7) \cdot P(3, 1) = 7! \cdot 3 = 15,120$ different ways.

29. By the multiplication principle, a person could schedule the evening of television viewing in $8 \cdot 5 \cdot 6 = 240$ different ways.

31. There is exactly one 3–letter subset of the letters A, B, and C, namely A, B, and C.

33. **(a)** initial

This word contains 3 i's, 1 n, 1 t, 1 a, and 1 ℓ. Use the formula for distinguishable permutations with $n = 7$, $n_1 = 3$, $n_2 = 1$, $n_3 = 1$, $n_4 = 1$ and $n_5 = 1$.

$$\frac{7!}{3!1!1!1!} = \frac{7 \cdot 6 \cdot 5 \cdot 4 \cdot 3!}{3!} = 840$$

There are 840 distinguishable permutations of the letters.

(b) little

Use the formula for distinguishable permutations with $n = 6$, $n_1 = 2$, $n_2 = 1$, $n_3 = 2$, and $n_4 = 1$.

$$\frac{6!}{2!1!2!1!} = \frac{6!}{2!2!}$$
$$= \frac{6 \cdot 5 \cdot 4 \cdot 3 \cdot 2 \cdot 1}{2 \cdot 1 \cdot 2 \cdot 1}$$

There are 180 distinguishable permutations.

(c) decreed

Use the formula for distinguishable permutations with $n = 7$, $n_1 = 2$, $n_2 = 3$, $n_3 = 1$, and $n_4 = 1$.

$$\frac{7!}{2!3!1!1!} = \frac{7!}{2!3!}$$
$$= \frac{7 \cdot 6 \cdot 5 \cdot 4 \cdot 3!}{2 \cdot 1 \cdot 3!}$$
$$= 420$$

There are 420 distinguishable permutations.

35. **(a)** The 9 books can be arranged in $P(9, 9) = 9! = 362,880$ ways.

(b) The blue books can be arranged in 4! ways, the green books can be arranged in 3! ways, and the red books can be arranged in 2! ways. There are 3! ways to choose the order of the 3 groups of books. Therefore, using the multiplication principle, the number of possible arrangements is

$$4!3!2!3! = 24 \cdot 6 \cdot 2 \cdot 6$$
$$= 1728.$$

(c) Use the formula for distinguishable permutations with $n = 9$, $n_1 = 4$, $n_2 = 3$, and $n_3 = 2$.

The number of distinguishable arrangements is

$$\frac{9!}{4!3!2!} = \frac{9 \cdot 8 \cdot 7 \cdot 6 \cdot 5 \cdot 4!}{4! \cdot 6 \cdot 2}$$
$$= 1260.$$

37. 5 of the 11 drugs can be administered in

$$P(11, 5) = \frac{11!}{(11 - 5)!} = \frac{11!}{6!}$$
$$= \frac{11 \cdot 10 \cdot 9 \cdot 8 \cdot 7 \cdot 6!}{6!}$$
$$= 55,440$$

different sequences.

39. (a) There are 4 tasks to be performed in selecting 4 letters for the call letters. The first task may be done in 2 ways, the second in 25, the third in 24, and the fourth in 23. By the multiplication principle, there will be $2 \cdot 25 \cdot 24 \cdot 23 = 27,600$ different call letter names possible.

(b) With repeats possible, there will be $2 \cdot 26 \cdot 26 \cdot 26 = 2 \cdot 26^3$ or 35,152 call letter names possible.

(c) To start with W or K, make no repeats, and end in R, there will be $2 \cdot 24 \cdot 23 \cdot 1 = 1104$ possible call letter names.

41. (a) Our number system has ten digits, which are 1 through 9 and 0.

There are 3 tasks to be performed in selecting 3 digits for the area code. The first task may be done in 8 ways, the second in 2, and the third in 10. By multiplication principle, there will be

$$8 \cdot 2 \cdot 10 = 160$$

different area codes possible.

There are 7 tasks to be performed in selecting 7 digits for the telephone number. The first task may be done in 8 ways, and the other 6 tasks may each be done in 10 ways. By the multiplication principle, there will be

$$8 \cdot 10^6 = 8,000,000$$

different telephone numbers possible within each area code.

(b) Some numbers, such as 911, 800, and 900 are reserved for special purposes and therefore unavailable for use as an area code.

43. There would be 4 tasks to be performed in selecting 4 digits for this new type of area code. The first task could be done in 8 ways, the second in 2, the third in 10, and the fourth in 10.
By the multiplication principle, there would be

$$8 \cdot 2 \cdot 10^2 = 1600$$

different area codes possible with this plan.

45. **(a)** There were $26^3 \cdot 10^3 = 17,576,000$ license plates possible that had 3 letters followed by 3 digits.

(b) There were $10^3 \cdot 26^3 = 17,576,000$ new license plates possible when plates were also issued having 3 digits followed by 3 numbers.

(c) There were $26 \cdot 10^3 \cdot 26^3 = 456,976,000$ new license plates possible when plates were also issued having 1 letter followed by 3 digits and then 3 letters.

47. If there are no restrictions on the digits used, there would be $10^5 = 100,000$ different 5-digit zip codes possible.
If the first digit is not allowed to be 0, there would be $9 \cdot 10^4 = 90,000$ zip codes possible.

Section 6.2

1. To evaluate $\binom{8}{3}$, use the formula $\binom{n}{r} = \dfrac{n!}{(n-r)!r!}$ with $n = 8$ and $r = 3$.

$$\binom{8}{3} = \frac{8!}{(8-3)!3!}$$
$$= \frac{8!}{5!3!}$$
$$= \frac{8 \cdot 7 \cdot 6 \cdot 5!}{5! \cdot 3 \cdot 2 \cdot 1} = 56$$

3. $\binom{12}{5} = \dfrac{12!}{(12-5)!5!} = \dfrac{12!}{7!5!}$
$$= \frac{12 \cdot 11 \cdot 10 \cdot 9 \cdot 8 \cdot 7!}{7! \cdot 5 \cdot 4 \cdot 3 \cdot 2 \cdot 1} = 792$$

5. $\binom{6}{0} = \dfrac{6!}{(6-0)!0!}$
$$= \frac{6!}{6!0!} = 1$$

7. $\binom{21}{10} = \dfrac{21!}{(21-10)!10!}$
$$= \frac{21!}{11!10!}$$
$$= 352,716$$

9. $\binom{25}{16} = \dfrac{25!}{(25-16)!16!}$
$$= \frac{25!}{9!16!}$$
$$= 2,042,975$$

11. $\binom{27}{10} = \dfrac{27!}{(27-10)!10!}$
$$= \frac{27!}{17!10!}$$
$$= 8,436,285$$

13. Pick 13 from 52. There are
$$\binom{52}{13} = \frac{52!}{13!39!}$$
$$= 635,013,559,600$$

different 13-card bridge hands possible.
In scientific notation, this answer would be written as 6.3501×10^{11}.

15. **(a)** There are
$$\binom{5}{2} = \frac{5!}{3!2!} = \frac{5 \cdot 4 \cdot 3!}{3! \cdot 2 \cdot 1} = 10$$

different 2-card combinations possible.

(b) The 10 possible hands are

$\{1, 2\}, \{2, 3\}, \{3, 4\}, \{4, 5\},$
$\{1, 3\}, \{2, 4\}, \{3, 5\}, \{1, 4\},$
$\{2, 5\}, \{1, 5\}.$

Of these, 7 contain a card numbered less than 3.

17. To have *at least* 2 good hitters among the 3 chosen, there will either be *exactly* 2 good hitters or 3 good hitters. The number of ways the coach can choose exactly 2 good hitters (and hence 1 poor hitter) is

$$\binom{5}{2} \cdot \binom{4}{1} = 10 \cdot 4 = 40.$$

The number of ways the coach can choose 3 good hitters is

$$\binom{5}{3} = 10.$$

The total number of ways to select at least 2 good hitters is

$$40 + 10 = 50.$$

19. Choose 2 letters from $\{L, M, N\}$; order is important.

(a)

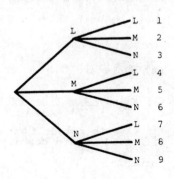

There are 9 ways to choose 2 letters if repetition is allowed.

(b)

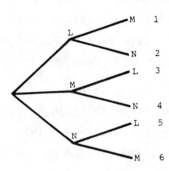

There are 6 ways to choose 2 letters if no repeats are allowed.

(c) $\binom{3}{2} = \dfrac{3!}{1!2!} = 3$

This answer differs from both (a) and (b).

23. Order does not matter in choosing members of a committee, so use combinations rather than permutations.

(a) The number of committees whose members are all men is

$$\binom{8}{5} = \frac{8!}{3!5!} = \frac{8 \cdot 7 \cdot 6 \cdot 5!}{3 \cdot 2 \cdot 1 \cdot 5!}$$
$$= 56.$$

(b) The number of committees whose members are all women is

$$\binom{11}{5} = \frac{11!}{6!5!} = \frac{11 \cdot 10 \cdot 9 \cdot 8 \cdot 7 \cdot 6!}{6! \cdot 5 \cdot 4 \cdot 3 \cdot 2 \cdot 1}$$
$$= 462.$$

(c) The 3 men can be chosen in

$$\binom{8}{3} = \frac{8!}{5!3!} = \frac{8 \cdot 7 \cdot 6 \cdot 5!}{5! \cdot 3 \cdot 2 \cdot 1}$$
$$= 56 \text{ ways.}$$

The 2 women can be chosen in

$$\binom{11}{2} = \frac{11!}{9!2!} = \frac{11 \cdot 10 \cdot 9!}{9! \cdot 2 \cdot 1}$$
$$= 55 \text{ ways.}$$

Using the multiplication principle, a committee of 3 men and 2 women can be chosen in

$$56 \cdot 55 = 3080 \text{ ways.}$$

25. Order does not matter, so use combinations.

(a) The 3 students who will take part in the course can be chosen in

$$\binom{12}{3} = \frac{12!}{9!3!}$$
$$= \frac{12 \cdot 11 \cdot 10 \cdot 9!}{9! \cdot 3 \cdot 2 \cdot 1}$$
$$= 220 \text{ ways.}$$

(b) The 9 students who will not take part in the course can be chosen in

$$\binom{12}{9} = \frac{12!}{3!9!} = 220 \text{ ways.}$$

27. Order is important, so use permutations.
The number of ways in which the children can find seats is

$$P(12, 11) = \frac{12!}{(12 - 11)!} = \frac{12!}{1!}$$
$$= 12!$$
$$= 479,001,600.$$

In scientific notation, this answer would be written as 4.79×10^8.

29. Since order is important, use a permutation. (Each secretary is being assigned to a manager, which is essentially the same as putting them in numbered slots.)
The secretaries can be selected in $P(7, 3) = 7 \cdot 6 \cdot 5 = 210$ different ways.

31. Since order does not matter, use combinations.

(a) There are $\binom{25}{3} = 2300$ possible samples of 3 apples.

(b) There are $\binom{5}{3} = 10$ possible samples of 3 rotten apples.

(c) There are $\binom{5}{1}\binom{20}{2} = 950$ possible samples with exactly 1 rotten apple.

33. Since order is important, use a permutation. The plants can be arranged in

$P(9, 5) = 9 \cdot 8 \cdot 7 \cdot 6 \cdot 5 = 15,120$ different ways.

35. Since order does not matter, use combinations.

(a) There are $\binom{20}{5} = 15,504$ different ways to select five of the orchids.

(b) If 2 special orchids must be included in the show, that leaves 18 other orchids from which the other 3 orchids for the show must be chosen. This can be done in

$\binom{18}{3}$ = 816 different ways.

37. $\binom{n}{n-1} = \dfrac{n!}{[n-(n-1)]!(n-1)!}$

$\qquad = \dfrac{n!}{1!(n-1)!}$

$\qquad = \dfrac{n \cdot (n-1)!}{1 \cdot (n-1)!}$

$\qquad = n$

39. There are 2 types of meat and 6 types of extras. Order does not matter here, so use combinations.

(a) There are $\binom{2}{1}$ ways to choose one type of meat and $\binom{6}{3}$ ways to choose exactly three extras. By the multiplication principle, there are

$$\binom{2}{1} \cdot \binom{6}{3} = 2 \cdot 20 = 40$$

different ways to order a hamburger with exactly three extras.

(b) There are $\binom{6}{3}$ = 20 different ways to choose exactly three extras.

(c) "At least five extras" means "5 extras or 6 extras." There are $\binom{6}{5}$ different ways to choose exactly 5 extras and $\binom{6}{6}$ ways to choose

exactly 6 extras, so there are

$\binom{6}{5} + \binom{6}{6}$ = 6 + 1 = 7 different ways to choose at least five extras.

41. Select 8 of the 16 smokers and 8 of the 20 nonsmokers; order does not matter in the group, so use combinations. There are

$$\binom{16}{8} \cdot \binom{20}{8} = 1,621,233,900$$

different ways to select the study group. In scientific notation, this answer would be written as 1.6212×10^9.

43. Order does not matter in choosing a delegation, so use combinations. This committee has 5 + 4 = 9 members.

(a) There are

$$\binom{9}{3} = \frac{9!}{6!3!}$$

$$= \frac{9 \cdot 8 \cdot 7 \cdot 6!}{6!3 \cdot 2 \cdot 1}$$

$$= 84 \text{ possible delegations.}$$

(b) To have all Democrats, the number of possible delegations is

$$\binom{5}{3} = 10.$$

(c) To have 2 Democrats and 1 Republican, the number of possible delegations is

$$\binom{5}{2} \cdot \binom{4}{1} = 10 \cdot 4 = 40.$$

(d) We have previously calculated that there are 84 possible delegations, of which 10 consist of all Democrats. Those 10 delegations are the only ones with no Republicans, so the remaining 84 − 10 = 74 delegations include at least one Republican.

45. In the lottery, 6 different numbers are to be chosen from the 99 numbers.

(a) There are

$$\binom{99}{6} = \frac{99!}{93!\,6!} = 1,120,529,256$$

different ways to choose 6 numbers if order is not important. In scientific notation, this answer would be written as 1.1205×10^9.

(b) There are

$$P(99,\ 6) = \frac{99!}{93!} = 806,781,064,320$$

different ways to choose 6 numbers if order matters. In scientific notation, this answer would be written as 8.0678×10^{11}.

Section 6.3

1. There are $\binom{10}{3}$ samples of 3 apples.

$$\binom{10}{3} = \frac{10 \cdot 9 \cdot 8}{3 \cdot 2 \cdot 1} = 120$$

There are $\binom{6}{3}$ samples of 3 red apples.

$$\binom{6}{3} = \frac{6 \cdot 5 \cdot 4}{3 \cdot 2 \cdot 1} = 20$$

Thus,

$$P(\text{all red apples}) = \frac{20}{120} = \frac{1}{6}.$$

3. There are $\binom{4}{2}$ samples of 2 yellow apples.

$$\binom{4}{2} = \frac{4 \cdot 3}{2 \cdot 1} = 6$$

There are $\binom{6}{1} = 6$ samples of 1 red apple. Thus, there are $6 \cdot 6 = 36$ samples of 3 in which 2 are yellow and 1 red. Thus,

$$P(\text{2 yellow and 1 red apple}) = \frac{36}{120}$$
$$= \frac{3}{10}.$$

5. The number of 2−card hands is

$$\binom{52}{2} = \frac{52 \cdot 51}{2 \cdot 1} = 1326.$$

7. There are $\binom{52}{2} = 1326$ different 2−card hands. The number of 2−card hands with exactly one ace is

$$\binom{4}{1}\binom{48}{1} = 4 \cdot 48 = 192.$$

The number of 2−card hands with two aces is

$$\binom{4}{2} = 6.$$

Thus there are 198 hands with at least one ace. Therefore,

P(the 2-card hand contains an ace)

$$= \frac{198}{1326} = \frac{33}{221} \approx .149.$$

9. There are $\binom{52}{2} = 1326$ different 2-card hands. There are $\binom{13}{2} = 78$ ways to get a 2-card hand where both cards are of a single named suit, but there are 4 suits to choose from. Thus,

P(two cards of same suit)

$$= \frac{4 \cdot \binom{13}{2}}{\binom{52}{2}} = \frac{312}{1326} = \frac{52}{221} \approx .235.$$

11. There are $\binom{52}{2} = 1326$ different 2-card hands. There are 12 face cards in a deck, so there are 40 cards that are not face cards.

Thus,

P(no face cards)

$$= \frac{\binom{40}{2}}{\binom{52}{2}} = \frac{780}{1326} = \frac{130}{221} \approx .588.$$

13. There are 26 choices for each slip pulled out, and there are 5 slips pulled out, so there are

$$26^5 = 11,881,376$$

different "words" that can be formed from the letters. If the "word" must be "chuck," there is only one choice for each of the 5 letters

(the first slip must contain a "c," the second an "h," and so on). Thus,

P(word is "chuck")

$$= \frac{1^5}{26^5} = \left(\frac{1}{26}\right)^5 \approx 8.42 \times 10^{-8}.$$

15. There are $26^5 = 11,881,376$ different "words" that can be formed. If the "word" is to have no repetition of letters, then there are 26 choices for the first letter, but only 25 choices for the second (since the letters must all be different), 24 choices for the third, and so on. Thus,

P(all different letters)

$$= \frac{26 \cdot 25 \cdot 24 \cdot 23 \cdot 22}{26^5}$$

$$= \frac{1 \cdot 25 \cdot 24 \cdot 23 \cdot 22}{26^4}$$

$$= \frac{303,600}{456,976}$$

$$= \frac{18,975}{28,561} \approx .664.$$

17. There are $\binom{52}{5}$ different 5-card poker hands. There are 4 royal flushes, one for each suit. Thus,

P(royal flush)

$$= \frac{4}{\binom{52}{5}} = \frac{4}{2,598,960}$$

$$= \frac{1}{649,740}$$

$$\approx 1.54 \times 10^{-6}.$$

19. The four of a kind can be chosen in 13 ways and then is matched with 1 of the remaining 48 cards to make a 5-card hand containing 4 of a kind, Thus there are $13 \cdot 48 = 624$ poker hands with 4 of a kind. It follows that

P(four of a kind)

$$= \frac{624}{\binom{52}{5}} = \frac{624}{2,598,960}$$

$$= \frac{1}{4165} \approx 2.40 \times 10^{-4}.$$

21. There are $\binom{52}{13}$ different 13-card bridge hands. Since there are only 13 hearts, there is exactly one way to get a bridge hand containing only hearts. Thus,

$$P(\text{only hearts}) = \frac{1}{\binom{52}{13}}.$$

23. There are $\binom{4}{3}$ ways to obtain 3 aces, $\binom{4}{3}$ ways to obtain 3 kings, and $\binom{44}{7}$ ways to obtain the 7 remaining cards. Thus,

P(exactly 3 aces and exactly 3 kings)

$$= \frac{\binom{4}{3}\binom{4}{3}\binom{44}{7}}{\binom{52}{13}}.$$

25. There are 21 books, so the number of selections of any 6 books is

$$\binom{21}{6} = 54,264.$$

(a) The probability that the selection consisted of 3 Hughes and 3 Morrison books is

$$\frac{\binom{9}{3}\binom{7}{3}}{\binom{21}{6}} = \frac{84 \cdot 35}{54,264} = \frac{2940}{54,264} \approx .054.$$

(b) A selection containing exactly 4 Baldwin books will contain 2 of the 16 books by the other authors, so the probability is

$$\frac{\binom{5}{4}\binom{16}{2}}{\binom{21}{6}} = \frac{5 \cdot 120}{54,264} = \frac{600}{54,264} \approx .011.$$

(c) The probability of a selection consisting of 2 Hughes, 3 Baldwin, and 1 Morrison book is

$$\frac{\binom{9}{2}\binom{5}{3}\binom{7}{1}}{\binom{21}{6}} = \frac{36 \cdot 10 \cdot 7}{54,264}$$

$$= \frac{2520}{54,274} \approx .046.$$

(d) A selection consisting of at least 4 Hughes books may contain 4, 5, or 6 Hughes books, with any remaining books by the other authors. Therefore, the probability is

$$\frac{\binom{9}{4}\binom{12}{2} + \binom{9}{5}\binom{12}{1} + \binom{9}{6}}{\binom{21}{6}}$$

$$= \frac{126 \cdot 66 + 126 \cdot 12 + 84}{54,264}$$

$$= \frac{8316 + 1512 + 84}{54,264}$$

$$= \frac{9912}{54,264} \approx .183.$$

(e) Since there are 9 Hughes books and 5 Baldwin books, there are 14 books written by males. The probability of a selection with exactly 4 books written by males is

$$\frac{\binom{14}{4}\binom{7}{2}}{\binom{21}{6}} = \frac{1001 \cdot 21}{54,264}$$

$$= \frac{21,021}{54,264} \approx .387.$$

(f) A selection with no more than 2 books written by Baldwin may contain 0, 1, or 2 books by Baldwin, with the remaining books by the other authors. Therefore, the probability is

$$\frac{\binom{16}{6} + \binom{5}{1}\binom{16}{5} + \binom{5}{2}\binom{16}{4}}{\binom{21}{6}}$$

$$= \frac{8008 + 5 \cdot 4368 + 10 \cdot 1820}{54,264}$$

$$= \frac{8008 + 21,840 + 18,200}{54,264}$$

$$= \frac{48,048}{54,264} \approx .885.$$

29.   P(at least 2 presidents have the same birthday)

$$= 1 - P(\text{no 2 presidents have the same birthday})$$

The number of ways that 41 people can have the same or different birthdays is $(365)^{41}$. The number of ways that 41 people can have all different birthdays is the number of permutations of 365 things taken 41 at a time or P(365, 41). Thus,

P(at least 2 presidents have the same birthday)

$$= 1 - \frac{P(365, 41)}{365^{41}}.$$

(Be careful to realize that the symbol P is sometimes used to indicate permutations and sometimes used to indicate probability; in this solution, the symbol is used both ways.)

31.   Since there are 435 members of the House of Representatives, and there are only 365 days in a year, it is a certain event that at least 2 people will have the same birthday. Thus,

P(at least 2 members have the same birthday) = 1.

33.   Each of the four people can choose to get off at any one of the seven floors, so there are 7^4 ways the four people can leave the elevator. The number of ways the people can leave at different floors is the number of permutations of 7 things (floors) taken 4 at a time or

$$P(7, 4) = 7 \cdot 6 \cdot 5 \cdot 4 = 840.$$

Thus, the probability that no two passengers leave at the same floor is

$$\frac{P(7, 4)}{7^4} = \frac{840}{2401} = \frac{120}{343} \approx .3499.$$

(Note the similarity of this problem and the "birthday problem.")

35. There are 9 ways to choose 1 typewriter from the shipment of 9.
Since 2 of the 9 are defective,
there are 7 ways to choose 1 non-
defective typewriter. Thus,

 P(1 drawn from the 9 is not defective)
 = 7/9.

37. There are $\binom{9}{3}$ ways to choose 3 typewriters.

$$\binom{9}{3} = \frac{9!}{3!6!} = \frac{9 \cdot 8 \cdot 7}{3 \cdot 2 \cdot 1} = 84$$

There are $\binom{7}{3}$ ways to choose 3 non-
defective typewriters.

$$\binom{7}{3} = \frac{7!}{3!4!} = \frac{7 \cdot 6 \cdot 5}{3 \cdot 2 \cdot 1} = 35$$

Thus,

 P(3 drawn from the 9 are non-defective)
 $= \frac{35}{84} = \frac{5}{12}$.

39. There are $\binom{12}{4} = 495$ different ways
to choose 4 engines for testing from
the crate of 12. A crate will not
be shipped if any of the 4 in the
sample is defective. If there are 2
defectives in the crate, then there
are $\binom{10}{4} = 210$ ways of choosing a
sample with no defectives. Thus,

 P(shipping a crate with 2 defectives)
 $= \frac{210}{495} \approx .424$.

41. There are $\binom{99}{6} = 1,120,529,256$
different ways to pick 6 numbers
from 1 to 99, but there is only 1
way to win; the 6 numbers you pick
must exactly match the 6 winning
numbers, without regard to order.
Thus,

 P(win the big prize)
 $$= \frac{1}{1,120,529,256}$$
 $\approx 8.9 \times 10^{-10}$.

43. **(a)** There were 28 games played in
the season, since the numbers in the
"Won" column have a sum of 28 (and
the numbers in the "Lost" column
have a sum of 28). Each of those 28
games had 2 possible outcomes;
either Team A won and Team B lost,
or else Team A lost and Team B won.
By the multiplication principle,
this means that there were 2^{28}
different win/lose progressions
possible. Any one of the 8 teams
could have been the one that won
all of its games, any one of the
remaining 7 teams could have been
the one that won all but one of its
games, and so on, until there is
only one team left and it is the one
that lost all of its games. By the
multiplication principle, this means
that there were

$$8 \cdot 7 \cdot 6 \cdot 5 \cdot 4 \cdot 3 \cdot 2 \cdot 1 = 8!$$

different "perfect progressions"
possible.

Thus,

P("perfect progression" in an
8-team league)

$$= \frac{8!}{2^{28}} \approx .000150.$$

(b) If there are n teams in the league, then the "Won" column will begin with n − 1, followed by n − 2, then n − 3, and so on down to 0. It can be shown that the sum of these n numbers is $\frac{n(n-1)}{2}$, so there are $2^{n(n-1)/2}$ different win/lose progressions possible. The n teams can be ordered in n! different ways, so there are n! different "perfect progressions" possible. Thus,

P("perfect progression" in an
n-team league)

$$= \frac{n!}{2^{n(n-1)/2}}.$$

45. This exercise should be solved by using the Monte Carlo method. The answers may vary from run to run since the random numbers will vary from run to run. By non-computer methods, the answers are as follows.

(a) P(no aces) $= \dfrac{\binom{48}{13}}{\binom{52}{13}} \approx .3038$

(b) P(2 kings and 2 aces)

$$= \frac{\binom{4}{2}\binom{4}{2}\binom{44}{9}}{\binom{52}{13}} \approx .0402$$

(c) P(only 3 suits are represented)

$$= \frac{4 \cdot \binom{39}{13}}{\binom{52}{13}} \approx .6651$$

Section 6.4

1. This is a Bernoulli trial problem with P(success) = P(girl) = 1/2. The probability of exactly x successes in n trials is

$$\binom{n}{x}p^x(1-p)^{n-x},$$

where p is the probability of success in a single trial. We have n = 5, x = 2, and p = 1/2. Note that $1 - p = 1 - \frac{1}{2} = \frac{1}{2}$.

P(exactly 2 girls)

$$= \binom{5}{2}\left(\frac{1}{2}\right)^2\left(\frac{1}{2}\right)^3$$

$$= \frac{10}{32} = \frac{5}{16} \approx .313$$

3. We have n = 5, x = 0, p = 1/2, and 1 − p = 1/2.

$$P(\text{no girls}) = \binom{5}{0}\left(\frac{1}{2}\right)^0\left(\frac{1}{2}\right)^5$$

$$= \frac{1}{32} \approx .031$$

5. "At least 4 girls" means either 4 or 5 girls.

P(at least 4 girls)

$$= \binom{5}{4}\left(\frac{1}{2}\right)^4\left(\frac{1}{2}\right)^1 + \binom{5}{5}\left(\frac{1}{2}\right)^5\left(\frac{1}{2}\right)^0$$

$$= \frac{5}{32} + \frac{1}{32} = \frac{6}{32} = \frac{3}{16} \approx .188$$

7. P(no more than 3 boys)

$$= 1 - P(\text{at least 4 boys})$$

$$= 1 - P(4 \text{ boys or 5 boys})$$

$$= 1 - [P(4 \text{ boys}) + P(5 \text{ boys})]$$

$$= 1 - \left(\frac{5}{32} + \frac{1}{32}\right)$$

$$= 1 - \frac{6}{32}$$

$$= 1 - \frac{3}{16} = \frac{13}{16} \approx .813$$

9. On one roll, $P(1) = \frac{1}{6}$. We have $n = 12$, $x = 12$, and $p = 1/6$. Note that $1 - p = 5/6$. Thus,

P(exactly twelve 1's)

$$= \binom{12}{12}\left(\frac{1}{6}\right)^{12}\left(\frac{5}{6}\right)^{0}$$

$$\approx 4.6 \times 10^{-10}.$$

11. P(exactly one 1)

$$= \binom{12}{1}\left(\frac{1}{6}\right)^{1}\left(\frac{5}{6}\right)^{11} \approx .269$$

13. "No more than three 1's" means 0, 1, 2, or 3 ones. Thus,

P(no more than three 1's)

$$= P(\text{zero 1's}) + P(\text{one 1})$$

$$+ P(\text{two 1's}) + P(\text{three 1's})$$

$$= \binom{12}{0}\left(\frac{1}{6}\right)^{0}\left(\frac{5}{6}\right)^{12} + \binom{12}{1}\left(\frac{1}{6}\right)^{1}\left(\frac{5}{6}\right)^{11}$$

$$+ \binom{12}{2}\left(\frac{1}{6}\right)^{2}\left(\frac{5}{6}\right)^{10} + \binom{12}{3}\left(\frac{1}{6}\right)^{3}\left(\frac{5}{6}\right)^{9}$$

$$\approx .875.$$

15. Each time the coin is tossed $P(\text{head}) = \frac{1}{2}$. We have $n = 5$, $x = 5$, $p = 1/2$, and $1 - p = 1/2$.

Thus,

P(all heads)

$$= \binom{5}{5}\left(\frac{1}{2}\right)^{5}\left(\frac{1}{2}\right)^{0}$$

$$= \frac{1}{32} \approx .031.$$

17. P(no more than 3 heads)

$$= P(0 \text{ heads}) + P(1 \text{ head})$$

$$+ P(2 \text{ heads}) + P(3 \text{ heads})$$

$$= \binom{5}{0}\left(\frac{1}{2}\right)^{0}\left(\frac{1}{2}\right)^{5} + \binom{5}{1}\left(\frac{1}{2}\right)^{1}\left(\frac{1}{2}\right)^{4}$$

$$+ \binom{5}{2}\left(\frac{1}{2}\right)^{2}\left(\frac{1}{2}\right)^{3} + \binom{5}{3}\left(\frac{1}{2}\right)^{3}\left(\frac{1}{2}\right)^{2}$$

$$= \frac{26}{32} = \frac{13}{16} \approx .813$$

21. We have $n = 10$, $x = 2$, $p = .2$, and $1 - p = .8$, so

$$P(x = 2) = \binom{10}{2}(.2)^{2}(.8)^{8} \approx .302.$$

23. We have

P(x ≤ 3)

$$= P(x = 0) + P(x = 1)$$

$$+ P(x = 2) + P(x = 3)$$

$$= \binom{10}{0}(.2)^{0}(.8)^{10} + \binom{10}{1}(.2)^{1}(.8)^{9}$$

$$+ \binom{10}{2}(.2)^{2}(.8)^{8} + \binom{10}{3}(.2)^{3}(.8)^{7}$$

$$\approx .879.$$

25. We have $n = 10$, $x = 10$, $p = .9$, and $1 - p = .1$, so

$$P(x = 10) = \binom{10}{10}(.9)^{10}(.1)^{0}$$

$$\approx .349.$$

27. We have

    $P(x \geq 9)$

      $= P(x = 9) + P(x = 10)$

      $= \binom{10}{9}(.9)^9(.1)^1 + \binom{10}{10}(.9)^{10}(.1)^0$

      $\approx .387 + .349$

      $= .736.$

29. We have $n = 6$, $x = 2$, $p = 1/5$, and
    $1 - p = 4/5$. Thus,

    P(exactly 2 correct)

      $= \binom{6}{2}\left(\frac{1}{5}\right)^2\left(\frac{4}{5}\right)^4$

      $\approx .246.$

31. We have

    P(at least 4 correct)

      $= P(4 \text{ correct}) + P(5 \text{ correct})$
      $+ P(6 \text{ correct})$

      $= \binom{6}{4}\left(\frac{1}{5}\right)^4\left(\frac{4}{5}\right)^2 + \binom{6}{5}\left(\frac{1}{5}\right)^5\left(\frac{4}{5}\right)^1$

      $+ \binom{6}{6}\left(\frac{1}{5}\right)^6\left(\frac{4}{5}\right)^0$

      $\approx .017.$

33. We have $n = 3$, $x = 1$, $p = .1$, and
    $1 - p = .9$. Thus,

    P(exactly 1 loses everything)

      $= \binom{3}{1}(.1)^1(.9)^2 \approx .243.$

35. We have $n = 20$, $p = .05$, and
    $1 - p = .95$. Thus,

    P(at most 2 defective transistors)

      $= P(x \leq 2)$

      $= P(x = 0) + P(x = 1) + P(x = 2)$

    $= \binom{20}{0}(.05)^0(.95)^{20} + \binom{20}{1}(.05)^1(.95)^{19}$

      $+ \binom{20}{2}(.05)^2(.95)^{18}$

    $\approx .925.$

37. We have $n = 20$, $x = 17$, $p = .7$, and
    $1 - p = .3$. Thus,

    P(exactly 17 cured)

      $= \binom{20}{17}(.7)^{17}(.3)^3$

      $= (1140)(.7)^{17}(.3)^3$

      $\approx .072.$

39. P(at least 18 cured)

      $= P(\text{exactly 18 cured})$
      $+ P(\text{exactly 19 cured})$
      $+ P(20 \text{ cured})$

      $= \binom{20}{18}(.7)^{18}(.3)^2 + \binom{20}{19}(.7)^{19}(.3)^1$

      $+ \binom{20}{20}(.7)^{20}(.3)^0$

      $= 190(.7)^{18}(.3)^2 + 20(.7)^{19}(.3)$
      $+ (.7)^{20}$

      $\approx .035$

41. We have $n = 100$, $p = .012$, and
    $1 - p = .988$. Thus,

    $P(x \leq 2)$

      $= P(x = 0) + P(x = 1) + P(x = 2)$

      $= \binom{100}{0}(.012)^0(.988)^{100}$

      $+ \binom{100}{1}(.012)^1(.988)^{99}$

      $+ \binom{100}{2}(.012)^2(.988)^{98}$

      $= (.012)^0(.988)^{100}$

      $+ 100(.012)^1(.988)^{99}$

      $+ 4950(.012)^2(.988)^{98}$

      $\approx .881.$

43. We have n = 6, x = 3, p = .70, and 1 − p = .30. Thus,

P(exactly 3 recover)

$$= \binom{6}{3}(.7)^3(.3)^3$$

$$\approx .185.$$

45. P(no more than 3 recover)

= P(0 recover) + P(1 recovers)

  + P(2 recover) + P(3 recover)

$$= \binom{6}{0}(.7)^0(.3)^6 + \binom{6}{1}(.7)^1(.3)^5$$

$$+ \binom{6}{2}(.7)^2(.3)^4 + \binom{6}{3}(.7)^3(.3)^3$$

$$\approx .256$$

47. We have n = 10,000,

p = 2.5 × 10⁻⁷ = .00000025, and 1 − p = .99999975. Thus,

P(at least 1 mutation occurs)

= 1 − P(none occur)

$$= 1 - \binom{10,000}{0}(p)^0(1-p)^{10,000}$$

$$= 1 - (.99999975)^{10,000} \approx .0025.$$

49. We have n = 5, p = 1/3, and 1 − p = 2/3. Thus,

P(x ≤ 3)

= P(x = 0) + P(x = 1) + P(x = 2)

  + P(x = 3)

$$= \binom{5}{0}\left(\tfrac{1}{3}\right)^0\left(\tfrac{2}{3}\right)^5 + \binom{5}{1}\left(\tfrac{1}{3}\right)^1\left(\tfrac{2}{3}\right)^4$$

$$+ \binom{5}{2}\left(\tfrac{1}{3}\right)^2\left(\tfrac{2}{3}\right)^3 + \binom{5}{3}\left(\tfrac{1}{3}\right)^3\left(\tfrac{2}{3}\right)^2$$

$$= \frac{232}{243} \approx .955.$$

51. We have n = 10, x = 7, p = .2, and 1 − p = .8. Thus,

$$P(x = 7) = \binom{10}{7}(.2)^7(.8)^3 \approx .00079.$$

53. We have

P(x < 8)

= 1 − P(x ≥ 8)

= 1 − [P(x = 8) + P(x = 9) + P(x = 10)]

= 1 − P(x = 8) − P(x = 9) − P(x = 10)

$$= 1 - \binom{10}{8}(.2)^8(.8)^2 - \binom{10}{9}(.2)^9(.8)^1$$

$$- \binom{10}{10}(.2)^{10}(.8)^0$$

= 1 − .000074 − .000004 − .0000001

= .999922.

55. Let success mean not getting the flu. Then we have n = 134, p = .80, and 1 − p = .20.

(a) If exactly 10 of the people get the flu, then x = 134 − 10 = 124 since we have let success mean *not* getting the flu. Thus,

$$P(x = 124) = \binom{134}{124}(.80)^{124}(.20)^{10}$$

$$\approx .000036.$$

(b) If no more than 10 people get the flu, then more than 10 people did *not* get the flu, so we are interested in x ≥ 10. Thus,

P(x ≥ 10)

= 1 − P(x < 10)

= 1 − P(x = 9) − P(x = 8) − P(x = 7)

  − P(x = 6) − P(x = 5) − P(x = 4)

  − P(x = 3) − P(x = 2) − P(x = 1)

  − P(x = 0)

$$\approx .000054.$$

(c) If none of the people get the flu, then x = 134. Thus,

$$P(x = 134)$$

$$= \binom{134}{134}(.80)^{134}(.20)^0$$

$$\approx .000000000000103$$

$$\approx 1.0 \times 10^{-13}.$$

57. Let success mean producing a defective item. Then we have n = 75, p = .05, and 1 − p = .95.

(a) If there are exactly 5 defective items, then x = 5. Thus,

$$P(x = 5) = \binom{75}{5}(.05)^5(.95)^{70}$$

$$\approx .148774.$$

(b) If there are no defective items, then x = 0. Thus,

$$P(x = 0) = \binom{75}{0}(.05)^0(.95)^{75}$$

$$\approx .021344.$$

(c) If there is at least 1 defective item, then we are interested in x ≥ 1. We have

$$P(x \geq 1) = 1 - P(x = 0)$$

$$\approx 1 - .021344$$

$$= .978656.$$

Section 6.5

1. (a) P(1 germinated) = $\frac{0}{10}$ = 0,

P(2 germinated) = $\frac{1}{10}$ = .1,

and so on. The probability distribution is as follows.

Number Germinated	0	1	2	3	4	5
Probability	0	0	.1	.3	.4	.2

(b) The histogram corresponding to this probability distribution consists of 6 rectangles, the first two of which have no height. See the histogram in the back of the textbook.

3. (a) P(0 bullseyes) = $\frac{0}{25}$ = 0,

P(1 bullseye) = $\frac{1}{25}$ = .04,

and so on.

Number of Bullseyes	0	1	2	3	4	5	6
Probability	0	.04	0	.16	.40	.32	.08

(b) See the histogram in the back of the textbook.

5. (a) P(0 with disease) = $\frac{3}{20}$ = .15,

P(1 with disease) = $\frac{5}{20}$ = .25,

and so on.

Number with Disease	0	1	2	3	4	5
Probability	.15	.25	.3	.15	.1	.05

(b) See the histogram in the back of the textbook.

7. Let x denote the number of heads observed. Then x can take on 0, 1, 2, 3, 4 as values. The probabilities are as follows.

$$P(x = 0) = \binom{4}{0}\left(\frac{1}{2}\right)^0\left(\frac{1}{2}\right)^4 = \frac{1}{16}$$

$$P(x = 1) = \binom{4}{1}\left(\frac{1}{2}\right)^1\left(\frac{1}{2}\right)^3 = \frac{4}{16} = \frac{1}{4}$$

$$P(x = 2) = \binom{4}{2}\left(\frac{1}{2}\right)^2\left(\frac{1}{2}\right)^2 = \frac{6}{16} = \frac{3}{8}$$

$$P(x = 3) = \binom{4}{3}\left(\frac{1}{2}\right)^3\left(\frac{1}{2}\right)^1 = \frac{4}{16} = \frac{1}{4}$$

$$P(x = 4) = \binom{4}{4}\left(\frac{1}{2}\right)^4\left(\frac{1}{2}\right)^0 = \frac{1}{16}$$

Therefore, the probability distribution is as follows.

Number of Heads	0	1	2	3	4
Probability	$\frac{1}{16}$	$\frac{1}{4}$	$\frac{3}{8}$	$\frac{1}{4}$	$\frac{1}{16}$

9. Let x denote the number of aces drawn. Then x can take on values 0, 1, 2, 3. The probabilities are as follows.

$$P(x = 0) = \binom{3}{0}\left(\frac{48}{52}\right)\left(\frac{47}{51}\right)\left(\frac{46}{50}\right) \approx .783$$

$$P(x = 1) = \binom{3}{1}\left(\frac{4}{52}\right)\left(\frac{48}{51}\right)\left(\frac{47}{50}\right) \approx .204$$

$$P(x = 2) = \binom{3}{2}\left(\frac{4}{52}\right)\left(\frac{3}{51}\right)\left(\frac{48}{50}\right) \approx .013$$

$$P(x = 3) = \binom{3}{3}\left(\frac{4}{52}\right)\left(\frac{3}{51}\right)\left(\frac{2}{50}\right) \approx .0002$$

Therefore, the probability distribution is as follows.

Number of Aces	0	1	2	3
Probability	.783	.204	.013	.0002

For Exercises 11–15, see the histograms in the back of the textbook.

11. Draw a histogram with 5 rectangles, corresponding to x = 0, x = 1, x = 2, x = 3, and x = 4. $P(x \leq 2)$ corresponds to

$$P(x = 0) + P(x = 1) + P(x = 2),$$

so shade the first 3 rectangles in the histogram.

13. Draw a histogram with 4 rectangles, corresponding to x = 0, x = 1, x = 2, and x = 3.

P(at least one ace) = $P(x \geq 1)$

corresponds to

$$P(x = 1) + P(x = 2) + P(x = 3),$$

so shade the last 3 rectangles.

15. Draw a histogram with 5 rectangles, the same histogram as in Exercise 11. $P(x = 2 \text{ or } x = 3)$ corresponds to $P(x = 2) + P(x = 3)$, so shade those 2 rectangles.

17. E(x) = 2(.1) + 3(.4) + 4(.3) + 5(.2)
    = 3.6

19. $E(z) = 9(.14) + 12(.22) + 15(.36)$

    $+ 18(.18) + 21(.10)$

    $= 14.64$

21. It is possible (but not necessary) to begin by writing the histogram's data as a probability distribution, which would look as follows.

x	1	2	3	4
P(x)	.2	.3	.1	.4

    The expected value of x is

    $E(x) = 1(.2) + 2(.3) + 3(.1) + 4(.4)$

    $= 2.7.$

23. The expected value of x is

    $E(x) = 6(.1) + 12(.2) + 18(.4)$

    $+ 24(.2) + 30(.1)$

    $= 18.$

25. Using the data from Example 6, the expected winnings for Mary are

    $E(x) = (-.4)\left(\frac{1}{4}\right) + .4\left(\frac{1}{4}\right)$

    $+ .4\left(\frac{1}{4}\right) + (-.4)\left(\frac{1}{4}\right)$

    $= 0.$

    Yes, it is still a fair game if Mary tosses and Donna calls.

27. Below is the probability distribution of x, which stands for the person's net winnings.

x	$99	$39	-$1
P(x)	$\frac{1}{500} = .002$	$\frac{2}{500} = .004$	$\frac{497}{500} = .994$

The expected value of the person's winnings is

$E(x) = 99(.002) + 39(.004)$

$+ (-1)(.994)$

$\approx -\$.64$ or $-64¢.$

Since the expected value of the winnings is not 0, this is not a fair game.

29. The number of possible samples is

$$\binom{7}{3} = \frac{7!}{3!4!} = \frac{7 \cdot 6 \cdot 5}{3 \cdot 2 \cdot 1} = 35.$$

The number of samples containing no yellows and therefore 3 whites is $\binom{4}{3} = 4$, so the probability of drawing a sample containing no yellow is 4/35.

The number of samples containing 1 yellow and therefore 2 whites is $\binom{3}{1}\binom{4}{2} = 3 \cdot 6 = 18$, so the probability of drawing a sample containing 1 yellow is 18/35.

Similarly, the probability of drawing a sample containing 2 yellows is

$$\frac{\binom{3}{2}\binom{4}{1}}{\binom{7}{3}} = \frac{12}{35},$$

and the probability of drawing a sample containing 3 yellows is

$$\frac{\binom{3}{3}}{\binom{7}{3}} = \frac{1}{35}.$$

Let x denote the number of yellow marbles drawn. The probability distribution of x is as follows.

x	0	1	2	3
P(x)	$\frac{4}{35}$	$\frac{18}{35}$	$\frac{12}{35}$	$\frac{1}{35}$

The expected value is

$$E(x) = 0\left(\frac{4}{35}\right) + 1\left(\frac{18}{35}\right) + 2\left(\frac{12}{35}\right) + 3\left(\frac{1}{35}\right)$$

$$= \frac{45}{35} = \frac{9}{7} \approx 1.3 \text{ yellow marbles.}$$

31. The probability that the delegation contains no liberals and 3 conservatives is

$$\frac{\binom{5}{0}\binom{4}{3}}{\binom{9}{3}} = \frac{1 \cdot 4}{84} = \frac{4}{84}.$$

Similarly, use combinations to calculate the remaining probabilities for the probability distribution.

(a) Let x denote the number of liberals on the delegation. The probability distribution of x is as follows.

x	0	1	2	3
P(x)	$\frac{4}{84}$	$\frac{30}{84}$	$\frac{40}{84}$	$\frac{10}{84}$

The expected value is

$$E(x) = 0\left(\frac{4}{84}\right) + 1\left(\frac{30}{84}\right) + 2\left(\frac{40}{84}\right) + 3\left(\frac{10}{84}\right)$$

$$= \frac{140}{84} = \frac{5}{3} \approx 1.67 \text{ liberals.}$$

(b) Let y denote the number of conservatives on the committee. The probability distribution of y is as follows.

y	0	1	2	3
P(y)	$\frac{10}{84}$	$\frac{40}{84}$	$\frac{30}{84}$	$\frac{4}{84}$

The expected value is

$$E(y) = 0\left(\frac{10}{84}\right) + 1\left(\frac{40}{84}\right) + 2\left(\frac{30}{84}\right) + 3\left(\frac{4}{84}\right)$$

$$= \frac{112}{84} = \frac{4}{3} \approx 1.33 \text{ conservatives.}$$

33. Let x represent the number of junior members on the committee. Use combinations to find the probabilities of 0, 1, 2, and 3 junior members. The probability distribution of x is as follows.

x	0	1	2	3
P(x)	$\frac{57}{203}$	$\frac{95}{203}$	$\frac{45}{203}$	$\frac{6}{203}$

The expected value is

$$E(x) = 0\left(\frac{57}{203}\right) + 1\left(\frac{95}{203}\right) + 2\left(\frac{45}{203}\right) + 3\left(\frac{6}{203}\right)$$

$$= 1 \text{ junior member.}$$

35. The probability of drawing 2 diamonds is

$$\frac{\binom{13}{2}}{\binom{52}{2}} = \frac{78}{1326},$$

and the probability of not drawing 2 diamonds is

$$1 - \frac{78}{1326} = \frac{1248}{1326}.$$

Let x denote your net winnings.   Then the expected value of the game is

$$E(x) = 4.5\left(\frac{78}{1326}\right) + (-.5)\left(\frac{1248}{1326}\right) = -\frac{273}{1326} \approx \$-.21 \quad \text{or} \quad -21\text{¢}.$$

The game is not fair since your expected winnings are not zero.

37. The probability of getting exactly 3 of the 4 selections correct and winning this game is $4\left(\frac{1}{13}\right)^3\left(\frac{12}{13}\right) \approx .001681$.   The probability of losing is .998319.   If you win, your winnings are $199.   Otherwise, you lose $1 (win -$1).   If x denotes your winnings, then the expected value is

$$E(x) = 199(.001681) + (-1)(.998319)$$

$$= .334519 - .998319 = -.6638$$

$$\approx -\$.66 \quad \text{or} \quad -66\text{¢}.$$

39.   In this form of roulette, $P(\text{even}) = \frac{18}{37}$ and $P(\text{noneven}) = \frac{19}{37}$.

If an even number comes up, you win $1.   Otherwise, you lose $1 (win -$1.)   If x denotes your winnings, then the expected value is

$$E(x) = 1\left(\frac{18}{37}\right) + (-1)\left(\frac{19}{37}\right) = -\frac{1}{37} \approx -2.7\text{¢}.$$

41.   In this form of the game Keno,

$$P(\text{your number comes up}) = \frac{20}{80} = \frac{1}{4} \text{ and } P(\text{your number doesn't come up}) = \frac{60}{80} = \frac{3}{4}.$$

If your number comes up, you win $2.20.   Otherwise, you lose $1 (win -$1).   If x denotes your winings, then the expected value is

$$E(x) = 2.20\left(\frac{1}{4}\right) - 1\left(\frac{3}{4}\right) = .55 - .75 = -\$.20 \quad \text{or} \quad -20\text{¢}.$$

45.   The expected value is

$$E(x) = 0(.01) + 1(.05) + 2(.15) \ \ + 3(.26) + 4(.33) + 5(.14) + 6(.06)$$

$$= 3.51 \text{ complaints.}$$

47.

Account number	Existing volume	Potential add volume	Probability of getting it	Expected value	Exist. vol. + exp. value	Class
1	$15,000	$10,000	.25	$2500	$17,500	C
2	$40,000	$0	---	-----	$40,000	C
3	$20,000	$10,000	.20	$2000	$22,000	C
4	$50,000	$10,000	.10	$1000	$51,000	B
5	$5000	$50,000	.50	$25,000	$30,000	C
6	$0	$100,000	.60	$60,000	$60,000	A
7	$30,000	$20,000	.80	$16,000	$46,000	B

49. **(a)** Let x denote the amount of damage in millions of dollars. For seeding, the expected value is

$$E(x) = (.038)(335.8) + (.143)(191.1) + (.392)(100) + (.255)(46.7) + (.172)(16.3)$$
$$\approx \$94.0 \text{ million}$$

For not seeding, the expected value is

$$E(x) = (.054)(335.8) + (.206)(191.1) + (.480)(100) + (.206)(46.7) + (.054)(16.3)$$
$$\approx \$116.0 \text{ million.}$$

(b) Seed, since the total expected damage is less with that option.

51. Let x denote the winnings. The probability distribution is as follows.

x	P(x)
100,000	1/2,000,000
40,000	2/2,000,000
10,000	2/2,000,000
0	1,999,995/2,000,000

The expected value is

$$E(x) = 100,000\left(\frac{1}{2,000,000}\right) + 40,000\left(\frac{2}{2,000,000}\right) + 10,000\left(\frac{2}{2,000,000}\right) + 0\left(\frac{1,999,995}{2,000,000}\right)$$
$$= .05 + .04 + .01 + 0 = \$.10 = 10¢$$

Since the expecting winnings are 10¢, if entering the contest costs 50¢ then it would not be worth it to enter.

Chapter 6   Review Exercises

1. 6 shuttle vans can line up at the airport in

$$P(6, 6) = 6! = 720$$

different ways.

3. 3 oranges can be taken from a bag of 12 in

$$\binom{12}{3} = \frac{12!}{9!3!} = \frac{12 \cdot 11 \cdot 10}{3 \cdot 2 \cdot 1} = 220$$

different ways.

5. 2 pictures from a group of 5 different pictures can be arranged in

$$P(5, 2) = 5 \cdot 4 = 20$$

different ways.

7. **(a)** There are 2! ways to arrange the landscapes, 3! ways to arrange the puppies, and 2 choices whether landscapes or puppies come first. Thus, the pictures can be arranged in

$$(2!)(3!) \cdot 2 = 24$$

different ways.

(b) The pictures must be arranged puppy, landscape, puppy, landscape, puppy. Arrange the puppies in 3! or 6 ways. Arrange the landscapes in 2! or 2 ways. In this scheme, the pictures can be arranged in $6 \cdot 2 = 12$ different ways.

9. **(a)** There are $7 \cdot 5 \cdot 4 = 140$ different groups of 3 representatives possible.

(b) $7 \cdot 5 \cdot 4 = 140$ is the number of groups with 3 representatives. For 2 representatives, the number of groups is

$$7 \cdot 5 + 7 \cdot 4 + 5 \cdot 4 = 83.$$

For 1 representative, the number of groups is

$$7 + 5 + 4 = 16.$$

The total number of these groups is

$$140 + 83 + 16 = 239$$

groups.

13. It is impossible to draw 3 blue balls, since there are only 2 blue balls in the basket; hence,

$$P(\text{all blue balls}) = 0.$$

15. P(exactly 2 black balls)

$$= \frac{\binom{4}{2}\binom{7}{1}}{\binom{11}{3}} = \frac{42}{165} = \frac{14}{55} \approx .255$$

17. P(2 green balls and 1 blue ball)

$$= \frac{\binom{5}{2}\binom{2}{1}}{\binom{11}{3}} = \frac{20}{165} = \frac{4}{33} \approx .121$$

19. Let x denote the number of girls. We have n = 6, x = 6, p = 1/2, and 1 − p = 1/2, so

$$P(\text{all girls}) = \binom{6}{6}\left(\frac{1}{2}\right)^6\left(\frac{1}{2}\right)^0$$

$$= \frac{1}{64} \approx .016.$$

21. Let x denote the number of boys, and then p = 1/2 and 1 − p = 1/2. We have

P(no more than 2 boys)

$= P(x \leq 2)$

$= P(x = 0) + P(x = 1) + P(x = 2)$

$= \binom{6}{0}\left(\frac{1}{2}\right)^0\left(\frac{1}{2}\right)^6 + \binom{6}{1}\left(\frac{1}{2}\right)^1\left(\frac{1}{2}\right)^5$

$\quad + \binom{6}{2}\left(\frac{1}{2}\right)^2\left(\frac{1}{2}\right)^4$

$= \frac{11}{32} \approx .344.$

23. $P(2 \text{ spades}) = \dfrac{\binom{13}{2}}{\binom{52}{2}} = \dfrac{78}{1326}$

$\qquad\qquad = \dfrac{1}{17} \approx .059$

25. P(exactly 1 face card)

$= \dfrac{\binom{12}{1}\binom{40}{1}}{\binom{52}{2}} = \dfrac{480}{1326} = \dfrac{80}{221} \approx .362$

27. P(at most 1 queen)

$= P(0 \text{ queens}) + P(1 \text{ queen})$

$= \dfrac{\binom{48}{2}}{\binom{52}{2}} + \dfrac{\binom{4}{1}\binom{48}{1}}{\binom{52}{2}}$

$= \dfrac{1128}{1326} + \dfrac{192}{1326}$

$= \dfrac{1320}{1326} = \dfrac{220}{221} \approx .9955$

29. Add up the frequencies to obtain n = 23. Divide the frequencies by n to obtain the probabilities.

(a) The probability distribution is as follows.

x	8	9	10	11	12	13	14
P(x)	$\frac{1}{23}$	0	$\frac{2}{23}$	$\frac{5}{23}$	$\frac{8}{23}$	$\frac{4}{23}$	$\frac{3}{23}$

(b) The histogram consists of 7 rectangles, the second of which has no height.

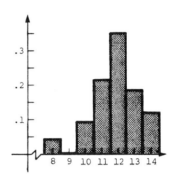

(c) The expected value is

$E(x) = 8\left(\frac{1}{23}\right) + 9(0) + 10\left(\frac{2}{23}\right) + 11\left(\frac{5}{23}\right)$

$\qquad + 12\left(\frac{8}{23}\right) + 13\left(\frac{4}{23}\right) + 14\left(\frac{3}{23}\right)$

$= \dfrac{273}{23} \approx 11.87.$

31. **(a)** There are n = 36 possible outcomes. Let x represent the sum of the dice, and note that the possible values of x are the whole numbers from 2 to 12. The probability distribution is as follows.

x	2	3	4	5	6
P(x)	$\frac{1}{36}$	$\frac{2}{36}$	$\frac{3}{36}$	$\frac{4}{36}$	$\frac{5}{36}$

x	7	8	9	10	11	12
P(x)	$\frac{6}{36}$	$\frac{5}{36}$	$\frac{4}{36}$	$\frac{3}{36}$	$\frac{2}{36}$	$\frac{1}{36}$

(b) The histogram consists of 11 rectangles.

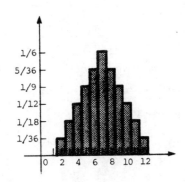

(c) The expected value is

$$E(x) = 2\left(\frac{1}{36}\right) + 3\left(\frac{2}{36}\right) + 4\left(\frac{3}{36}\right) + 5\left(\frac{4}{36}\right)$$

$$+ 6\left(\frac{5}{36}\right) + 7\left(\frac{6}{36}\right) + 8\left(\frac{5}{36}\right) + 9\left(\frac{4}{36}\right)$$

$$+ 10\left(\frac{3}{36}\right) + 11\left(\frac{2}{36}\right) + 12\left(\frac{1}{36}\right)$$

$$= \frac{252}{36} = 7.$$

33. The probability that corresponds to the shaded region of the histogram is the total of the shaded areas, which is

$$1(.3) + 1(.2) + 1(.1) = .6.$$

35. The probability of rolling a 6 is 1/6, and your net winnings would be $2. The probability of rolling a 5 is 1/6, and your net winnings would be $1. The probability of rolling something else is 4/6, and your net winnings would be -$2. Let x represent your winnings. The expected value is

$$E(x) = 2\left(\frac{1}{6}\right) + 1\left(\frac{1}{6}\right) + (-2)\left(\frac{4}{6}\right)$$

$$= -\frac{5}{6} \approx -\$.833 \quad \text{or} \quad -83.3¢.$$

This is not a fair game since the expected value is not 0.

37. Let x denote the number of girls. The probability distribution is as follows.

x	0	1	2	3	4	5
P(x)	$\frac{1}{32}$	$\frac{5}{32}$	$\frac{10}{32}$	$\frac{10}{32}$	$\frac{5}{32}$	$\frac{1}{32}$

The expected value is

$$E(x) = 0\left(\frac{1}{32}\right) + 1\left(\frac{5}{32}\right) + 2\left(\frac{10}{32}\right) + 3\left(\frac{10}{32}\right)$$

$$+ 4\left(\frac{5}{32}\right) + 5\left(\frac{1}{32}\right)$$

$$= \frac{80}{32} = 2.5 \text{ girls.}$$

39. $P(3 \text{ clubs}) = \dfrac{\binom{13}{3}}{\binom{52}{3}} = \dfrac{286}{22,100} \approx .0129$

Thus,

P(win) = .0129 and
P(lose) = 1 - .0129 = .9871.

Let x represent the amount you should pay.

Your net winnings are 100 - x if you win and -x if you lose.

If it is a fair game, your expected winnings will be 0. Thus, E(x) = 0 becomes

$$.0129(100 - x) + .9871(-x) = 0$$
$$1.29 - .0129x - .9871x = 0$$
$$1.29 - x = 0$$
$$x = 1.29.$$

You should pay $1.29.

41. We have $n = 20$, $x = 3$, $p = .01$, and $1 - p = .99$, so

$$P(x = 3) = \binom{20}{3}(.01)^3(.99)^{17}$$

$$\approx .00096.$$

43. $P(x \geq 12)$

$$= P(x = 12) + P(x = 13) + \cdots$$

$$+ P(x = 20)$$

$$= \binom{20}{12}(.01)^{12}(.99)^8$$

$$+ \binom{20}{13}(.01)^{13}(.99)^7$$

$$+ \cdots + \binom{20}{20}(.01)^{20}(.99)^0$$

45. Let x denote your winnings. Reduce the $1000 prize by 62¢ for the cost of entering. (This cost covers the stamped envelope for mailing the entry and the self-addressed stamped envelope enclosed with the entry.) Then, the expected value is

$$E(x) = 999.38\left(\frac{1}{4000}\right) + (-.62)\left(\frac{3999}{4000}\right)$$

$$\approx -.37,$$

which is a loss of 37¢.

47. If a box is good (probability .9) and the merchant samples an excellent piece of fruit from that box (probability .80), then he will accept the box and earn a $200 profit on it.

If a box is bad (probability .1) and he samples an excellent piece of fruit from the box (probability .30), then he will accept the box and earn a −$1000 profit on it.

If the merchant ever samples a non-excellent piece of fruit, he will not accept the box. In this case he pays nothing and earns nothing, so the profit will be $0.

Let x denote the merchant's earnings. Note that $.9(.80) = .72$,

$$.1(.30) = .03,$$

and $1 - (.72 + .03) = .25$.

The probability distribution is as follows.

x	200	−1000	0
P(x)	.72	.03	.25

The expected value when the merchant samples the fruit is

$$E(x) = 200(.72) + (-1000)(.03) + 0(.25)$$

$$= 144 - 30 + 0$$

$$= \$114.$$

We must also consider the case in which the merchant does not sample the fruit. Let x again denote the merchant's earnings. The probability distribution is as follows.

x	200	−1000
P(x)	.9	.1

The expected value when the merchant does not sample the fruit is

$$E(x) = 200(.9) + (-1000)(.1)$$
$$= 180 - 100$$
$$= \$80.$$

Combining these two results, the expected value of the right to sample is \$114 − \$80 = \$34, which corresponds to choice (c).

Extended Application

1. **(a)** $C(M_0)$
   $$= NL[1 - (1 - p_1)(1 - p_2)(1 - p_3)]$$
   $$= 3(54)[1 - (.91)(.76)(.83)]$$
   $$= \$69.01$$

   (b) $C(M_2)$
   $$= H_2 + NL[1 - (1 - p_1)(1 - p_3)]$$
   $$= 40 + 3(54)[1 - (.91)(.83)]$$
   $$= \$79.64$$

   (c) $C(M_3)$
   $$= H_3 + NL[1 - (1 - p_1)(1 - p_2)]$$
   $$= 9 + 3(54)[1 - (.91)(.76)]$$
   $$= \$58.96$$

   (d) $C(M_{12})$
   $$= H_1 + H_2 + NL[1 - (1 - p_3)]$$
   $$= 15 + 40 + 3(54)[1 - .83]$$
   $$= \$82.54$$

   (e) $C(M_{13})$
   $$= H_1 + H_3 + NL[1 - (1 - p_2)]$$
   $$= 15 + 9 + 3(54)[1 - .76]$$
   $$= \$62.88$$

(f) $C(M_{123})$
   $$= H_1 + H_2 + H_3 + NL[1 - 1]$$
   $$= 15 + 40 + 9$$
   $$= \$64.00$$

2. Policy M_3, stocking only part 3 on the truck, leads to the lowest expected cost.

3. It is not necessary for the probabilities to add up to 1 because it is possible that no parts will be needed. That is, the events of needing parts 1, 2, and 3 are not the only events in the sample space.

4. For 3 different parts we have 8 different policies: 1 with no parts, 3 with 1 part, 3 with 2 parts, and 1 with 3 parts. The number of different policies, $8 = 2^3$, is the number of subsets of a set containing 3 distinct elements. If there are n different parts, the number of policies is the number of subsets of a set containing n distinct elements which we showed in Chapter 5 to be 2^n.

CHAPTER 6 TEST

1. Evaluate the following factorials, permutations, and combinations.

   (a) 7! **(b)** $\binom{8}{3}$ **(c)** P(6, 4)

2. The 24 members of the 3rd grade Mitey Mites hockey team must select a head basher, a second basher, and a designated tripper for their team. How many ways can this be done?

3. A basketball team consists of 7 good shooters and 5 poor shooters.

   (a) In how many ways can a 5-person team be selected?

   (b) In how many ways can a team consisting of only good shooters be selected?

   (c) In how many ways can a team consisting of 3 good and 2 poor shooters be selected?

4. In Ohio, most license plates consist of three letters followed by three digits. How many different plates are possible in the following situations?

   (a) Repeats of letters and digits are allowed.

   (b) Repeats of letters, but not digits are allowed.

   (c) The first two letters must be AR, and repeats are allowed.

5. A coin and a die are tossed. List the sample space for this experiment.

6. A three-card hand is drawn from a standard deck of 52 cards. Set up the probability of each of the following events.

   (a) The hand contains exactly 2 hearts.

   (b) The hand contains fewer than 2 hearts.

7. A shipment of bolts has 20% of the bolts defective.

   (a) What is the probability of finding 2 or fewer defective bolts in a sample of 15?

   (b) What is the probability that 9 bolts will have to be sampled before 2 defectives are found?

   (c) How many bolts must be sampled to ensure a probability of at least .6 that a defective bolt is found?

8. A nickel, a dime, and a quarter are tossed simultaneously. Let the random variable x denote the number of heads observed in the experiment, and prepare a probability distribution.

9. Two dice are rolled, and the total number of points is recoreded. Find the expected value.

10. There is a game called Double or Nothing, and it costs $1 to play. If you draw the ace of spades, you are paid $2, but you are paid nothing if you draw any other card. Is this a fair game?

CHAPTER 6   TEST ANSWERS

1. **(a)** 5040 **(b)** 56 **(c)** 360   2. 12,144 ways   3. **(a)** 792 ways

(b) 21 ways (c) 350 ways   4. **(a)** $26 \cdot 26 \cdot 26 \cdot 10 \cdot 10 \cdot 10 \cdot = 17{,}576{,}000$

(b) $26 \cdot 26 \cdot 26 \cdot 10 \cdot 9 \cdot 8 = 12{,}654{,}720$ **(c)** $1 \cdot 1 \cdot 26 \cdot 10 \cdot 10 \cdot 10 = 26{,}000$

5. $\{(h, 1), (h, 2), (h, 3), (h, 4), (h, 5), (h, 6),$
   $(t, 1), (t, 2), (t, 3), (t, 4), (t, 5), (t, 6)\}$

6. **(a)** $\dfrac{\binom{13}{2}\binom{39}{1}}{\binom{52}{3}}$ **(b)** $\dfrac{\binom{13}{0}\binom{39}{3} + \binom{13}{1}\binom{39}{2}}{\binom{52}{3}}$

7. **(a)** $\binom{15}{0}(.2)^0(.8)^{15} + \binom{15}{1}(.2)^1(.8)^{14} + \binom{15}{2}(.2)^2(.8)^{13} \approx .3980$

(b) $\binom{8}{1}(.2)^2(.8)^7 \approx .0671$ **(c)** 5

8.

x	0	1	2	3
P(x)	$\frac{1}{8}$	$\frac{3}{8}$	$\frac{3}{8}$	$\frac{1}{8}$

9. 7

10. No, it is not a fair game.

CHAPTER 7   STATISTICS

Section 7.1

1. **(a)** – **(b)** Since 0 – 24 is to be the first interval and there are 25 numbers between 0 and 24 inclusive, we will let all six intervals be of size 25. The other five intervals are 25 – 49, 50 – 74, 75 – 99, 100 – 124, and 125 – 149. Keeping a tally of how many data values lie in each interval leads to the following frequency distribution.

Interval	Frequency
0 – 24	4
25 – 49	3
50 – 74	6
75 – 99	3
100 – 124	5
125 – 149	9

(c) Draw the histogram. It consists of 6 bars of equal width and having heights as determined by the frequency of each interval. See the histogram that appears in the back of the textbook.

(d) To construct the frequency polygon, join consecutive midpoints of the tops of the histogram bars with straight line segments. See the back of the textbook.

3. **(a)** – **(b)** Since 70 – 74 is to be the first interval, we let all the intervals be of size 5. The largest data value is 111, so the last interval that will be needed is 110 – 114. The frequency distribution is as follows.

Interval	Frequency
70 – 74	2
75 – 79	1
80 – 84	3
85 – 89	2
90 – 94	6
95 – 99	5
100 – 104	6
105 – 109	4
110 – 114	2

(c) Draw the histogram. It consists of 7 bars of equal width and having heights as determined by the frequency of each interval. See the histogram that appears in the back of the textbook.

(d) Construct the frequency polygon by joining consecutive midpoints of the tops of the histogram bars with straight line segments. See the back of the textbook.

7. The mean of the 5 numbers is

$$\bar{x} = \frac{\Sigma x}{5}$$

$$= \frac{8 + 10 + 16 + 21 + 25}{5}$$

$$= \frac{80}{5} = 16.$$

9. $\Sigma x = 21,900 + 22,850 + 24,930$

$+ \ 29,710 + 28,340 + 40,000$

$= \ 167,730$

The mean of the 6 numbers is

$\bar{x} = \dfrac{\Sigma x}{6} = \dfrac{167,730}{6} = 27,955.$

11. $\Sigma x = 9.4 + 11.3 + 10.5 + 7.4 + 9.1$

$+ \ 8.4 + 9.7 + 5.2 + 1.1 + 4.7$

$= \ 76.8$

The mean of the 10 numbers is

$\bar{x} = \dfrac{\Sigma x}{10} = \dfrac{76.8}{10} = 7.68 \approx 7.7.$

13. Add to the frequency distribution a new column, "Value × Frequency."

Value	Frequency	Value × Frequency
3	4	3 · 4 = 12
5	2	5 · 2 = 10
9	1	9 · 1 = 9
12	3	12 · 3 = 36
	Total: 10	Total: 67

The means is $\bar{x} = \dfrac{67}{10} = 6.7.$

15.

Value	Frequency	Value × Frequency
12	4	12 · 4 = 48
13	2	13 · 2 = 26
15	5	15 · 5 = 75
19	3	19 · 3 = 57
22	1	22 · 1 = 22
23	5	23 · 5 = 115
	Total: 20	Total: 343

The mean is $\bar{x} = \dfrac{343}{20} = 17.15 \approx 17.2.$

17. 12, 18, 32, 51, 58, 92, 106

The median is the middle number, 51.

19. 100, 114, 125, 135, 150, 172

The median is the mean of the 2 middle numbers, which is

$\dfrac{125 + 135}{2} = \dfrac{260}{2} = 130.$

21. First arrange the numbers in numerical order, from smallest to largest.

3.4, 9.1, 22.6, 28.4, 29.8,
32.1, 47.6, 59.8

There are 8 numbers here; the median is the mean of the 2 middle numbers, which is

$\dfrac{28.4 + 29.8}{2} = \dfrac{58.2}{2} = 29.1.$

23. 4, 9, 8, 6, 9, 2, 1, 3

The mode is the number that occurs most often. Here, the mode is 9.

25. 74, 68, 68, 68, 75, 75, 74, 74, 70

The mode is the number that occurs most often. Here, there are two modes, 68 and 74, since they both appear three times.

27. 6.8, 6.3, 6.3, 6.9, 6.7, 6.4, 6.1, 6.0

The mode is 6.3.

31.

Interval	Midpoint, x	Frequency, f	Product, xf
70 – 74	72	2	144
75 – 79	77	1	77
80 – 84	82	3	246
85 – 89	87	2	174
90 – 94	92	6	552
95 – 99	97	5	485
100 – 104	102	6	612
105 – 109	107	4	428
110 – 114	112	2	224
		Total: 31	Total: 2942

The mean of this collection of grouped data is

$$\bar{x} = \frac{2942}{31} \approx 94.9.$$

The intervals 90 – 94 and 100 – 104 each contain the most data values, 6, so they are the modal classes.

33. Draw the histogram. It consists of 5 bars of equal width and having heights as determined by the frequency of each interval (the fourth bar has no height). Construct the frequency polygon by joining the consecutive midpoints of the tops of the histogram bars with straight line segments. See the answer graph in the back of the textbook.

35. $\Sigma x = 2200 + 2000 + 1750 + 2200 + 2400 + 2800 + 2800 + 2450 + 2600 + 2750$

   $= 23,950$

The mean of the wheat production is

$$\bar{x} = \frac{\Sigma x}{10} = \frac{23,950}{10} = 2395 \text{ million bushels.}$$

37. **(a)** The height of the 10 – 19 histogram bar looks like about 17.5, so about 17.5% of the population was between 10 and 19 years old.

(b) The 60 – 69 bar has a height of about 8, so about 8% of the population was between 60 and 69 years old.

(c) The 20 – 29 bar appears to have the greatest height, so that age group had the largest percentage of the population.

39. **(a)** $\bar{x} = \dfrac{\Sigma x}{n} = \dfrac{666}{12} = 55.5$ **(b)** $\bar{x} = \dfrac{\Sigma x}{n} = \dfrac{347}{12} \approx 28.9$

The mean of the maximum tempera-   The mean of the minimum tempera-

tures is 55.5°F.   tures is about 28.9°F.

Section 7.2

1. The standard deviation of a sample of numbers is the square root of the variance of the sample.

3. The range is $74 - 29 = 53$, the difference of the highest and lowest numbers in the set.

The mean is $\bar{x} = \dfrac{1}{7}(42 + 38 + 29 + 74 + 82 + 71 + 35) = \dfrac{371}{7} = 53$.

To prepare for calculating the standard deviation, construct a table of the set of numbers, their deviations from the mean $\bar{x}$, and the squares of these deviations.

x	$x - \bar{x}$	$(x - \bar{x})^2$
42	−11	121
38	−15	225
29	−24	576
74	21	441
82	29	841
71	18	324
35	−18	324
		Total: 2852

The standard deviation is $s = \sqrt{\dfrac{2852}{7 - 1}} \approx \sqrt{475.3} \approx 21.8$.

5.   The range is $287 - 241 = 46$.   The mean is

$$\bar{x} = \frac{1}{6}(241 + 248 + 251 + 257 + 252 + 287) = 256.$$

x	x − $\bar{x}$	(x − $\bar{x}$)²
241	−15	225
248	−8	64
251	−5	25
257	1	1
252	−4	16
287	31	961
		Total: 1292

The standard deviation is $s = \sqrt{\dfrac{1292}{6 - 1}} = \sqrt{258.4} \approx 16.1$.

7.   The range is $27 - 3 = 24$.   The mean is

$$\bar{x} = \frac{1}{10}(3 + 7 + 4 + 12 + 15 + 18 + 19 + 27 + 24 + 11) = 14.$$

x	x − $\bar{x}$	(x − $\bar{x}$)²
3	−11	121
7	−7	49
4	−10	100
12	−2	4
15	1	1
18	4	16
19	5	25
27	13	169
24	10	100
11	−3	9
		Total: 594

The standard deviation is $s = \sqrt{\dfrac{594}{10 - 1}} = \sqrt{66} \approx 8.1$.

9. Expand the table to include columns for the midpoint x of each interval, and for
    fx, x², and fx².

Interval	f	x	fx	x²	fx²
0 – 24	4	12	48	144	576
25 – 49	3	37	111	1369	4107
50 – 74	6	62	372	3844	23,064
75 – 99	3	87	261	7569	22,707
100 – 124	5	112	560	12,544	62,720
125 – 149	9	137	1233	18,769	168,921
Totals:	30		2585		282,095

    The mean of the grouped data is

    $$\bar{x} = \frac{\Sigma fx}{n} = \frac{2585}{30} \approx 86.2.$$

    The standard deviation for the grouped data is

    $$s = \sqrt{\frac{\Sigma fx^2 - n\bar{x}^2}{n-1}} = \sqrt{\frac{282,095 - 30(86.2)^2}{30-1}} \approx \sqrt{2046.7} \approx 45.2.$$

11. Start with the frequency distribution that was the answer to Exercise 3 of
    Section 7.1, and expand the table to include columns for the midpoint x of each
    interval, and for fx, x², and fx².

Interval	f	x	fx	x²	fx²
70 – 74	2	72	144	5184	10,368
75 – 79	1	77	77	5929	5929
80 – 84	3	82	246	6724	20,172
85 – 89	2	87	174	7569	15,138
90 – 94	6	92	552	8464	50,784
95 – 99	5	97	485	9409	47,045
100 – 104	6	102	612	10,404	62,424
105 – 109	4	107	428	11,449	45,796
110 – 114	2	112	224	12,544	25,088
Totals:	31		2942		282,744

The mean of the grouped data is

$$\bar{x} = \frac{\Sigma fx}{n} = \frac{2942}{31} \approx 94.9.$$

The standard deviation for the grouped data is

$$s = \sqrt{\frac{\Sigma fx^2 - n\bar{x}^2}{n - 1}}$$

$$= \sqrt{\frac{282,744 - 31(94.9)^2}{31 - 1}}$$

$$\approx \sqrt{118.0} \approx 10.9.$$

13. Use k = 2 in Chebyshev's theorem.

$$P(\mu - 2\sigma \le x \le \mu + 2\sigma) \ge 1 - \frac{1}{2^2} = \frac{3}{4},$$

so at least 3/4 of the distribution is within 2 standard deviations of the mean.

15. Use k = 5 in Chebyshev's theorem.

$$P(\mu - 5\sigma \le x \le \mu + 5\sigma) \ge 1 - \frac{1}{5^2} = \frac{24}{25},$$

so at least 24/25 of the distribution is within 5 standard deviations of the mean.

17. Here $32 = 50 - 3 \cdot 6 = \mu - 3\sigma$ and $68 = 50 + 3 \cdot 6 = \mu + 3\sigma$, so Chebyshev's theorem applies with k = 3; hence at least

$$1 - \frac{1}{k^2} = \frac{8}{9} \approx 88.9\%$$

of the numbers lie between 32 and 68.

19. The answer here is the complement of the answer to Exercise 16. It was found that at least 75% of the distribution of numbers are between 38 and 62, so at most 100% - 75% = 25% of the numbers are less than 38 or more than 62.

21. 300, 320, 380, 420, 500, 2000

(a) $\mu = \frac{1}{6}(300 + 320 + 380 + 420$

$+ 500 + 2000)$

$= \frac{1}{6}(3920) \approx 653.33.$

The mean of the salaries is $653.33.

x	x²
300	90,000
320	102,400
380	144,400
420	176,400
500	250,000
2000	4,000,000
	Total: 4,763,200

$$\sigma = \sqrt{\frac{\Sigma x^2 - n\bar{\mu}^2}{n - 1}}$$

$$= \sqrt{\frac{4,763,200 - 6(653.33)^2}{6 - 1}}$$

$$= \sqrt{440,426.67} \approx 663.65$$

The standard deviation of the salaries is $663.65.

(b) $653.33 - 663.65 = -10.32$ is 1 standard deviation below the mean, and $653.33 + 663.65 = 1316.98$ is 1 standard deviation above the mean.

5 of the data values lie between these two numbers, so 5 of the workers earn wages within 1 standard deviation of the mean.

(c) 653.33 − 2(663.65) = −673.97 is 2 standard deviations below the mean, and 653.33 + 2(663.65) = 1980.63 is 2 standard deviations above the mean.

5 of the data values lie between these two numbers, so 5 of the workers earn wages within 2 standard deviations of the mean.

(d) Use k = 2 in Chebyshev's theorem.

$$P(\mu - 2\sigma \leq x \leq \mu + 2\sigma) \geq 1 - \frac{1}{2^2} = \frac{3}{4}$$

Chebyshev's theorem guarantees that at least 3/4 of a distribution will lie within 2 standard deviations of the mean, which agrees with our findings in part (c) above.

23. 15, 18, 19, 23, 25, 25, 28, 30, 34, 38

(a) $\bar{x} = \frac{1}{10}(15 + 18 + 19 + 23 + 25 + 25 + 28 + 30 + 34 + 38) = \frac{1}{10}(255) = 25.5$

The mean life of the sample of Brand X batteries is 25.5 hr.

x	$x - \bar{x}$	$(x - \bar{x})^2$
15	−10.5	110.25
18	−7.5	56.25
19	−6.5	42.25
23	−2.5	6.25
25	−.5	.25
25	−.5	.25
28	2.5	6.25
30	4.5	20.25
34	8.5	72.25
38	12.5	156.25
		Total: 470.50

$$s = \sqrt{\frac{470.50}{10 - 1}} \approx \sqrt{52.28} \approx 7.2$$

The standard deviation of the Brand X battery lives is 7.2 hr.

(b) Forever Power has a smaller standard deviation (4.1 hr, as opposed to 7.2 hr for Brand X), which indicates a more uniform life.

(c) Forever Power has a higher mean (26.2 hr, as opposed to 25.5 hr for Brand X), which indicates a longer average life.

25.

					Sample Number					
	1	2	3	4	5	6	7	8	9	10
	2	3	-2	-3	-1	3	0	-1	2	0
	-2	-1	0	1	2	2	1	2	3	0
	1	4	1	2	4	2	2	3	2	2
(a) $\bar{x}$	$\frac{1}{3}$	2	$-\frac{1}{3}$	0	$\frac{5}{3}$	$\frac{7}{3}$	1	$\frac{4}{3}$	$\frac{7}{3}$	$\frac{2}{3}$
(b) s	2.1	2.6	1.5	2.6	2.5	.6	1	2.1	.6	1.2

(c) $\bar{X} = \dfrac{\Sigma \bar{x}}{n} \approx \dfrac{11.3}{10} = 1.13$

(d) $\bar{s} = \dfrac{\Sigma s}{n} = \dfrac{16.8}{10} = 1.68$

(e) The upper control limit for the sample means is

$$\mu + k_1 \bar{s} = 1.13 + (1.954)(1.68) \approx 4.41.$$

The lower control limit for the sample means is

$$\mu - k_1 s = 1.13 - (1.954)(1.68) \approx -2.15.$$

(f) The upper control limit for the sample standard deviations is

$$k_2 \bar{s} = 2.568(1.68) \approx 4.31.$$

The lower control limit for the sample standard deviations is

$$k_3 \bar{s} = 0(1.68) = 0.$$

27. 8.54, 8.31, 8.24, 7.43,
6.70, 6.53, 6.87

x	x^2
8.54	72.9316
8.31	69.0561
8.24	67.8976
7.43	55.2049
6.70	44.89
6.53	42.6409
6.87	47.1969
Totals: 52.62	399.818

(a) The mean is

$$\mu = \frac{\Sigma x}{n} = \frac{52.62}{7} \approx 7.5171 \approx 7.52.$$

1987 with 7.43 thousand unemployed
is the year that has unemployment
closest to the mean.

(b) The standard deviation is

$$\sigma = \sqrt{\frac{\Sigma x^2 - n\mu^2}{n - 1}}$$

$$= \sqrt{\frac{399.818 - 7(7.5171)^2}{7 - 1}}$$

$$\approx \sqrt{.7117} \approx .84.$$

(c) $7.52 - .84 = 6.68$ is 1 standard deviation below the mean, and
$7.52 + .84 = 8.36$ is 1 standard deviation above the mean.
8.31, 8.24, 7.43, 6.70, and 6.87 are
all between 6.68 and 8.36, so unemployment is within 1 standard
deviation of the mean in 5 of the
years.

29. This exercise should be solved using
a computer or a calculator with a
standard deviation key. The answers
are 1.8158 mm for the mean and
.4451 mm for the standard deviation.

Section 7.3

1. The peak in a normal curve occurs
directly above the mean.

3. For normal distributions where $\mu \neq 0$
or $\sigma \neq 1$, z-scores are found by using
the formula $z = \frac{x - \mu}{\sigma}$.

5. Use the table, "Area Under a Normal
Curve to the Left of z", in the
Appendix. To find the percent of
the area under a normal curve between the mean and 2.50 standard
deviations from the mean, subtract
the table entry for $z = 0$ (rep-
resenting the mean) from the table
entry for $z = 2.5$.

$$.9938 - .5000 = .4938$$

Therefore, 49.38% of the area lies
between μ and $\mu + 2.5\sigma$.

7. Subtract the table entry for
$z = -1.71$ from the table entry
for $z = 0$.

$$.5000 - .0436 = .4564$$

45.64% of the area lies between
μ and $\mu - 1.71\sigma$.

9. $P(1.41 \leq z \leq 2.83)$

    $= P(z \leq 2.83) - P(z \leq 1.41)$

    $= .9977 - .9207$

    $= .077$

    7.7% of the total area under the standard normal curve lies between $z = 1.41$ and $z = 2.83$.

11. $P(-2.48 \leq z \leq -.05)$

    $= P(z \leq -.05) - P(z \leq -2.48)$

    $= .4801 - .0066$

    $= .4735$

    47.35% of the area lies between $z = -2.48$ and $z = -.05$.

13. $P(-3.11 \leq z \leq 1.44)$

    $= P(z \leq 1.44) - P(z \leq -3.11)$

    $= .9251 - .0009$

    $= .9242$

    92.42% of the area lies between $z = -3.11$ and $z = 1.44$.

15. $P(-.42 \leq z \leq .42)$

    $= P(z \leq .42) - P(z \leq -.42)$

    $= .6628 - .3372$

    $= .3256$

    32.56% of the area lies between $z = -.42$ and $z = .42$.

17. 5% of the total area is to the left of z.

    Use the table backwards.
    Look in the body of the table for an area of .05, and find the corresponding z using the left column and top column of the table.

The closest values to .05 in the body of the table are .0505, which corresponds to $z = -1.64$, and .0495, which corresponds to $z = -1.65$.

19. 15% of the area is to the right of z. If 15% of the area is to the right of z, then 85% of the area is to the left of z. The closest value to .85 in the body of the table is .8504, which corresponds to $z = 1.04$.

21. Let x represent the bolt diameter.

    $$\mu = .25, \sigma = .02$$

    First, find the probability that a bolt has a diameter less than or equal to .3 in, that is, $P(x \leq .3)$. The z-score corresponding to $x = .3$ is

    $$z = \frac{x - \mu}{\sigma}$$

    $$= \frac{.3 - .25}{.02} = 2.5.$$

    Using the table, find the area to the left of $z = 2.5$. This gives us

    $$P(x \leq .3) = P(z \leq 2.5) = .9938.$$

    Then

    $$P(x > .3) = 1 - P(x \leq .3)$$

    $$= 1 - .9938$$

    $$= .0062.$$

23. Let x represent a grocery bill.

    $$\mu = 52.25, \sigma = 19.50$$

    The middle 50% of the grocery bills have cutoffs at 25% below the mean and 25% above the mean.

At 25% below the mean, the area to the left is .2500, which corresponds to about z = −.67. At 25% above the mean, the area to the left is .7500, which corresponds to about z = .67. Find the x-value that corresponds to each z-score.

For z = −.67,

$$-.67 = \frac{x - 52.25}{19.50} \quad \text{or} \quad x \approx 39.19.$$

For z = .67,

$$.67 = \frac{x - 52.25}{19.50} \quad \text{or} \quad x \approx 65.32.$$

The middle 50% of the customers spend between $39.19 and $65.32.

25. Let x represent the length of a fish.

$$\mu = 12.3, \ \sigma = 4.1$$

Find the z-score for x = 18.

$$z = \frac{18 - 12.3}{4.1} \approx 1.39$$

$$\begin{aligned}
P(x > 18) &= 1 - P(x \leq 18) \\
&= 1 - P(z \leq 1.39) \\
&= 1 - .9177 = .0823
\end{aligned}$$

27. Let x represent the life of a light bulb.

$$\mu = 500, \ \sigma = 100$$

At least 500 hr means x ≥ 500.

$$P(x \geq 500) = 1 - P(x < 500)$$

Find the z-score that corresponds to x = 500.

$$z = \frac{500 - 500}{100} = 0$$

$$\begin{aligned}
P(x \geq 500) &= P(z \geq 0) \\
&= 1 - .5000 = .5000
\end{aligned}$$

Thus, 10,000(.5) = 5000 bulbs can be expected to last at least 500 hr.

29. Let x represent the life of a light bulb.

$$\mu = 500, \ \sigma = 100$$

Between 650 and 780 hr means the following.

For x = 650, $z = \frac{650 - 500}{100} = 1.5$.

For x = 780, $z = \frac{780 - 500}{100} = 2.8$.

$$\begin{aligned}
P(650 &\leq x \leq 780) \\
&= P(1.5 \leq z \leq 2.8) \\
&= P(z \leq 2.8) - P(z \leq 1.5) \\
&= .9974 - .9332 \\
&= .0642
\end{aligned}$$

Thus, 10,000(.0642) = 642 bulbs can be expected to last between 650 and 780 hr.

31. Let x represent the life of a light bulb.

$$\mu = 500, \ \sigma = 100$$

Less than 740 hr means x < 740.

For x = 740, $z = \frac{740 - 500}{100} = 2.4$.

$$P(x < 740) = P(z < 2.4) = .9918$$

Thus, 10,000(.9918) = 9918 bulbs can be expected to last at least 740 hr.

33. Let x represent the weight of a package.

$$\mu = 16.5, \ \sigma = .5$$

For $x = 16$, $z = \dfrac{16 - 16.5}{.5} = -1$.

$P(x < 16) = P(z < -1) = .1587$

is the fraction of the boxes that are underweight.

35. Let x represent the weight of a package.

$$\mu = 16.5, \ \sigma = .2$$

For $x = 16$, $z = \dfrac{16 - 16.5}{.2} = -2.5$

$P(x < 16) = P(z < -2.5) = .0062$

is the fraction of the boxes that are underweight.

37. Let x represent the weight of a chicken.

$$\mu = 1850, \ \sigma = 150$$

More than 1700 g means x > 1700.

For $x = 1700$, $z = \dfrac{1700 - 1850}{150} = -1.0$.

$$
\begin{aligned}
P(x > 1700) &= 1 - P(x \le 1700) \\
&= 1 - P(z \le -1.0) \\
&= 1 - .1587 \\
&= .8413
\end{aligned}
$$

Thus, 84.13% of the chickens will weigh more than 1700 g.

39. Let x represent the weight of a chicken.

$$\mu = 1850, \ \sigma = 150$$

Between 1750 and 1900 g means $1750 \le x \le 1900$.

For $x = 1750$, $z = \dfrac{1750 - 1850}{150} = -.67$.

For $x = 1900$, $z = \dfrac{1900 - 1850}{150} = -.33$.

$$
\begin{aligned}
P(1750 &\le x \le 1900) \\
&= P(-.67 \le z \le -.33) \\
&= P(z \le -.33) - P(z \le -.67) \\
&= .6293 - .2514 \\
&= .3779
\end{aligned}
$$

Thus, 37.79% of the chicken will weigh between 1750 and 1900 g.

41. Let x represent the weight of a chicken.

$$\mu = 1850, \ \sigma = 150$$

Less than 1550 g means x < 1550.

For $x = 1550$, $z = \dfrac{1550 - 1850}{150} = -2.0$.

$$
\begin{aligned}
P(x < 1550) &= P(z < -2.0) \\
&= .0228
\end{aligned}
$$

Thus, 2.28% of the chickens will weigh less than 1550 g.

43. Let x represent the amount of vitamins a person needs. Then

$$
\begin{aligned}
P(x \le \mu + 2.5\sigma) &= P(z \le 2.5) \\
&= .9938.
\end{aligned}
$$

99.38% of the people will receive adequate amounts of vitamins.

45. The Recommended Daily Allowance is

$$\mu + 2.5\sigma = 159 + (2.5)(12)$$
$$= 189 \text{ units.}$$

47. Let x represent a driving speed.

$$\mu = 50, \ \sigma = 10$$

At the 85th percentile, the area to the left is .8500, which corresponds to about z = 1.04.
Find the x-value that corresponds to this z-score.

$$z = \frac{x - \mu}{\sigma}$$
$$1.04 = \frac{x - 50}{10}$$
$$10.4 = x - 50$$
$$60.4 = x$$

The 85th percentile speed for this road is 60.4 mph.

49. Let x stand for a student's total points.

$$P\left(x \ge \mu + \frac{3}{2}\sigma\right) = P(z \ge 1.5)$$
$$= 1 - P(z \le 1.5)$$
$$= 1 - .9332$$
$$= .0668$$

Thus, 6.68% of the students receive A's.

51. Let x stand for a student's total points.

$$P\left(\mu - \frac{1}{2}\sigma \le x \le \mu + \frac{1}{2}\sigma\right)$$
$$= P(-.5 \le z \le .5)$$
$$= P(z \le .5) - P(z \le -.5)$$
$$= .6915 - .3085$$
$$= .383$$

Thus, 38.3% of the students receive C's.

53. Let x represent a student's test score.

$$\mu = 74, \ \sigma = 6$$

Since the top 8% get A's, we want to find the number a for which

$$P(x \ge a) = .08,$$
or　　$$P(x \le a) = .92.$$

Read the table backwards to find the z-score for an area of .92, which is 1.41. Find the value of x that corresponds to z = 1.41.

$$z = \frac{x - \mu}{\sigma}$$
$$1.41 = \frac{x - 74}{6}$$
$$8.46 = x - 74$$
$$82.46 = x$$

The bottom cutoff score for an A is 82.

55. Let x represent a student's test score.

$$\mu = 74, \ \sigma = 6$$

23% of the students will receive D's and F's, so to find the bottom cut-off score for a C we need to find the number c for which

$$P(x \le c) = .23.$$

Read the table backwards to find the z-score for an area of .23, which is -.74. Find the value of x that corresponds to z = -.74.

$$-.74 = \frac{x - 74}{6}$$

$$-4.44 = x - 74$$

$$69.56 = x$$

The bottom cutoff score for a C is 70.

Exercises 57–63 should be completed using a computer. The solution may vary according to the computer program that is used. The answers to these exercises are given below.

57. $P(1.372 \le z \le 2.548) = .0796$

59. $P(z > -2.476) = .9933$

61. $P(12.275 < x < 28.432) = .1527$

63. $P(x < 17.462) = .0051$

Section 7.4

3. (a) Use the formula

$$P(x) = \binom{n}{x} p^x (1 - p)^{n-x}$$

to calculate the probabilities.

$$P(x) = \binom{6}{x} \left(\frac{1}{6}\right)^x \left(\frac{5}{6}\right)^{6-x};$$

$$x = 0, 1, 2, \ldots, 6$$

$$P(0) = \binom{6}{0} \left(\frac{1}{6}\right)^0 \left(\frac{5}{6}\right)^6 \approx .335$$

$$P(6) = \binom{6}{6} \left(\frac{1}{6}\right)^6 \left(\frac{5}{6}\right)^0 \approx .00002 \approx .000$$

Find P(1) through P(5) similarly.

The distribution is as follows.

x	0	1	2	3
P(x)	.335	.402	.201	.054

x	4	5	6
P(x)	.008	.001	.000

(b) The mean is $\mu = np = (6)\left(\frac{1}{6}\right)$

$$= 1.00.$$

(c) The standard deviation is

$$\sigma = \sqrt{np(1 - p)}$$

$$= \sqrt{6\left(\frac{1}{6}\right)\left(\frac{5}{6}\right)}$$

$$= \sqrt{\frac{5}{6}} \approx .91.$$

5. (a)

$$P(x) = \binom{3}{x} (.02)^x (.98)^{3-x};$$

$$x = 0, 1, 2, 3$$

$$P(0) = \binom{3}{0} (.02)^0 (.98)^3 \approx .941$$

$$P(3) = \binom{3}{3} (.02)^3 (.98)^0$$

$$= .000008 \approx .000$$

x	0	1	2	3
P(x)	.941	.058	.001	.000

(b) $\mu = np = 3(.02) = .06$

(c) $\sigma = \sqrt{np(1 - p)}$

$$= \sqrt{3(.02)(.98)} \approx .24$$

7. **(a)**

$$P(x) = \binom{4}{x}(.7)^x(.3)^{4-x};$$
$$x = 0, 1, 2, 3, 4$$

$$P(0) = \binom{4}{0}(.7)^0(.3)^4 = .0081$$

$$P(4) = \binom{4}{4}(.7)^4(.3)^0 = .2401$$

x	0	1	2	3	4
P(x)	.0081	.0756	.2646	.4116	.2401

(b) $\mu = np = (4)(.7) = 2.8$

(c) $\sigma = \sqrt{np(1-p)}$
$$= \sqrt{4(.7)(.3)}$$
$$\approx .92$$

9. $n = 500$, $p = .025$

$\mu = np = (500)(.025) = 12.5$
$\sigma = \sqrt{np(1-p)} = \sqrt{(500)(.025)(.975)}$
$$= \sqrt{12.1875} \approx 3.49$$

11. $n = 64$, $p = .80$
$\mu = np = 64(.80) = 51.2$

$\sigma = \sqrt{np(1-p)} = \sqrt{64(.80)(.20)}$
$$= \sqrt{10.24} \approx 3.2$$

13. The normal distribution can be used to approximate a binomial distribution as long as $np \geq 5$ and $n(1-p) \geq 5$.

15. 16 coins are tossed, which means $n = 16$ and $p = 1/2$. To approximate the binomial distribution, use a normal distribution with

$\mu = np = 16\left(\frac{1}{2}\right) = 8$ and

$\sigma = \sqrt{np(1-p)} = \sqrt{16\left(\frac{1}{2}\right)\left(\frac{1}{2}\right)} = 2.$

Exactly 7 heads corresponds to the area under the normal curve between $x = 6.5$ and $x = 7.5$. The corresponding z-scores are

$$z = \frac{6.5 - 8}{2} = -.75$$

and $\quad z = \frac{7.5 - 8}{2} = -.25.$

P(exactly 7 heads)
$$= P(6.5 \leq x \leq 7.5)$$
$$= P(-.75 \leq z \leq -.25)$$
$$= P(z \leq -.25) - P(z \leq -.75)$$
$$= .4013 - .2266$$
$$= .1747$$

17. $n = 16$, $p = \frac{1}{2}$

As in Exercise 15, $\mu = 8$ and $\sigma = 2$. Less than 12 tails corresponds to the area under the normal curve to the left of $x = 11.5$. The corresponding z-score is

$$z = \frac{11.5 - 8}{2} = 1.75.$$

P(less than 12 tails)
$$= P(x \leq 11.5)$$
$$= P(z \leq 1.75)$$
$$= .9599$$

19. $n = 1000$, $p = \frac{1}{2}$

We have

$\mu = np = 1000\left(\frac{1}{2}\right) = 500$ and

$\sigma = \sqrt{np(1-p)} = \sqrt{1000\left(\frac{1}{2}\right)\left(\frac{1}{2}\right)}$
$$= \sqrt{250} \approx 15.81.$$

Exactly 510 heads corresponds to the area under the normal curve between $x = 509.5$ and $x = 510.5$. The corresponding z-scores are

$$z = \frac{509.5 - 500}{15.81} \approx .60$$

and

$$z = \frac{510.5 - 500}{15.81} \approx .66.$$

P(exactly 510 heads)

  $= P(509.5 \le x \le 510.5)$

  $= P(.60 \le z \le .66)$

  $= P(z \le .66) - P(z \le .60)$

  $= .7454 - .7257$

  $= .0197$

21. $n = 1000$, $p = \frac{1}{2}$

As in Exercise 19, $\mu = 500$ and $\sigma = 15.81$.

Less than 470 tails corresponds to the area under the normal curve to the left of $x = 469.5$. The corresponding z-score is

$$z = \frac{469.5 - 500}{15.8} = -1.93.$$

P(less than 470 tails)

  $= P(x \le 469.5)$

  $= P(z \le -1.93)$

  $= .0268$

23. $n = 120$, $p = \frac{1}{6}$

We have $\mu = np = 120\left(\frac{1}{6}\right) = 20$ and

$\sigma = \sqrt{np(1 - p)}$

  $= \sqrt{120\left(\frac{1}{6}\right)\left(\frac{5}{6}\right)}$

  $\approx 4.08.$

We want the area between $x = 23.5$ and $x = 24.5$. The corresponding z-scores are .86 and 1.103.

P(exactly twenty-four 6's)

  $= P(23.5 \le x \le 24.5)$

  $= P(.86 \le z \le 1.103)$

  $= P(z \le 1.103) - P(z \le .86)$

  $= .8643 - .8051$

  $= .0592$

25. $n = 120$, $p = \frac{1}{6}$

As in Exercise 23, $\mu = 20$ and $\sigma = 4.08$.

We want the area to the left of $x = 21.5$. The corresponding z-score is .37.

P(fewer than twenty-two 6's)

  $= P(x \le 21.5)$

  $= P(z \le .37)$

  $= .6443$

27. We have $n = 10,000$ and $p = .02$, so $\mu = np = 200$ and $\sigma = \sqrt{np(1 - p)} = 14$. We want the area to the right of $x = 222.5$. The corresponding z-score value is 1.61.

P(more than 222 defects)

  $= P(x \ge 222.5)$

  $= P(z \ge 1.61)$

  $= 1 - P(z \le 1.61)$

  $= 1 - .9463$

  $= .0537$

29. n = 120, p = .6

We have $\mu = np = 120(.6) = 72$ and
$\sigma = \sqrt{np(1 - p)} = \sqrt{120(.6)(.4)} \approx 5.4$.

(a) P(exactly 80 units)

   $= P(79.5 \leq x \leq 80.5)$

   $= P(1.39 \leq z \leq 1.57)$

   $= P(z \leq 1.57) - P(z \leq 1.39)$

   $= .9418 - .9177$

   $= .0241$

(b) P(at least 70 units)

   $= P(x \geq 69.5)$

   $= P(z \geq -.46)$

   $= 1 - P(z \leq -.46)$

   $= 1 - .3228$

   $= .6772$

31. n = 25, p = .80

We have $\mu = np = 25(.80) = 20$ and
$\sigma = \sqrt{np(1 - p)} = \sqrt{25(.80)(.20)} = 2$.

P(exactly 20 cured)

   $= P(19.5 \leq x \leq 20.5)$

   $= P(-.25 \leq z \leq .25)$

   $= P(z \leq .25) - P(z \leq -.25)$

   $= .5987 - .4013$

   $= .1974$

33. $P(x = 0) = \binom{25}{0}(.80)^0(.20)^{25}$

   $= (.20)^{25}$

   $= 3.36 \times 10^{-18}$

   ≈ 0

35. n = 1400, p = .55

We have $\mu = np = 1400(.55) = 770$ and
$\sigma = \sqrt{np(1 - p)} = \sqrt{1400(.55)(.45)}$

   ≈ 18.6.

P(at least 750 people)

   $= P(x \geq 749.5)$

   $= P(z \geq -1.10)$

   $= 1 - P(z \leq -1.10)$

   $= 1 - .1357$

   $= .8643$

Exercises 37 and 39 should be completed using a computer. The solution may vary according to the computer program that is used. The answers to these exercises are given below.

37.   **(a)** P(exactly 5 are color–blind)

   $= .0472$

   (b) P(no more than 5 are color–blind)

   $= .9875$

   (c) P(at least 1 is color–blind)

   $= .8814$

39.   **(a)** P(all 58 like it)

   $= 1.04 \times 10^{-9} \approx 0$

   (b) P(exactly 28, 29, or 30 like it)

   $= .0018$

Chapter 7 Review Exercises

3. **(a)** Since 450 – 474 is to be the first interval, we will let all the intervals be of size 25. The largest data value is 566, so the last interval that will be needed is 550 – 574. The frequency distribution is as follows.

Interval	Frequency
450 – 474	5
475 – 499	6
500 – 524	5
525 – 549	2
550 – 574	2

(b) Draw the histogram. It consists of 5 bars of equal width and having heights as determined by the frequency of each interval. See the histogram that appears in the back of the textbook.

(c) Construct the frequency polygon by joining consecutive midpoints of the tops of the histogram bars with straight line segments. See the back of the textbook.

5. $\Sigma x = 41 + 60 + 67 + 68 + 72 + 74 + 78 + 83 + 90 + 97 = 730$

The mean of the 10 numbers is

$$\bar{x} = \frac{\Sigma x}{10} = \frac{730}{10} = 73.$$

7.

Interval	Midpoint, x	Frequency, f	Product, xf
10 – 19	14.5	6	87
20 – 29	24.5	12	294
30 – 39	34.5	14	483
40 – 49	44.5	10	445
50 – 59	54.5	8	436
		Total: 50	Total: 1745

The mean of this collection of grouped data is

$$\bar{x} = \frac{1745}{50} = 34.9.$$

11. Arrange the numbers in numerical order, from smallest to largest.

$$35, \ 36, \ 36, \ 38, \ 38, \ 42, \ 44, \ 48$$

There are 8 numbers here; the median is the mean of the 2 middle numbers, which is

$$\frac{38 + 38}{2} = \frac{76}{2} = 38.$$

The mode is the number that occurs most often. Here, there are two modes, 36 and 38, since they both appear twice.

13. The modal class for the distribution of Exercise 8 is the interval 55 - 59, since it contains more data values than any of the other intervals.

17. The range is 93 - 26 = 67, the difference of the highest and lowest numbers in the distribution.

The mean is $\bar{x} = \frac{1}{10}(26 + 43 + 51 + 29 + 37 + 56 + 29 + 82 + 74 + 93) = \frac{520}{10} = 52.$

Construct a table with the values of x, $x - \bar{x}$, and $(x - \bar{x})^2$.

x	$x - \bar{x}$	$(x - \bar{x})^2$
26	-26	676
43	-9	81
51	-1	1
29	-23	529
37	-15	225
56	4	16
29	-23	529
82	30	900
74	22	484
93	41	1681
	Total:	5122

The standard deviation is $s = \sqrt{\frac{5122}{10 - 1}} \approx \sqrt{569.1} \approx 23.9.$

19. Start with the frequency distribution that was the answer to Exercise 8, and expand the table to include columns for the midpoint x of each interval, and for fx, x^2, and fx^2.

Interval	f	x	fx	x^2	fx^2
40 – 44	2	42	84	1764	3528
45 – 49	5	47	235	2209	11,045
50 – 54	7	52	364	2704	18,928
55 – 59	10	57	570	3249	32,490
60 – 64	4	62	248	3844	15,376
65 – 69	1	67	67	4489	4489
Totals:	29		1568		85,856

The mean of the grouped data is

$$\bar{x} = \frac{\Sigma fx}{n} = \frac{1568}{29} \approx 54.07.$$

The standard deviation for the grouped data is

$$s = \sqrt{\frac{\Sigma fx^2 - n\bar{x}^2}{n - 1}} = \sqrt{\frac{85,856 - 29(54.07)^2}{29 - 1}} \approx \sqrt{38.3} \approx 6.2.$$

21. A skewed distribution has the largest frequency at one end rather than in the middle.

23. (a) Using the standard normal curve table,

$$P(-2.5 \le z \le 2.5) = .9938 - .0062 = .9876,$$

so 98.76% of the distribution is within 2.5 standard deviations of the mean.

(b) According to Chebyshev's theorem,

$$P(-2.5 \le z \le 2.5) \ge 1 - \frac{1}{(2.5)^2} \approx .84,$$

so at least 84% of the distribution is within 2.5 standard deviations of the mean.

25. Using the standard normal curve table,

$$P(z < .41) = .6591.$$

27. $P(1.53 \leq z \leq 2.82)$

$= P(z \leq 2.82) - P(z \leq 1.53)$

$= .9976 - .9370$

$= .0606$

29. The normal distribution is not a good approximation of a binomial distribution that has a value of p close to 0 or 1 because the histogram of such a binomial distribution is skewed and therefore not close to the shape of a normal distribution.

31. $P(x < 32)$

$= P\left(z < \dfrac{32 - 32.1}{.1}\right)$

$= P(z < -1)$

$= .1587$ or 15.87%

33. $n = 500$, $p = .06$

To approximate the binomial distribution, use a normal distribution with

$\mu = np = 500(.06) = 30$ and
$\sigma = \sqrt{np(1 - p)}$
$= \sqrt{28.2}$
$\approx 5.31.$

(a) 25 or fewer overstuffed corresponds to the area under the normal curve to the left of $x = 25.5$. The corresponding z-score is

$$z = \dfrac{25.5 - 30}{5.3} = -.85.$$

$P(25$ or fewer overstuffed$)$

$= P(x \leq 25.5)$

$= P(z < -.85)$

$= .1977$

(b) $P($exactly 30 overstuffed$)$

$= P(29.5 \leq x \leq 30.5)$

$= P(-.09 \leq z \leq .09)$

$= .5359 - .4641$

$= .0718$

(c) $P($more than 40 overstuffed$)$

$= P(x \geq 40.5)$

$= P(z \geq 1.98)$

$= 1 - .9761$

$= .0239$

35. The table below records the mean and standard deviation for diet A and for diet B.

	μ	σ
diet A	2.7	2.14
diet B	1.3	.9

(a) Diet A had the greater mean gain, since the mean for diet A is larger.

(b) Diet B had a more consistent gain, since diet B has a smaller standard deviation.

37. $n = 100$, $p = .98$

We have $\mu = np = 100(.98) = 98$ and

$\sigma = \sqrt{np(1 - p)} = \sqrt{100(.98)(.02)}$
$= 1.4.$

At least 95% or .95(100) = 95 of the flies corresponds to the area under the normal curve to the right of x = 94.5.   The corresponding z-score is

$$z = \frac{94.5 - 98}{1.4} = -2.5.$$

$$
\begin{aligned}
P(\text{at least 95 flies are killed}) &= P(x \geq 94.5) \\
&= P(z \geq -2.5) \\
&= 1 - P(z \leq -2.5) \\
&= 1 - .0062 \\
&= .9938
\end{aligned}
$$

39.   n = 100, p = .98

As in Exercise 37, we have $\mu = 98$ and $\sigma = 1.4$.   All the flies, or 100 flies, corresponds to the area under the normal curve between x = 99.5 and x = 100.5. The corresponding z-scores are

$$z = \frac{99.5 - 98}{1.4} \approx 1.07 \quad \text{and}$$

$$z = \frac{100.5 - 98}{1.4} \approx 1.79.$$

$$
\begin{aligned}
P(\text{all the flies are killed}) &= P(99.5 \leq x \leq 100.5) \\
&= P(1.07 \leq z \leq 1.79) \\
&= P(z \leq 1.79) - P(z \leq 1.07) \\
&= .9633 - .8577 \\
&= .1056
\end{aligned}
$$

41. $\mu = 42$, $\sigma = 12$

Find the z-score for z = 35.

$$z = \frac{35 - 42}{12} \approx -.58$$

$$P(x \leq 35) = P(z \leq -.58) = .2810$$

28.10% of the residents commute no more than 35 min per day.

43. $\mu = 42$, $\sigma = 12$

For x = 38, z ≈ −.33.

For x = 60, z = 1.5.

$$
\begin{aligned}
P(38 \leq x \leq 60) &= P(-.33 \leq z \leq 1.5) \\
&= P(z \leq 1.5) - P(z \leq -.33) \\
&= .9332 - .3707 \\
&= .5625
\end{aligned}
$$

56.25% of the residents commute between 38 and 60 min per day.

45.

Interval	x	Tally	f	x · f
1 – 3	2	卌 I	6	12
4 – 6	5	卌	5	25
7 – 9	8	卌 卌 I	11	88
10 – 12	11	卌 卌 卌 卌	20	220
13 – 15	14	卌 I	6	84
16 – 18	17	II	2	34
Totals			50	463

x	P(x)	x · p(x)
0	.001	0
1	.010	.010
2	.044	.088
3	.117	.351
4	.205	.820
5	.246	1.230
6	.205	1.230
7	.117	.819
8	.044	.352
9	.010	.090
10	.001	.010
Totals:	1.000	5.000

(a) The mean of the frequency distribution is

$$\overline{x} = \frac{\Sigma(x \cdot f)}{n} = \frac{463}{50} = 9.26 \approx 9.3.$$

The expected value of the probability distribution is

$$E(x) = \Sigma(x \cdot P(x)) = 5.$$

(b) The standard deviation of the frequency distribution is

$$s = \sqrt{\frac{\Sigma fx^2 - n\overline{x}^2}{n - 1}} = \sqrt{\frac{5027 - 50(9.26)^2}{50 - 1}} \approx 3.9.$$

The standard deviation of the probability distribution is

$$s = \sqrt{np(1 - p)} = \sqrt{10(.5)(.5)} \approx 1.58.$$

(c) Use k = 2 in Chebyshev's theorem since $1 - \frac{1}{2^2} = \frac{3}{4} = 75\%$. For the frequency distribution, the interval $\overline{x} - 2s \le x \le \overline{x} + 2s$ becomes

$$9.3 - 2(3.9) \le x \le 9.3 + 2(3.9)$$

$$\text{or} \qquad\qquad 1.5 \le x \le 17.1.$$

For the probability distribution, the interval is

$$5 - 2(1.58) \le x \le 5 + 2(1.58) \quad \text{or} \quad 1.84 \le x \le 8.16.$$

(d) 95.44% of the area under the normal approximation of the binomial probability distribution will lie between z = -2 and z = 2.

$$z = -2 \text{ means } -2 = \frac{x - 5}{1.58} \quad \text{or} \quad 1.84 = x.$$

$$z = 2 \text{ means } 2 = \frac{x - 5}{1.58} \quad \text{or} \quad 8.16 = x.$$

The interval is $1.84 \le x \le 8.16$.

(e) The normal distribution can't be used to answer probability questions about the frequency distribution because the histogram of the frequency distribution is not close enough to the shape of a normal curve.

CHAPTER 7   TEST

1. A certain species of fruit fly is 40% red–winged and 60% black–winged. Let x represent the number of red–winged flies. Compute the probability distribution for x if 4 flies are chosen at random.

2. The following table gives the probability distribution for a random variable x. Find the expected value for x.

x	3	4	5	7
P(x)	.1	.3	.5	.1

3. Find the mean and standard deviation for the set of numbers given below. Also find the median and the mode.

   3, 3, 4, 4, 4, 5, 6, 6, 6, 6

4. Find the mean and standard deviation of a binomial distribution with n = 56 and p = .4.

5. A distribution of 100 incomes has μ = $10,500 and σ = $1000. Use Chebyshev's theorem to find:

   (a) A range of incomes which would include at least 75 of the 100 incomes.

   (b) The minimum number of incomes we would expect to find in the interval from $7500 to $13,500.

6. Using the normal curve table, find the percent of the total area under the normal curve in each case:

   (a) z between –1.13 and 2.14.   (b) z greater than 2.1.

7. Al's Quick Photo store develops an average roll of film in 2.3 min. The standard deviation in time is .6 min. Assume a normal distribution.

   (a) Out of 1000 rolls of film, how many will be developed in less than 3.5 min?

   (b) What percentage of film is developed in between 2 and 3 min?

8. A loaded coin with P(head) = .6 is tossed 100 times. Using the normal curve approximation to the binomial distribution, find the probability of getting each of the following:

   (a) at least 55 heads.   **(b)** exactly 61 heads.

   (c) between 55 and 65 heads (inclusive).

CHAPTER 7 TEST ANSWERS

1.

x	0	1	2	3	4
P(x)	.1296	.3456	.3456	.1536	.0256

2. E(x) = 4.7   3. $\mu = 4.7$, $\sigma \approx 1.25$ 4. $\mu = 22.4$, $\sigma \approx 3.67$

Median = 4.5, mode = 6

5. **(a)** $8500 to $12,500   **(b)** 88   6. **(a)** .8546   **(b)** .0179

7. **(a)** 977   **(b)** 57%   8. **(a)** .8686   **(b)** .0819   **(c)** .7372

CHAPTER 8   MARKOV CHAINS

Section 8.1

1. $\begin{bmatrix} \frac{2}{3} & \frac{1}{2} \end{bmatrix}$ could not be a probability vector because the sum of the entries in the row is not equal to 1.

3. [0   1] could be a probability vector since it is a matrix of only one row, having nonnegative entries whose sum is 1.

5. [.4   .2   0] could not be a probability vector because the sum of the entries in the row is not equal to 1.

7. [.07 .04 .37 .52] could be a probability vector. It is a matrix of only one row, having nonnegative entries whose sum is 1.

9. [0, −.2 .6 .6] could not be a probability vector because it has a negative entry.

11.
$$\begin{array}{cc} & \begin{array}{cc} A & B \end{array} \\ \begin{array}{c} A \\ B \end{array} & \begin{bmatrix} \frac{2}{3} & \frac{1}{3} \\ 1 & 0 \end{bmatrix} \end{array}$$

This could be a transition matrix since it is a square matrix, all entries are between 0 and 1, inclusive, and the sum of the entries in each row is 1.

To draw the transition diagram, give names to the two states (such as A and B) and label the probabilities of going from one state to another. In this case,

$$P_{AA} = \frac{2}{3}, \qquad P_{AB} = \frac{1}{3},$$

$$P_{BA} = 1, \text{ and } P_{BB} = 0.$$

See the transition diagram in the back of the textbook.

13. $\begin{bmatrix} \frac{1}{4} & \frac{3}{4} & 0 \\ 2 & 0 & 1 \\ 1 & \frac{2}{3} & 3 \end{bmatrix}$

This could not be a transition matrix because it has entries that are greater than 1.

15. $\begin{bmatrix} \frac{1}{3} & \frac{1}{2} & 1 \\ 0 & 1 & 0 \\ \frac{1}{2} & \frac{1}{2} & 1 \end{bmatrix}$

This could not be a transition matrix because the sum of the entries in the first row and in the third row is more than 1.

17. The transition diagram provides the information

$$P_{AA} = .9, \ P_{AB} = .1, \ P_{AC} = 0,$$

$$P_{BA} = .1, \ P_{BB} = .6, \ P_{BC} = .3,$$

$$P_{CA} = 0, \ P_{CB} = .3, \text{ and } P_{CC} = .7.$$

The transition matrix associated with this diagram is

$$
\begin{array}{c} \\ A \\ B \\ C \end{array}
\begin{array}{ccc} A & B & C \end{array}
\left[\begin{array}{ccc} .9 & .1 & 0 \\ .1 & .6 & .3 \\ 0 & .3 & .7 \end{array}\right].
$$

19. $A = \left[\begin{array}{cc} 1 & 0 \\ .8 & .2 \end{array}\right]$

$$
A^2 = \left[\begin{array}{cc} 1 & 0 \\ .8 & .2 \end{array}\right]\left[\begin{array}{cc} 1 & 0 \\ .8 & .2 \end{array}\right] = \left[\begin{array}{cc} 1 & 0 \\ .96 & .04 \end{array}\right]
$$

$$
A^3 = \left[\begin{array}{cc} 1 & 0 \\ .8 & .2 \end{array}\right]\left[\begin{array}{cc} 1 & 0 \\ .96 & .04 \end{array}\right]
$$

$$
= \left[\begin{array}{cc} 1 & 0 \\ .992 & .008 \end{array}\right]
$$

The entry in row 1, column 2 of A^3 gives the probability that state 1 changes to state 2 after 3 repetitions of the experiment. This probability is 0.

21. $C = \left[\begin{array}{cc} .5 & .5 \\ .72 & .28 \end{array}\right]$

$$
C^2 = \left[\begin{array}{cc} .5 & .5 \\ .72 & .28 \end{array}\right]\left[\begin{array}{cc} .5 & .5 \\ .72 & .28 \end{array}\right]
$$

$$
= \left[\begin{array}{cc} .61 & .39 \\ .5616 & .4384 \end{array}\right]
$$

$$
C^3 = \left[\begin{array}{cc} .5 & .5 \\ .72 & .28 \end{array}\right]\left[\begin{array}{cc} .61 & .39 \\ .5616 & .4384 \end{array}\right]
$$

$$
= \left[\begin{array}{cc} .5858 & .4142 \\ .596448 & .403552 \end{array}\right]
$$

The probability that state 1 changes to state 2 after 3 repetitions is .4142, since that is the entry in row 1, column 2 of C^3.

23. $E = \left[\begin{array}{ccc} .8 & .1 & .1 \\ .3 & .6 & .1 \\ 0 & 1 & 0 \end{array}\right]$

$$
E^2 = \left[\begin{array}{ccc} .8 & .1 & .1 \\ .3 & .6 & .1 \\ 0 & 1 & 0 \end{array}\right]\left[\begin{array}{ccc} .8 & .1 & .1 \\ .3 & .6 & .1 \\ 0 & 1 & 0 \end{array}\right]
$$

$$
= \left[\begin{array}{ccc} .67 & .24 & .09 \\ .42 & .49 & .09 \\ .3 & .6 & .1 \end{array}\right]
$$

$$
E^3 = \left[\begin{array}{ccc} .8 & .1 & .1 \\ .3 & .6 & .1 \\ 0 & 1 & 0 \end{array}\right]\left[\begin{array}{ccc} .67 & .24 & .09 \\ .42 & .49 & .09 \\ .3 & .6 & .1 \end{array}\right]
$$

$$
= \left[\begin{array}{ccc} .608 & .301 & .091 \\ .483 & .426 & .091 \\ .42 & .49 & .09 \end{array}\right]
$$

The probability that state 1 changes to state 2 after 3 repetitions is .301, since that is the entry in row 1, column 2 of E^3.

25. **(a)** We are asked to show that $I(P^n) = (\cdots(((IP)P)P)\cdots P)$ is true for any natural number n, where the expression on the right side of the equation has a total of n factors of P. This may be proven by mathematical induction on n.

When n = 1, the statement becomes $I(P) = (IP)$, which is obviously true. When n = 2, the statement becomes $(P^2) = (IP)P$, or $I(PP) = (IP)P$, which is true since matrix multiplication is associative. Next, assume the nth statement is true in order to show that (n + 1)st statement is true. That is, assume that

$$I(P^n) = (\cdots(((IP)P)P)\cdots P)$$

is true. Associativity plays a role here also.

$$I(P^{n+1}) = I(P^n \cdot P)$$
$$= (IP^n)P$$
$$= (\cdots(((IP)P)P)\cdots P)P$$

Conclude that

$$I(P^n) = (\cdots(((IP)P)P)\cdots P)$$

is true for any natural number n.

27. The probability vector is [.4   .6].

(a) $[.4 \quad .6]\begin{bmatrix} .8 & .2 \\ .35 & .65 \end{bmatrix}$

$= [.53 \quad .47]$

Thus, after 1 week Johnson has 53% market share, and Northclean has a 47% share.

(b) $C^2 = \begin{bmatrix} .8 & .2 \\ .35 & .65 \end{bmatrix}\begin{bmatrix} .8 & .2 \\ .35 & .65 \end{bmatrix}$

$= \begin{bmatrix} .71 & .29 \\ .5075 & .4925 \end{bmatrix}$

$[.4 \quad .6]\begin{bmatrix} .71 & .29 \\ .5075 & .4925 \end{bmatrix}$

$= [.5885 \quad .4115]$

After 2 weeks, Johnson has a 58.85% market share, and Northclean has a 41.15% share.

(c) $C^3 = \begin{bmatrix} .8 & .2 \\ .35 & .65 \end{bmatrix}\begin{bmatrix} .71 & .29 \\ .5075 & .4925 \end{bmatrix}$

$= \begin{bmatrix} .6695 & .3305 \\ .5784 & .4216 \end{bmatrix}$

$[.4 \quad .6]\begin{bmatrix} .6695 & .3305 \\ .5784 & .4216 \end{bmatrix}$

$= [.6148 \quad .3852]$

After 3 weeks, the shares are 61.48% and 38.52%, respectively.

(d) $C^4 = \begin{bmatrix} .8 & .2 \\ .35 & .65 \end{bmatrix}\begin{bmatrix} .6695 & .3305 \\ .5784 & .4216 \end{bmatrix}$

$= \begin{bmatrix} .6513 & .3487 \\ .6103 & .3897 \end{bmatrix}$

$[.4 \quad .6]\begin{bmatrix} .6513 & .3487 \\ .6103 & .3897 \end{bmatrix}$

$= [.62666 \quad .37334]$

After 4 weeks the shares are 62.666% and 37.334%, respectively.

29. The transition matrix P is

$$\begin{array}{c} \\ G_0 \\ G_1 \\ G_2 \end{array} \begin{array}{ccc} G_0 & G_1 & G_2 \end{array} \\ \begin{bmatrix} .85 & .1 & .05 \\ 0 & .8 & .2 \\ 0 & 0 & 1 \end{bmatrix}.$$

We have 50,000 new policy holders, all in G_0. The probability vector for these people is

$$\begin{array}{ccc} G_0 & G_1 & G_2 \end{array} \\ [1 \quad 0 \quad 0].$$

(a) After 1 yr, the distribution of people in each group is

$$[1 \quad 0 \quad 0]\begin{bmatrix} .85 & .1 & .05 \\ 0 & .8 & .2 \\ 0 & 0 & 1 \end{bmatrix}$$

$= [.85 \quad .1 \quad .05].$

There are

$(.85)(50,000) = 42,500$ people in G_0,

$(.1)(50,000) = 5000$ people in G_1, and

$(.05)(50,000) = 2500$ people in G_2.

(b)

$$P^2 = \begin{bmatrix} .85 & .1 & .05 \\ 0 & .8 & .2 \\ 0 & 0 & 1 \end{bmatrix}\begin{bmatrix} .85 & .1 & .05 \\ 0 & .8 & .2 \\ 0 & 0 & 1 \end{bmatrix}$$

$$= \begin{bmatrix} .7225 & .165 & .1125 \\ 0 & .64 & .36 \\ 0 & 0 & 1 \end{bmatrix}$$

After 2 yr, the distribution of people in each group is

$$[1 \quad 0 \quad 0]\begin{bmatrix} .7225 & .165 & .1125 \\ 0 & .64 & .36 \\ 0 & 0 & 1 \end{bmatrix}$$

$$= [.7225 \quad .165 \quad .1125].$$

There are

$(.7225)(50,000) = 36,125$ in G_0,

$(.165)(50,000) = 8250$ in G_1, and

$(.1125)(50,000) = 5625$ in G_2.

(c)

$$P^3 = \begin{bmatrix} .85 & .1 & .05 \\ 0 & .8 & .2 \\ 0 & 0 & 1 \end{bmatrix}\begin{bmatrix} .7225 & .165 & .1125 \\ 0 & .64 & .36 \\ 0 & 0 & 1 \end{bmatrix}$$

$$= \begin{bmatrix} .61413 & .20425 & .18163 \\ 0 & .512 & .488 \\ 0 & 0 & 1 \end{bmatrix}$$

After 3 yr, the distribution is

$$[1 \quad 0 \quad 0]\begin{bmatrix} .61413 & .20425 & .18163 \\ 0 & .512 & .488 \\ 0 & 0 & 1 \end{bmatrix}$$

$$= [.61413 \quad .20425 \quad .18163].$$

There are

$(50,000)(.61413) = 30,706$ in G_0,

$(50,000)(.20425) = 10,213$ in G_1, and

$(50,000)(.18163) = 9081$ in G_2.

(d)

$$P^4 = \begin{bmatrix} .85 & .1 & .05 \\ 0 & .8 & .2 \\ 0 & 0 & 1 \end{bmatrix}\begin{bmatrix} .6141 & .2043 & .1816 \\ 0 & .512 & .488 \\ 0 & 0 & 1 \end{bmatrix}$$

$$\begin{bmatrix} .52201 & .22481 & .25318 \\ 0 & .4096 & .5904 \\ 0 & 0 & 1 \end{bmatrix}$$

The probabilities are

$$[1 \quad 0 \quad 0]\begin{bmatrix} .52201 & .22481 & .25318 \\ 0 & .4096 & .5904 \\ 0 & 0 & 1 \end{bmatrix}$$

$$= [.52201 \quad .22481 \quad .25318].$$

There are

$(50,000)(.52201) = 26,100$ in G_0,

$(50,000)(.22481) = 11,241$ in G_1, and

$(50,000)(.25318) = 12,659$ in G_2.

31. $P = \begin{bmatrix} .825 & .175 & 0 \\ .060 & .919 & .021 \\ .049 & 0 & .951 \end{bmatrix}$

The initial probability vector is

$$[.26 \quad .6 \quad .14].$$

(a) The share held by each type is given by

$$[.26 \quad .6 \quad .14]\begin{bmatrix} .825 & .175 & 0 \\ .060 & .919 & .021 \\ .049 & 0 & .951 \end{bmatrix}$$

$$= [.257 \quad .597 \quad .146].$$

(b)

$$P^2 = \begin{bmatrix} .825 & .175 & 0 \\ .060 & .919 & .021 \\ .049 & 0 & .951 \end{bmatrix}\begin{bmatrix} .825 & .175 & 0 \\ .060 & .919 & .021 \\ .049 & 0 & .951 \end{bmatrix}$$

$$= \begin{bmatrix} .691 & .305 & .004 \\ .106 & .855 & .039 \\ .087 & .009 & .904 \end{bmatrix}$$

The share held by each is

$$[.26 \quad .6 \quad .14] \begin{bmatrix} .691 & .305 & .004 \\ .106 & .855 & .039 \\ .087 & .009 & .904 \end{bmatrix}$$

$$= [.255 \quad .594 \quad .151].$$

(c)

$$P^3 = \begin{bmatrix} .825 & .175 & 0 \\ .060 & .919 & .021 \\ .049 & 0 & .951 \end{bmatrix} \begin{bmatrix} .691 & .305 & .004 \\ .106 & .855 & .039 \\ .087 & .009 & .904 \end{bmatrix}$$

$$= \begin{bmatrix} .589 & .401 & .01 \\ .141 & .804 & .055 \\ .117 & .023 & .860 \end{bmatrix}$$

The share held by each after 3 yr is

$$[.26 \quad .6 \quad .14] \begin{bmatrix} .589 & .401 & .01 \\ .141 & .804 & .055 \\ .117 & .023 & .860 \end{bmatrix}$$

$$= [.254 \quad .590 \quad .156].$$

33. **(a)** If there is no one in line, then after 1 min there will be either 0, 1, or 2 people in line with probabilities $p_{00} = 1/2$, $p_{01} = 1/3$, and $p_{02} = 1/6$. If there is one person in line, then that person will be served and either 0, 1, or 2 new people will join the line, with probabilities $p_{10} = 1/2$, $p_{11} = 1/3$, and $p_{12} = 1/6$. If there are two people in line, then one of them will be served and either 1 or 2 new people will join the line, with probabilities $p_{21} = 1/2$ and $p_{22} = 1/2$; it is impossible for both people in line to be served, so $p_{20} = 0$. Therefore,

the transition matrix is

$$A = \begin{bmatrix} \frac{1}{2} & \frac{1}{3} & \frac{1}{6} \\ \frac{1}{2} & \frac{1}{3} & \frac{1}{6} \\ 0 & \frac{1}{2} & \frac{1}{2} \end{bmatrix}.$$

(b) The transition matrix for a two-minute period is

$$A^2 = \begin{bmatrix} \frac{1}{2} & \frac{1}{3} & \frac{1}{6} \\ \frac{1}{2} & \frac{1}{3} & \frac{1}{6} \\ 0 & \frac{1}{2} & \frac{1}{2} \end{bmatrix} \begin{bmatrix} \frac{1}{2} & \frac{1}{3} & \frac{1}{6} \\ \frac{1}{2} & \frac{1}{3} & \frac{1}{6} \\ 0 & \frac{1}{2} & \frac{1}{2} \end{bmatrix}$$

$$= \begin{bmatrix} \frac{5}{12} & \frac{13}{36} & \frac{2}{9} \\ \frac{5}{12} & \frac{13}{36} & \frac{2}{9} \\ \frac{1}{4} & \frac{5}{12} & \frac{1}{3} \end{bmatrix}.$$

(c) The probability that a queue with no one in line has two people in line 2 min later is 2/9, since that is the entry in row 1, column 3 of A^2.

35. **(a)** We use L for liberal, C for conservative, and I for independent. The transition matrix P is give by

$$\begin{array}{c} \\ L \\ C \\ I \end{array} \begin{array}{ccc} L & C & I \\ \begin{bmatrix} .80 & .15 & .05 \\ .20 & .70 & .10 \\ .20 & .20 & .60 \end{bmatrix} \end{array}.$$

(b) The initial distribution is given by

$$I = [.40 \quad .45 \quad .15].$$

(c) On month later in July,

$$IP = [.44 \quad .405 \quad .155],$$

so there will be 44% liberals, 40.5% conservatives, and 15.5% independents.

(d) Two months later in August,

$$IP^2 = (IP)P = [.464 \quad .3805 \quad .1555],$$

so there will be 46.4% liberals, 38.05% conservatives, and 15.55% independents.

(e) Three months later in September,

$$IP^3 = (IP^2)P = [.4784 \quad .36705 \quad .15455],$$

so there will be 47.84% liberals, 36.705% conservatives, and 15.455% independents.

(f) Four months later in October,

$$IP^4 = (IP^3)P = [.48704 \quad .359605 \quad .153355],$$

so there will be 48.704% liberals, 35.9605% conservatives, and 15.3355% independents.

37. This exercise should be solved by computer methods. The solution will vary according to the computer program that is used. The answers are as follows. The first five powers of the transition matrix are

$$A = \begin{bmatrix} .1 & .2 & .2 & .3 & .2 \\ .2 & .1 & .1 & .2 & .4 \\ .2 & .1 & .4 & .2 & .1 \\ .3 & .1 & .1 & .2 & .3 \\ .1 & .3 & .1 & .1 & .4 \end{bmatrix}, \quad A^2 = \begin{bmatrix} .2 & .15 & .17 & .19 & .29 \\ .16 & .2 & .15 & .18 & .31 \\ .19 & .14 & .24 & .21 & .22 \\ .16 & .19 & .16 & .2 & .29 \\ .16 & .19 & .14 & .17 & .34 \end{bmatrix},$$

$$A^3 = \begin{bmatrix} .17 & .178 & .171 & .191 & .29 \\ .171 & .178 & .161 & .185 & .305 \\ .18 & .163 & .191 & .197 & .269 \\ .175 & .174 & .164 & .187 & .3 \\ .167 & .184 & .158 & .182 & .309 \end{bmatrix}, \quad A^4 = \begin{bmatrix} .1731 & .175 & .1683 & .188 & .2956 \\ .1709 & .1781 & .1654 & .1866 & .299 \\ .1748 & .1718 & .1753 & .1911 & .287 \\ .1712 & .1775 & .1667 & .1875 & .2971 \\ .1706 & .1785 & .1641 & .1858 & .301 \end{bmatrix},$$

and

$$A^5 = \begin{bmatrix} .17193 & .17643 & .1678 & .18775 & .29609 \\ .17167 & .17689 & .16671 & .18719 & .29754 \\ .17293 & .17488 & .17007 & .18878 & .29334 \\ .17192 & .17654 & .16713 & .18741 & .297 \\ .17142 & .17726 & .16629 & .18696 & .29807 \end{bmatrix}.$$

The probability that state 2 changes to state 4 after 5 repetitions of the experiment is .18719, since that is the entry in row 2, column 4 of A^5.

39. This exercise should be solved by computer methods. The solution will vary according to the computer program that is used. The answers are as follows.

   (a) The proportion is .847423 or about 85%.

   (b) The proportions are as listed in the probability vector

$$[.0128 \quad .0513 \quad .0962 \quad .8397].$$

Section 8.2

1. Let $A = \begin{bmatrix} .2 & .8 \\ .9 & .1 \end{bmatrix}$.

A is a regular transition matrix since $A^1 = A$ contains all positive entries.

3. Let $B = \begin{bmatrix} 1 & 0 \\ .6 & .4 \end{bmatrix}$.

$$B^2 = \begin{bmatrix} 1 & 0 \\ .6 & .4 \end{bmatrix}\begin{bmatrix} 1 & 0 \\ .6 & .4 \end{bmatrix}$$

$$= \begin{bmatrix} 1 & 0 \\ .84 & .16 \end{bmatrix}$$

B is not regular since any power of B will have [1   0] as its first row and thus cannot have all positive entries.

5. Let $P = \begin{bmatrix} 0 & 1 & 0 \\ .4 & .2 & .4 \\ 1 & 0 & 0 \end{bmatrix}$.

$$P^2 = \begin{bmatrix} 0 & 1 & 0 \\ .4 & .2 & .4 \\ 1 & 0 & 0 \end{bmatrix}\begin{bmatrix} 0 & 1 & 0 \\ .4 & .2 & .4 \\ 1 & 0 & 0 \end{bmatrix}$$

$$= \begin{bmatrix} .4 & .2 & .4 \\ .48 & .44 & .08 \\ 0 & 1 & 0 \end{bmatrix}$$

$$P^3 = \begin{bmatrix} 0 & 1 & 0 \\ .4 & .2 & .4 \\ 1 & 0 & 0 \end{bmatrix}\begin{bmatrix} .4 & .2 & .4 \\ .48 & .44 & .08 \\ 0 & 1 & 0 \end{bmatrix}$$

$$= \begin{bmatrix} .48 & .44 & .08 \\ .256 & .568 & .176 \\ .4 & .2 & .4 \end{bmatrix}$$

P is a regular transition matrix since P^3 contains all positive entries.

7. Let $P = \begin{bmatrix} \frac{1}{4} & \frac{3}{4} \\ \frac{1}{2} & \frac{1}{2} \end{bmatrix}$ and let V be the probability vector $[v_1 \quad v_2]$. We want to find V such that

$$VP = V,$$

or $[v_1 \quad v_2]\begin{bmatrix} \frac{1}{4} & \frac{3}{4} \\ \frac{1}{2} & \frac{1}{2} \end{bmatrix} = [v_1 \quad v_2].$

Use matrix multiplication on the left and obtain

$$\left[\frac{1}{4}v_1 + \frac{1}{2}v_2 \quad \frac{3}{4}v_1 + \frac{1}{2}v_2\right] = [v_1 \quad v_2].$$

Set corresponding entries from the two matrices equal to get

$$\frac{1}{4}v_1 + \frac{1}{2}v_2 = v_1$$

$$\frac{3}{4}v_1 + \frac{1}{2}v_2 = v_2.$$

Multiply both equations by 4 to eliminate fractions.

$$v_1 + 2v_2 = 4v_1$$
$$3v_1 + 2v_2 = 4v_2$$

Simplify both equations.

$$-3v_1 + 2v_2 = 0$$
$$3v_1 - 2v_2 = 0$$

This is a dependent system. To find the values of v_1 and v_2, an additional equation is needed. Since $V = [v_1 \quad v_2]$ is a probability vector,

$$v_1 + v_2 = 1.$$

To find v_1 and v_2, solve the system

$$-3v_1 + 2v_2 = 0 \quad (1)$$
$$v_1 + v_2 = 1. \quad (2)$$

From equation (2), $v_1 = 1 - v_2$. Substitute $1 - v_2$ for v_1 in equation (1) to obtain

$$-3(1 - v_2) + 2v_2 = 0$$
$$-3 + 3v_2 + 2v_2 = 0$$
$$-3 + 5v_2 = 0$$
$$v_2 = \frac{3}{5}.$$

Since $v_1 = 1 - v_2$, $v_1 = 2/5$, and the equilibrium vector is

$$V = [\tfrac{2}{5} \quad \tfrac{3}{5}].$$

9. Let $P = \begin{bmatrix} .3 & .7 \\ .4 & .6 \end{bmatrix}$ and let V be the probability vector $[v_1 \quad v_2]$. We want to find V such that

$$VP = V,$$

or $[v_1 \quad v_2]\begin{bmatrix} .3 & .7 \\ .4 & .6 \end{bmatrix} = [v_1 \quad v_2].$

By matrix multiplication and equality of matrices,

$$.3v_1 + .4v_2 = v_1$$
$$.7v_1 + .6v_2 = v_2.$$

Simplify these equations to get the dependent system

$$-.7v_1 + .4v_2 = 0$$
$$.7v_1 - .4v_2 = 0.$$

Since V is a probability vector,

$$v_1 + v_2 = 1.$$

To find v_1 and v_2, solve the system

$$.7v_1 - .4v_2 = 0$$
$$v_1 + v_2 = 0$$

by the substitution method. Observe that $v_2 = 1 - v_1$.

$$.7v_1 - .4(1 - v_1) = 0$$
$$1.1v_1 - .4 = 0$$
$$v_1 = \frac{.4}{1.1} = \frac{4}{11}$$
$$v_2 = 1 - \frac{4}{11} = \frac{7}{11}$$

The equilibrium vector is

$$[\tfrac{4}{11} \quad \tfrac{7}{11}].$$

11. Let V be the probability vector
$[v_1 \quad v_2 \quad v_3]$.

$$[v_1 \quad v_2 \quad v_3]\begin{bmatrix} .1 & .1 & .8 \\ .4 & .4 & .2 \\ .1 & .2 & .7 \end{bmatrix}$$

$$= [v_1 \quad v_2 \quad v_3]$$

$$.1v_1 + .4v_2 + .1v_3 = v_1$$
$$.1v_1 + .4v_2 + .2v_3 = v_2$$
$$.8v_1 + .2v_2 + .7v_3 = v_3$$

Simplify these equations to get the dependent system

$$-.9v_1 + .4v_2 + .1v_3 = 0$$
$$.1v_1 - .6v_2 + .2v_3 = 0$$
$$.8v_1 + .2v_2 - .3v_3 = 0.$$

Since V is a probability vector,

$$v_1 + v_2 + v_3 = 1.$$

Solving the above system of 4 equations using the Gauss–Jordan method, we obtain

$$v_1 = \frac{14}{83}, \quad v_2 = \frac{19}{83}, \quad v_3 = \frac{50}{83}.$$

Thus, the equilibrium vector is

$$\left[\frac{14}{83} \quad \frac{19}{83} \quad \frac{50}{83}\right].$$

13. Let V be the probability vector
$[v_1 \quad v_2 \quad v_3]$.

$$[v_1 \quad v_2 \quad v_3]\begin{bmatrix} .25 & .35 & .4 \\ .1 & .3 & .6 \\ .55 & .4 & .05 \end{bmatrix}$$

$$= [v_1 \quad v_2 \quad v_3]$$

$$.25v_1 + .1v_2 + .55v_3 = v_1$$
$$.35v_1 + .3v_2 + .4v_3 = v_2$$
$$.4v_1 + .6v_2 + .05v_3 = v_3$$

Simplify these equations to get the dependent system

$$-.75v_1 + .1v_2 + .55v_3 = 0$$
$$.35v_1 - .7v_2 + .4v_3 = 0$$
$$.4v_1 + .6v_2 - .95v_3 = 0.$$

Since V is a probability vector,

$$v_1 + v_2 + v_3 = 1.$$

Solving this system we obtain

$$v_1 = \frac{170}{563}, \quad v_2 = \frac{197}{563}, \quad v_3 = \frac{196}{563}.$$

Thus, the equilibrium vector is

$$\left[\frac{170}{563} \quad \frac{197}{563} \quad \frac{196}{563}\right].$$

15. $$[v_1 \quad v_2 \quad v_3]\begin{bmatrix} .85 & .1 & .05 \\ 0 & .8 & .2 \\ 0 & 0 & 1 \end{bmatrix}$$

$$= [v_1 \quad v_2 \quad v_3]$$

$$.85v_1 = v_1$$
$$.1v_1 + .8v_2 = v_2$$
$$.05v_1 + .2v_2 + v_3 = v_3$$

We also have $v_1 + v_2 + v_3 = 1$.
Solving this system, we obtain

$$v_1 = 0, \quad v_2 = 0, \quad v_3 = 1.$$

The equilibrium vector is

$$[0 \quad 0 \quad 1].$$

17. $$[v_1 \quad v_2 \quad v_3]\begin{bmatrix} .825 & .175 & 0 \\ .060 & .919 & .021 \\ .049 & 0 & .951 \end{bmatrix}$$

$$= [v_1 \quad v_2 \quad v_3]$$

$$.825v_1 + .060v_2 + .049v_3 = v_1$$
$$.175v_1 + .919v_2 = v_2$$
$$.021v_2 + .951v_3 = v_3$$

Also, $v_1 + v_2 + v_3 = 1$.

Solving this system, we obtain

$$v_1 = .244, \quad v_2 = .529, \quad v_3 = .227.$$

The equilibrium vector is

$$[.244 \quad .529 \quad .227].$$

19. $[v_1 \quad v_2 \quad v_3] \begin{bmatrix} .80 & .15 & .05 \\ 0 & .90 & .10 \\ .10 & .20 & .70 \end{bmatrix}$

$= [v_1 \quad v_2 \quad v_3]$

$.80v_2 \qquad + .10v_3 = v_1$

$.15v_1 + .90v_2 + .20v_3 = v_2$

$.05v_1 + .10v_2 + .70v_3 = v_3$

Also, $v_1 + v_2 + v_3 = 1$.

Solving this system, we obtain

$$v_1 = \frac{1}{2}, \quad v_2 = \frac{7}{20}, \quad v_3 = \frac{3}{20}.$$

The equilibrium vector is

$$\left[\frac{1}{2} \quad \frac{7}{20} \quad \frac{3}{20}\right].$$

21. Let V be the probability vector $[x_1 \quad x_2]$. We want to find V such that

$$V\begin{bmatrix} p & 1 - p \\ 1 - q & q \end{bmatrix} = V.$$

The system of equations is

$$px_1 + (1 - q)x_2 = x_1$$
$$(1 - p)x_1 + qx_2 = x_2.$$

Collecting like terms and simplifying leads to

$$(p - 1)x_1 + (1 - q)x_2 = 0,$$

so $$x_1 = \frac{1 - q}{1 - p}x_2.$$

Substituting this into $x_1 + x_2 = 1$, we obtain

$$\frac{1 - q}{1 - p}x_2 + x_2 = 1$$

or $$\frac{2 - p - q}{1 - p}x_2 = 1;$$

therefore,

$$x_2 = \frac{1 - p}{2 - p - q}$$

and $$x_1 = \frac{1 - q}{2 - p - q},$$

so

$$V = \left[\frac{1 - q}{2 - p - q} \quad \frac{1 - p}{2 - p - q}\right].$$

Since $0 < p < 1$ and $0 < q < 1$, the matrix is always regular.

23. Let V be the probability vector $[x_1 \quad x_2]$.

We have $P = \begin{bmatrix} a_{11} & a_{12} \\ a_{21} & a_{22} \end{bmatrix}$,

where $a_{11} + a_{21} = 1$
and $a_{12} + a_{22} = 1$.

The resulting equations are

$$a_{11}x_1 + a_{21}x_2 = x_1$$
$$a_{12}x_1 + a_{22}x_2 = x_2,$$

which we simplify to

$$(a_{11} - 1)x_1 + a_{21}x_2 = 0$$
$$a_{12}x_1 + (a_{22} - 1)x_2 = 0.$$

Hence, $x_1 = \frac{a_{21}}{1 - a_{11}}x_2$

$$= \frac{a_{21}}{a_{21}}x_2 = x_2,$$

which we substitute into $x_1 + x_2 = 1$, obtaining

$$x_2 + x_2 = 1$$
$$2x_2 = 1$$
$$x_2 = \frac{1}{2}$$

and therefore $x_1 = 1 - \frac{1}{2} = \frac{1}{2}$.

The equilibrium vector is $[\frac{1}{2} \quad \frac{1}{2}]$.

25. The transition matrix for the given information is

	Works	Doesn't work
Works	.9	.1
Dosen't Work	.7	.3

Let V be the probability vector $[v_1 \quad v_2]$.

$$[v_1 \quad v_2]\begin{bmatrix} .9 & .1 \\ .7 & .3 \end{bmatrix} = [v_1 \quad v_2]$$

$$.9v_1 + .7v_2 = v_1$$
$$.1v_1 + .3v_2 = v_2$$

Simplify these equations to get the dependent system

$$-.1v_1 + .7v_2 = 0$$
$$.1v_1 - .7v_2 = 0.$$

Also, $v_1 + v_2 = 1$, so $v_1 = 1 - v_2$.

$$.1(1 - v_2) - .7v_2 = 0$$
$$.1 - .8v_2 = 0$$

$$v_2 = 1/8, \quad v_1 = 7/8$$

The equilibrium vector is

$$[\frac{7}{8} \quad \frac{1}{8}].$$

The long-range probability that the line will work correctly is 7/8.

27. The transition matrix is

	Low	Medium	High
Low	.5	.4	.1
Medium	.25	.45	.3
High	.05	.4	.55

Let V be the probability vector $[v_1 \quad v_2 \quad v_3]$.

$$[v_1 \quad v_2 \quad v_3]\begin{bmatrix} .5 & .4 & .1 \\ .25 & .45 & .3 \\ .05 & .4 & .55 \end{bmatrix}$$

$$= [v_1 \quad v_2 \quad v_3]$$

$$.5v_1 + .25v_2 + .05v_3 = v_1$$
$$.4v_1 + .45v_2 + .4v_3 = v_2$$
$$.1v_1 + .3v_2 + .55v_3 = v_3$$

Also, $v_1 + v_2 + v_3 = 1$.
Solving this system, we obtain

$$v_1 = \frac{51}{209}, \quad v_2 = \frac{88}{209}, \quad v_3 = \frac{70}{209}.$$

The equilibrium vector is

$$[\frac{51}{209} \quad \frac{88}{209} \quad \frac{70}{209}].$$

Thus, the long-range proportions for low, medium, and high producers are 51/209, 88/209, and 70/209, respectively.

29. Let F, C, and R represent fair, cloudy, and rainy, respectively, and let $[x_1 \quad x_2 \quad x_3]$ be a probability vector.
The transition matrix is

	F	C	R
F	.60	.25	.15
C	.40	.35	.25
R	.35	.40	.25

and the resulting system of equations is

$$.60x_1 + .40x_2 + .35x_3 = x_1$$
$$.25x_1 + .35x_2 + .40x_3 = x_2$$
$$.15x_1 + .25x_2 + .25x_3 = x_3.$$

Also, $x_1 + x_2 + x_3 = 1.$
Solving this system, we obtain

$$x_1 = \frac{155}{318}, \quad x_2 = \frac{99}{318}, \quad x_3 = \frac{32}{159}.$$

The equilibrium vector is

$$\left[\frac{155}{318} \quad \frac{99}{318} \quad \frac{32}{159}\right].$$

Over the long term, the proportion of days that are expected to be fair, cloudy, and rainy are $\frac{155}{318} \approx 48.7\%$, $\frac{99}{318} \approx 31.1\%$, and $\frac{32}{159} \approx 20.1\%$, respectively.

31. The transition matrix is

$$\begin{bmatrix} p & 1-p \\ 1-p & p \end{bmatrix}.$$

The columns sum to $p + (1-p) = 1$, so by Exercise 23 equilibrium vector is $[\frac{1}{2} \quad \frac{1}{2}].$

The long-range prediction for the fraction of the people who will hear the decision correctly is 1/2.

33. The transition matrix is

$$P = \begin{bmatrix} .12 & .88 \\ .54 & .46 \end{bmatrix}. \text{ Let V be the}$$

probability vector $[v_1 \quad v_2].$

$$[v_1 \quad v_2]\begin{bmatrix} .12 & .88 \\ .54 & .46 \end{bmatrix} = [v_1 \quad v_2]$$

$$.12v_1 + .54v_2 = v_1$$
$$.88v_1 + .46v_2 = v_2$$

Simplify these equations to get the dependent system

$$-.88v_1 + .54v_2 = 0$$
$$.88v_1 - .54v_2 = 0.$$

Also, $v_1 + v_2 = 1.$
Solving this system, we obtain

$$v_1 = \frac{27}{71} \text{ and } v_2 = \frac{44}{71}, \text{ and note}$$

that $\frac{27}{71} \approx .38 = 38\%.$

About 38% of letters in English text are expected to be vowels.

35. This exercise should be solved by computer methods. The solution will vary according to the computer program that is used. The answer is

V = [.171898 .176519 .167414
     .187526 .296644]

37. This exercise should be solved by computer methods. The solution will vary according to the computer program that is used. The answers are as follows.

(a)

[.4 .6]; [.53 .47]; [.5885 .4115];
[.614825 .385175]; [.626671 .373329];
[.632002 .367998]; [.634401 .365599];
[.635480 .364520]; [.635966 .364034];
[.636185 .363815]

(b) .24; .11; .048; .022; .0097;
.0044; .0020; .00088; .00040; .00018

(c) The ratio is roughly .45 for each week.

(d) Each week, the difference between the probability vector and the equilibrium vector is slightly less than half of what it was the previous week.

(e) [.75 .25]; [.6875 .3125]; [.659375 .340625]; [.641023 .358977]; [.638461 .361539]; [.637307 .362693]; [.636788 .363212]; [.636555 .363445]; [.636450 .363550];

.11; .051; .023; .010; .0047; .0021; .00094; .00042; .00019; .000086

The ratio is roughly .45 for each week, which is the same conclusion as before.

Section 8.3

1.
$$\begin{array}{c c c c} & 1 & 2 & 3 \\ 1 & \left[.15 \right. & .05 & .8 \\ 2 & 0 & 1 & 0 \\ 3 & \left. .4 \right. & .6 & 0 \end{array}$$

Since $p_{22} = 1$, state 2 is absorbing. There is a probability of .05 of going from state 1 to state 2 and a probability of .6 of going from state 3 to state 2, so it is possible to go from each nonabsorbing state to the absorbing state. Thus, this is the transition matrix of an absorbing Markov chain.

3.
$$\begin{array}{c c c c} & 1 & 2 & 3 \\ 1 & \left[.4 \right. & 0 & .6 \\ 2 & 0 & 1 & 0 \\ 3 & \left. .9 \right. & 0 & .1 \end{array}$$

Since $p_{22} = 1$, state 2 is absorbing. Since $p_{12} = 0$ and $p_{32} = 0$, it is not possible to go from either of the nonabsorbing states (state 1 and state 3) to the absorbing state (state 2). Thus, this is not the transition matrix of an absorbing Markov chain.

5.
$$\begin{array}{c c c c c} & 1 & 2 & 3 & 4 \\ 1 & \left[.2 \right. & .5 & .1 & .2 \\ 2 & 0 & 1 & 0 & 0 \\ 3 & .9 & .02 & .04 & .04 \\ 4 & \left. 0 \right. & 0 & 0 & 1 \end{array}$$

Since $p_{22} = 1$ and $p_{44} = 1$, states 2 and 4 are absorbing. It is possible to get from state 1 to states 2 and 4, and from state 3 to states 2 and 4. Thus, this is the transition matrix of an absorbing Markov chain.

7.
$$\begin{array}{c c c c c} & 1 & 2 & 3 & 4 \\ 1 & \left[.1 \right. & .8 & 0 & .1 \\ 2 & 0 & 1 & 0 & 0 \\ 3 & 1 & 0 & 0 & 0 \\ 4 & \left. 0 \right. & 0 & 0 & 1 \end{array}$$

Since $p_{22} = 1$ and $p_{44} = 1$, states 2 and 4 are absorbing. It is possible to go from state 1 to states 2 and 4 and from state 3 to states 2 and 4 through state 1. Thus, this is the transition matrix of an absorbing Markov chain.

9. $P = \begin{bmatrix} 1 & 0 & 0 \\ 0 & 1 & 0 \\ \hline .2 & .3 & .5 \end{bmatrix}$

Here R = [.2   .3] and Q = [.5].
Find the fundamental matrix F.

$$F = [I - Q]^{-1}$$
$$= [1 - .5]^{-1} = [.5]^{-1}$$
$$= [2]$$

The product FR is

$$FR = [2][.2 \quad .3] = [.4 \quad .6].$$

$$\begin{array}{c} & 1 & 2 & 3 \\ 1 \\ 11. \quad 2 \\ 3 \end{array} \begin{bmatrix} .8 & .15 & .05 \\ 0 & 1 & 0 \\ 0 & 0 & 1 \end{bmatrix} = P$$

Rearrange the rows and columns of P
so that the absorbing states come
first.

$$\begin{array}{c} & 2 & 3 & 1 \\ 2 \\ 3 \\ \hline 1 \end{array} \begin{bmatrix} 1 & 0 & 0 \\ 0 & 1 & 0 \\ \hline .15 & .05 & .8 \end{bmatrix}$$

R = [.15   .05], Q = [.8]
$$F = [I - Q]^{-1} = [1 - .8]^{-1}$$
$$= [.2]^{-1} = [5]$$
$$FR = [5][.15 \quad .05] = [.75 \quad .25]$$

13. $\begin{bmatrix} 1 & 0 & 0 \\ 0 & 1 & 0 \\ \hline \frac{1}{3} & \frac{1}{3} & \frac{1}{3} \end{bmatrix}$

$$R = \begin{bmatrix} \frac{1}{3} & \frac{1}{3} \end{bmatrix}, \quad Q = \begin{bmatrix} \frac{1}{3} \end{bmatrix}$$

$$F = [I - Q]^{-1} = \left[1 - \frac{1}{3}\right]^{-1} = \left[\frac{3}{2}\right]$$

$$FR = \left[\frac{3}{2}\right]\left[\frac{1}{3} \quad \frac{1}{3}\right] = \left[\frac{1}{2} \quad \frac{1}{2}\right]$$

$$\begin{array}{c} & 1 & 2 & 3 & 4 \\ 1 \\ 2 \\ 15. \quad 3 \\ 4 \end{array} \begin{bmatrix} 1 & 0 & 0 & 0 \\ \frac{1}{3} & 0 & \frac{2}{3} & 0 \\ 0 & 0 & 1 & 0 \\ \frac{1}{4} & \frac{1}{4} & \frac{1}{4} & \frac{1}{4} \end{bmatrix} = P$$

Rearrange the rows and columns of P
so that the absorbing states come
first.

$$\begin{array}{c} & 1 & 3 & 2 & 4 \\ 1 \\ 3 \\ \hline 2 \\ 4 \end{array} \begin{bmatrix} 1 & 0 & 0 & 0 \\ 0 & 1 & 0 & 0 \\ \hline \frac{1}{3} & \frac{2}{3} & 0 & 0 \\ \frac{1}{4} & \frac{1}{4} & \frac{1}{4} & \frac{1}{4} \end{bmatrix}$$

$$R = \begin{bmatrix} \frac{1}{3} & \frac{2}{3} \\ \frac{1}{4} & \frac{1}{4} \end{bmatrix}$$

$$Q = \begin{bmatrix} 0 & 0 \\ \frac{1}{4} & \frac{1}{4} \end{bmatrix}$$

$$F = [I - Q]^{-1}$$

$$= \left(\begin{bmatrix} 1 & 0 \\ 0 & 1 \end{bmatrix} - \begin{bmatrix} 0 & 0 \\ \frac{1}{4} & \frac{1}{4} \end{bmatrix}\right)^{-1}$$

$$= \begin{bmatrix} 1 & 0 \\ -\frac{1}{4} & \frac{3}{4} \end{bmatrix}^{-1} = \begin{bmatrix} 1 & 0 \\ \frac{1}{3} & \frac{4}{3} \end{bmatrix}$$

$$FR = \begin{bmatrix} 1 & 0 \\ \frac{1}{3} & \frac{4}{3} \end{bmatrix}\begin{bmatrix} \frac{1}{3} & \frac{2}{3} \\ \frac{1}{4} & \frac{1}{4} \end{bmatrix} = \begin{bmatrix} \frac{1}{3} & \frac{2}{3} \\ \frac{4}{9} & \frac{5}{9} \end{bmatrix}$$

$$
\begin{array}{c}
\begin{array}{ccccc} 1 & 2 & 3 & 4 & 5 \end{array}\\
17.\quad \begin{array}{c} 1\\2\\3\\4\\5 \end{array}
\begin{bmatrix}
1 & 0 & 0 & 0 & 0\\
0 & 1 & 0 & 0 & 0\\
.1 & .2 & .3 & .2 & .2\\
.3 & .5 & .1 & 0 & .1\\
0 & 0 & 0 & 0 & 1
\end{bmatrix} = P
\end{array}
$$

Rearranging, we obtain the matrix

$$
\begin{array}{c}
\begin{array}{ccccc} 1 & 2 & 5 & 3 & 4 \end{array}\\
\begin{array}{c} 1\\2\\5\\ \\3\\4 \end{array}
\left[\begin{array}{ccc|cc}
1 & 0 & 0 & 0 & 0\\
0 & 1 & 0 & 0 & 0\\
0 & 0 & 1 & 0 & 0\\
\hline
.1 & .2 & .2 & .3 & .2\\
.3 & .5 & .1 & .1 & 0
\end{array}\right].
\end{array}
$$

$$R = \begin{bmatrix} .1 & .2 & .2\\ .3 & .5 & .1 \end{bmatrix}$$

$$Q = \begin{bmatrix} .3 & .2\\ .1 & 0 \end{bmatrix}$$

$$
\begin{aligned}
F &= [I - Q]^{-1}\\
&= \left(\begin{bmatrix} 1 & 0\\ 0 & 1 \end{bmatrix} - \begin{bmatrix} .3 & .2\\ .1 & 0 \end{bmatrix}\right)^{-1}\\
&= \begin{bmatrix} .7 & -.2\\ -.1 & 1 \end{bmatrix}^{-1}\\
&= \begin{bmatrix} \dfrac{25}{17} & \dfrac{5}{17}\\[2mm] \dfrac{5}{34} & \dfrac{35}{34} \end{bmatrix}
\end{aligned}
$$

$$
\begin{aligned}
FR &= \begin{bmatrix} \dfrac{25}{17} & \dfrac{5}{17}\\[2mm] \dfrac{5}{34} & \dfrac{35}{34} \end{bmatrix}\begin{bmatrix} \dfrac{1}{10} & \dfrac{2}{10} & \dfrac{2}{10}\\[2mm] \dfrac{3}{10} & \dfrac{5}{10} & \dfrac{1}{10} \end{bmatrix}\\
&= \begin{bmatrix} \dfrac{4}{17} & \dfrac{15}{34} & \dfrac{11}{34}\\[2mm] \dfrac{11}{34} & \dfrac{37}{68} & \dfrac{9}{68} \end{bmatrix}
\end{aligned}
$$

19. (a) The transition matrix is

$$
\begin{array}{c}
\begin{array}{ccccc} 0 & 1 & 2 & 3 & 4 \end{array}\\
\begin{array}{c} 0\\1\\2\\3\\4 \end{array}
\begin{bmatrix}
1 & 0 & 0 & 0 & 0\\
\frac{1}{2} & 0 & \frac{1}{2} & 0 & 0\\
0 & \frac{1}{2} & 0 & \frac{1}{2} & 0\\
0 & 0 & \frac{1}{2} & 0 & \frac{1}{2}\\
0 & 0 & 0 & 0 & 1
\end{bmatrix}.
\end{array}
$$

Rearranging, we have

$$
\begin{array}{c}
\begin{array}{ccccc} 0 & 4 & 1 & 2 & 3 \end{array}\\
\begin{array}{c} 0\\4\\ \\1\\2\\3 \end{array}
\left[\begin{array}{cc|ccc}
1 & 0 & 0 & 0 & 0\\
0 & 1 & 0 & 0 & 0\\
\hline
\frac{1}{2} & 0 & 0 & \frac{1}{2} & 0\\
0 & 0 & \frac{1}{2} & 0 & \frac{1}{2}\\
0 & \frac{1}{2} & 0 & \frac{1}{2} & 0
\end{array}\right].
\end{array}
$$

$$R = \begin{bmatrix} \frac{1}{2} & 0\\ 0 & 0\\ 0 & \frac{1}{2} \end{bmatrix}$$

$$Q = \begin{bmatrix} 0 & \frac{1}{2} & 0\\ \frac{1}{2} & 0 & \frac{1}{2}\\ 0 & \frac{1}{2} & 0 \end{bmatrix}$$

$$
\begin{aligned}
F &= [I - Q]^{-1}\\
&= \left(\begin{bmatrix} 1 & 0 & 0\\ 0 & 1 & 0\\ 0 & 0 & 1 \end{bmatrix} - \begin{bmatrix} 0 & \frac{1}{2} & 0\\ \frac{1}{2} & 0 & \frac{1}{2}\\ 0 & \frac{1}{2} & 0 \end{bmatrix}\right)^{-1}\\
&= \begin{bmatrix} 1 & -\frac{1}{2} & 0\\ -\frac{1}{2} & 1 & -\frac{1}{2}\\ 0 & -\frac{1}{2} & 1 \end{bmatrix}^{-1} = \begin{bmatrix} \frac{3}{2} & 1 & \frac{1}{2}\\ 1 & 2 & 1\\ \frac{1}{2} & 1 & \frac{3}{2} \end{bmatrix}
\end{aligned}
$$

$$FR = \begin{bmatrix} \frac{3}{2} & 1 & \frac{1}{2} \\ 1 & 2 & 1 \\ \frac{1}{2} & 1 & \frac{3}{2} \end{bmatrix} \begin{bmatrix} \frac{1}{2} & 0 \\ 0 & 0 \\ 0 & \frac{1}{2} \end{bmatrix}$$

$$= \begin{bmatrix} \frac{3}{4} & \frac{1}{4} \\ \frac{1}{2} & \frac{1}{2} \\ \frac{1}{4} & \frac{3}{4} \end{bmatrix}$$

(b) If player A starts with \$1, the probability of ruin for A is 3/4, since that is the entry in row 1, column 1 of FR. The 3/4 is the probability that the nonabsorbing state of starting with \$1 will lead to the absorbing state of ruin.

(c) If player A starts with \$3, the probability of ruin for A is 1/4, since that is the entry in row 3, column 1 of FR.

21. Use the formulas given in the textbook to calculate r and then x_a.

$$r = \frac{1 - .49}{.49} = 1.0408$$

The probability that A will be ruined in this situation is

$$x_a = \frac{(1.0408)^{10} - (1.0408)^{40}}{1 - (1.0408)^{40}}$$

$$\approx .8756.$$

23. a = 10, b = 10

Complete the chart by using the formulas given in the textbook for r and x_a.

p	r	x_a
.1	9	.9999999997
.2	4	.99999905
.3	$\frac{7}{3}$	.99979
.4	1.5	.98295
.5	1	.5
.6	$\frac{2}{3}$	.017046
.7	$\frac{3}{7}$	.000209
.8	.25	.00000095
.9	$\frac{1}{9}$	.0000000003

25. If an absorbing Markov chain has only one absorbing state, then the equilibrium vector has 1 in the position corresponding to the absorbing state and 0 in all of the other positions. Regardless of the initial state, the long-term trend will be for all states to end up in the single absorbing state.

27. **(a)** The transition matrix is

	Automobile	Light Rail	Stay Out
Automobile	.80	.15	.05
Light Rail	.05	.80	.15
Stay Out	0	0	1

Rearranging, we obtain the matrix

$$
\begin{array}{c}
\quad S \quad\ A \quad\ L \\
\begin{array}{c} S \\ A \\ L \end{array}
\left[\begin{array}{c|cc}
1 & 0 & 0 \\
\hline
.05 & .8 & .15 \\
.15 & .05 & .80
\end{array}\right].
\end{array}
$$

$$R = \begin{bmatrix} .05 \\ .15 \end{bmatrix}$$

$$Q = \begin{bmatrix} .8 & .15 \\ .05 & .80 \end{bmatrix}$$

$$F = [I - Q]^{-1}$$

$$= \left(\begin{bmatrix} 1 & 0 \\ 0 & 1 \end{bmatrix} - \begin{bmatrix} .8 & .15 \\ .05 & .80 \end{bmatrix} \right)^{-1}$$

$$= \begin{bmatrix} .2 & -.15 \\ -.05 & .2 \end{bmatrix}^{-1}$$

$$= \begin{bmatrix} 6.154 & 4.615 \\ 1.538 & 6.154 \end{bmatrix}$$

$$FR = \begin{bmatrix} 6.154 & 4.615 \\ 1.538 & 6.154 \end{bmatrix} \begin{bmatrix} .05 \\ .15 \end{bmatrix} = \begin{bmatrix} 1.000 \\ 1.000 \end{bmatrix}$$

(b) The probability that a person who commuted by car ends up avoiding the downtown area is 1, since that is the entry in row 1, column 1 of FR.

(c) The expected number of years until a person who commutes by auto-mobile this year ends up avoiding the downtown area is $10.769 \approx 10.77$ since that is the sum of the entries in row 1 of F.

29. **(a)** The transition matrix is

$$
\begin{array}{c}
\quad 1 \quad\ \ 2 \quad\ \ 3 \\
\begin{array}{c} 1 \\ 2 \\ 3 \end{array}
\left[\begin{array}{ccc}
.05 & .15 & .8 \\
.05 & .15 & .8 \\
0 & 0 & 1
\end{array}\right].
\end{array}
$$

Rearranging, we obtain the matrix

$$
\begin{array}{c}
\quad 3 \quad\ \ 1 \quad\ \ 2 \\
\begin{array}{c} 3 \\ 1 \\ 2 \end{array}
\left[\begin{array}{c|cc}
1 & 0 & 0 \\
\hline
.8 & .05 & .15 \\
.8 & .05 & .15
\end{array}\right].
\end{array}
$$

$$R = \begin{bmatrix} .8 \\ .8 \end{bmatrix}$$

$$Q = \begin{bmatrix} .05 & .15 \\ .05 & .15 \end{bmatrix}$$

(b) $F = [I - Q]^{-1}$

$$= \left(\begin{bmatrix} 1 & 0 \\ 0 & 1 \end{bmatrix} - \begin{bmatrix} .05 & .15 \\ .05 & .15 \end{bmatrix} \right)^{-1}$$

$$= \begin{bmatrix} .95 & -.15 \\ -.05 & .85 \end{bmatrix}^{-1}$$

$$= \begin{bmatrix} 1.0625 & .1875 \\ .0625 & 1.1875 \end{bmatrix}$$

$$FR = \begin{bmatrix} 1.0625 & .1875 \\ .0625 & 1.1875 \end{bmatrix} \begin{bmatrix} .8 \\ .8 \end{bmatrix} = \begin{bmatrix} 1 \\ 1 \end{bmatrix}$$

(c) The probability that the disease eventually disappears is 1, since that is the entry in row 2, column 1 of FR.

(d) The expected number of people is 1.25, since that is the sum of the entries in row 2 of F.

31. This exercise should be solved by computer methods. The solution will vary according to the computer pro-gram that is used. The answers are as follows.

(a) .6   **(b)** .5

Chapter 8   Review Exercises

3. $\begin{bmatrix} .4 & .6 \\ 1 & 0 \end{bmatrix}$

This could be a transition matrix since it is a square matrix, all entries are between 0 and 1, inclusive, and the sum of the entries in each row is 1.

5. $\begin{bmatrix} .8 & .2 & 0 \\ 0 & 1 & 0 \\ .1 & .4 & .5 \end{bmatrix}$

This could be a transition matrix.

7. (a) $C = \begin{bmatrix} .6 & .4 \\ 1 & 0 \end{bmatrix}$

$$C^2 = \begin{bmatrix} .6 & .4 \\ 1 & 0 \end{bmatrix}\begin{bmatrix} .6 & .4 \\ 1 & 0 \end{bmatrix}$$

$$= \begin{bmatrix} .76 & .24 \\ .6 & .4 \end{bmatrix}$$

$$C^3 = \begin{bmatrix} .6 & .4 \\ 1 & 0 \end{bmatrix}\begin{bmatrix} .76 & .24 \\ .6 & .4 \end{bmatrix}$$

$$= \begin{bmatrix} .696 & .304 \\ .76 & .24 \end{bmatrix}$$

(b) The probability that state 2 changes to state 1 after 3 repetitions is .76, since that is the entry in row 2, column 1 of C^3.

9. (a) $E = \begin{bmatrix} .2 & .5 & .3 \\ .1 & .8 & .1 \\ 0 & 1 & 0 \end{bmatrix}$

$$E^2 = \begin{bmatrix} .2 & .5 & .3 \\ .1 & .8 & .1 \\ 0 & 1 & 0 \end{bmatrix}\begin{bmatrix} .2 & .5 & .3 \\ .1 & .8 & .1 \\ 0 & 1 & 0 \end{bmatrix}$$

$$= \begin{bmatrix} .09 & .8 & .11 \\ .1 & .79 & .11 \\ .1 & .8 & .1 \end{bmatrix}$$

$$E^3 = \begin{bmatrix} .2 & .5 & .3 \\ .1 & .8 & .1 \\ 0 & 1 & 0 \end{bmatrix}\begin{bmatrix} .09 & .8 & .11 \\ .1 & .79 & .11 \\ .1 & .8 & .1 \end{bmatrix}$$

$$= \begin{bmatrix} .098 & .795 & .107 \\ .099 & .792 & .109 \\ .1 & .79 & .11 \end{bmatrix}$$

(b) The probability that state 2 changes to state 1 after 3 repetitions is .099, since that is the entry in row 2, column 1 of E^3.

11. $T^2 = \begin{bmatrix} .4 & .6 \\ .5 & .5 \end{bmatrix}\begin{bmatrix} .4 & .6 \\ .5 & .5 \end{bmatrix}$

$$= \begin{bmatrix} .46 & .54 \\ .45 & .55 \end{bmatrix}$$

The distribution after 2 repetitions is

$$[.3 \quad .7]\begin{bmatrix} .46 & .54 \\ .45 & .55 \end{bmatrix} = [.453 \quad .547].$$

To predict the long-range distribution, let V be the probability vector $[v_1 \quad v_2]$.

$$[v_1 \quad v_2]\begin{bmatrix} .46 & .54 \\ .45 & .55 \end{bmatrix} = [v_1 \quad v_2]$$

$$.46v_1 + .45v_2 = v_1$$
$$.54v_1 + .55v_2 = v_2$$
$$v_1 + \quad v_2 = 1$$

$$-.54v_1 + .45v_2 = 0$$

$$.54v_1 - .45v_2 = 0$$

$$v_2 = 1 - v_1$$

$$.54v_1 - .45(1 - v_1) = 0$$

$$.99v_1 = .45$$

$$v_1 = \frac{45}{99} = \frac{5}{11}$$

$$\approx .455$$

$$v_2 = \frac{54}{99} = \frac{6}{11}$$

$$\approx .545$$

The long-range distribution is

$$\begin{bmatrix} \frac{5}{11} & \frac{6}{11} \end{bmatrix} \quad \text{or} \quad [.455 \quad .545].$$

13. $T^2 = \begin{bmatrix} .6 & .2 & .2 \\ .3 & .3 & .4 \\ .5 & .4 & .1 \end{bmatrix} \begin{bmatrix} .6 & .2 & .2 \\ .3 & .3 & .4 \\ .5 & .4 & .1 \end{bmatrix}$

$$= \begin{bmatrix} .52 & .26 & .22 \\ .47 & .31 & .22 \\ .47 & .26 & .27 \end{bmatrix}$$

The distribution after 2 repetitions is

$$[.2 \quad .4 \quad .4] \begin{bmatrix} .52 & .26 & .22 \\ .47 & .31 & .22 \\ .47 & .26 & .27 \end{bmatrix}$$

$$= [.48 \quad .28 \quad .24].$$

To predict the long-range distribution, let V be the probability vector $[v_1 \quad v_2 \quad v_3]$.

$$[v_1 \quad v_2 \quad v_3] \begin{bmatrix} .6 & .2 & .2 \\ .3 & .3 & .4 \\ .5 & .4 & .1 \end{bmatrix}$$

$$= [v_1 \quad v_2 \quad v_3]$$

$$.6v_1 + .3v_2 + .5v_3 = v_1$$

$$.2v_1 + .3v_2 + .4v_3 = v_2$$

$$.2v_1 + .4v_2 + .1v_3 = v_3$$

Also, $v_1 + v_2 + v_3 = 1$.
Solving this system by the Gauss-Jordan method gives

$$v_1 = \frac{47}{95} \approx .495$$

$$v_2 = \frac{26}{95} \approx .274$$

$$v_3 = \frac{22}{95} \approx .232.$$

The long-range distribution is

$$\begin{bmatrix} \frac{47}{95} & \frac{26}{95} & \frac{22}{95} \end{bmatrix} \quad \text{or} \quad [.495 \quad .274 \quad .232].$$

17. $A = \begin{bmatrix} .4 & .2 & .4 \\ 0 & 1 & 0 \\ .6 & .3 & .1 \end{bmatrix}$

$A^2 = \begin{bmatrix} .4 & .4 & .2 \\ 0 & 1 & 0 \\ .3 & .45 & .25 \end{bmatrix}$

$A^3 = \begin{bmatrix} .28 & .54 & .18 \\ 0 & 1 & 0 \\ .27 & .585 & .145 \end{bmatrix}$

Note that the second row will always have zeros; hence, the matrix is not regular.

23.

$$\begin{array}{c c} & \begin{array}{c c c} 1 & 2 & 3 \end{array} \\ \begin{array}{c} 1 \\ 2 \\ 3 \end{array} & \begin{bmatrix} .2 & 0 & .8 \\ 0 & 1 & 0 \\ .7 & 0 & .3 \end{bmatrix} \end{array}$$

Since $p_{22} = 1$, state 2 is absorbing. Since $p_{12} = 0$ and $p_{32} = 0$, it is not possible to go from either of the nonabsorbing states to the absorbing state. Thus, this is not the transition matrix of an absorbing Markov chain.

$$\begin{array}{c} \\ \\ \text{25.} \\ \\ \end{array} \begin{array}{c} \\ 1 \\ 2 \\ 3 \end{array} \overset{\begin{array}{ccc} 1 & 2 & 3 \end{array}}{\begin{bmatrix} .2 & .5 & .3 \\ 0 & 1 & 0 \\ 0 & 0 & 1 \end{bmatrix}} = P$$

Rearranging, we have

$$\begin{array}{c} 2 \\ 3 \\ \\ 1 \end{array} \overset{\begin{array}{ccc} 2 & 3 & 1 \end{array}}{\left[\begin{array}{cc|c} 1 & 0 & 0 \\ 0 & 1 & 0 \\ \hline .5 & .3 & .2 \end{array} \right]}.$$

$$R = [.5 \quad .3], \; Q = [.2]$$
$$F = [I - Q]^{-1} = [1 - .2]^{-1} = [.8]^{-1}$$
$$= \left[\frac{10}{8}\right] = \left[\frac{5}{4}\right] \;\; \text{or} \;\; [1.25]$$
$$FR = [1.25][.5 \quad .3] = [.625 \quad .375]$$
$$\text{or} \; \left[\frac{5}{8} \quad \frac{3}{8}\right]$$

$$\begin{array}{c} \\ \\ \text{27.} \\ \\ \\ \end{array} \begin{array}{c} \\ 1 \\ 2 \\ 3 \\ 4 \end{array} \overset{\begin{array}{cccc} 1 & 2 & 3 & 4 \end{array}}{\begin{bmatrix} \frac{1}{5} & \frac{1}{5} & \frac{2}{5} & \frac{1}{5} \\ 0 & 1 & 0 & 0 \\ \frac{1}{2} & \frac{1}{4} & \frac{1}{8} & \frac{1}{8} \\ 0 & 0 & 0 & 1 \end{bmatrix}} = P$$

Rearranging, we have

$$\begin{array}{c} 2 \\ 4 \\ \\ 1 \\ 3 \end{array} \overset{\begin{array}{cccc} 2 & 4 & 1 & 3 \end{array}}{\left[\begin{array}{cc|cc} 1 & 0 & 0 & 0 \\ 0 & 1 & 0 & 0 \\ \hline \frac{1}{5} & \frac{1}{5} & \frac{1}{5} & \frac{2}{5} \\ \frac{1}{4} & \frac{1}{8} & \frac{1}{2} & \frac{1}{8} \end{array} \right]}.$$

$$R = \begin{bmatrix} \frac{1}{5} & \frac{1}{5} \\ \frac{1}{4} & \frac{1}{8} \end{bmatrix}$$

$$Q = \begin{bmatrix} \frac{1}{5} & \frac{2}{5} \\ \frac{1}{2} & \frac{1}{8} \end{bmatrix}$$

$$F = [I - Q]^{-1}$$

$$= \left(\begin{bmatrix} 1 & 0 \\ 0 & 1 \end{bmatrix} - \begin{bmatrix} \frac{1}{5} & \frac{2}{5} \\ \frac{1}{2} & \frac{1}{8} \end{bmatrix} \right)^{-1}$$

$$= \begin{bmatrix} \frac{4}{5} & -\frac{2}{5} \\ -\frac{1}{2} & \frac{7}{8} \end{bmatrix}^{-1}$$

$$= \begin{bmatrix} \frac{7}{4} & \frac{4}{5} \\ 1 & \frac{8}{5} \end{bmatrix}$$

$$FR = \begin{bmatrix} \frac{7}{4} & \frac{4}{5} \\ 1 & \frac{8}{5} \end{bmatrix} \begin{bmatrix} \frac{1}{5} & \frac{1}{5} \\ \frac{1}{4} & \frac{1}{8} \end{bmatrix}$$

$$= \begin{bmatrix} \frac{11}{20} & \frac{9}{20} \\ \frac{3}{5} & \frac{2}{5} \end{bmatrix} \;\; \text{or} \;\; \begin{bmatrix} .55 & .45 \\ .6 & .4 \end{bmatrix}$$

29. (a) The distribution after the campaign is

$$[.35 \quad .65] \begin{bmatrix} .8 & .2 \\ .4 & .6 \end{bmatrix} = [.54 \quad .46].$$

(b) $P^3 = \begin{bmatrix} .688 & .312 \\ .624 & .376 \end{bmatrix}$

The distribution after 3 campaigns is

$$[.35 \quad .65] \begin{bmatrix} .688 & .312 \\ .624 & .376 \end{bmatrix}$$
$$= [.6464 \quad .3536].$$

31. The distribution after one month is

$$[.4 \quad .4 \quad .2] \begin{bmatrix} .8 & .15 & .05 \\ .25 & .55 & .2 \\ .04 & .21 & .75 \end{bmatrix}$$

$$= [.428 \quad .332 \quad .25].$$

33. $P^2 = \begin{bmatrix} .8 & .15 & .05 \\ .25 & .55 & .2 \\ .04 & .21 & .75 \end{bmatrix} \begin{bmatrix} .8 & .15 & .05 \\ .25 & .55 & .2 \\ .04 & .21 & .75 \end{bmatrix}$

$$= \begin{bmatrix} .6795 & .213 & .1075 \\ .3455 & .382 & .2725 \\ .1145 & .279 & .6065 \end{bmatrix}$$

$P^3 = \begin{bmatrix} .8 & .15 & .05 \\ .25 & .55 & .2 \\ .04 & .21 & .75 \end{bmatrix} \begin{bmatrix} .6795 & .213 & .1075 \\ .3455 & .382 & .2725 \\ .1145 & .279 & .6065 \end{bmatrix}$

$$= \begin{bmatrix} .60115 & .24165 & .1572 \\ .3828 & .31915 & .29805 \\ .18561 & .29799 & .5164 \end{bmatrix}$$

The distribution after 3 months is

$$[.4 \quad .4 \quad .2] \begin{bmatrix} .60115 & .24165 & .1572 \\ .3828 & .31915 & .29805 \\ .18561 & .29799 & .5164 \end{bmatrix}$$

$$= [.431 \quad .284 \quad .285].$$

35. Let $P = \begin{bmatrix} .3 & .5 & .2 \\ .2 & .6 & .2 \\ .1 & .5 & .4 \end{bmatrix}$.

The probability that a man of normal weight will have a thin son is given by the entry in row 2, column 1 of P, which is .2.

37. $P^2 = \begin{bmatrix} .3 & .5 & .2 \\ .2 & .6 & .2 \\ .1 & .5 & .4 \end{bmatrix} \begin{bmatrix} .3 & .5 & .2 \\ .2 & .6 & .2 \\ .1 & .5 & .4 \end{bmatrix}$

$$= \begin{bmatrix} .21 & .55 & .24 \\ .2 & .56 & .24 \\ .17 & .55 & .28 \end{bmatrix}$$

$P^3 = \begin{bmatrix} .3 & .5 & .2 \\ .2 & .6 & .2 \\ .1 & .5 & .4 \end{bmatrix} \begin{bmatrix} .21 & .55 & .24 \\ .2 & .56 & .24 \\ .17 & .55 & .28 \end{bmatrix}$

$$= \begin{bmatrix} .197 & .555 & .248 \\ .196 & .556 & .248 \\ .189 & .555 & .256 \end{bmatrix}$$

The probability that a man of normal weight will have a thin great-grandson is given by the entry in row 2, column 1 of P^3, which is .196.

39. $P^2 = \begin{bmatrix} .21 & .55 & .24 \\ .2 & .56 & .24 \\ .17 & .55 & .28 \end{bmatrix}$

The probability that an overweight man will have an overweight grandson is given by the entry in row 3, column 3 of P^2, which is .28.

41. The distribution of men by weight after 1 generation is

$$[.2 \quad .55 \quad .25] \begin{bmatrix} .3 & .5 & .2 \\ .2 & .6 & .2 \\ .1 & .5 & .4 \end{bmatrix}$$

$$= [.195 \quad .555 \quad .25].$$

43. The distribution of men by weight after 3 generations is

$$[.2 \quad .55 \quad .25] \begin{bmatrix} .197 & .555 & .248 \\ .196 & .556 & .248 \\ .189 & .555 & .256 \end{bmatrix}$$

$$= [.194 \quad .556 \quad .25].$$

45. If the offspring both carry genes AA, then so must their offspring; hence, state 1 ends up in state 1 with probability 1. If the offspring both carry genes aa, then so must their offspring; hence, state 6 ends up in state 6 with probability 1. If AA mates with aa, then the offspring will carry genes Aa; hence, state 3 ends up in state 4 with probability 1. If AA mates with Aa, there are four possible outcomes for a pair of offspring; AA and AA is one of the outcomes, so state 2 ends up in state 1 with probability 1/4, AA and Aa can happen two ways, so state 2 ends up in state 2 with probability 2/4 or 1/2, and Aa and Aa is the last possible outcome, so state 2 ends up in state 4 with probability 1/4. If Aa mates with Aa, there are sixteen possible outcomes for a pair of offspring; state 4 ends up in states 1, 2, 3, 4, 5, 6 with respective probabilities 1/16, 1/4, 1/8, 1/4, 1/4, and 1/16. If Aa mates with aa, there are four possible outcomes for a pair of offspring, corresponding to three of the possible states; state 5 ends up in states 4, 5, 6 with respective probabilities 1/4, 1/2, and 1/4. This verifies that the transition matrix for this mating experiment is

$$
\begin{array}{c}
\\ 1 \\ 2 \\ 3 \\ 4 \\ 5 \\ 6
\end{array}
\begin{array}{cccccc}
1 & 2 & 3 & 4 & 5 & 6 \\
\end{array}
\begin{bmatrix}
1 & 0 & 0 & 0 & 0 & 0 \\
\frac{1}{4} & \frac{1}{2} & 0 & \frac{1}{4} & 0 & 0 \\
0 & 0 & 0 & 1 & 0 & 0 \\
\frac{1}{16} & \frac{1}{4} & \frac{1}{8} & \frac{1}{4} & \frac{1}{4} & \frac{1}{16} \\
0 & 0 & 0 & \frac{1}{4} & \frac{1}{2} & \frac{1}{4} \\
0 & 0 & 0 & 0 & 0 & 1
\end{bmatrix}.
$$

47. Rearrange the rows and columns of the transition matrix so that the absorbing states come first.

$$
\begin{array}{c}
\\ 1 \\ 6 \\ 2 \\ 3 \\ 4 \\ 5
\end{array}
\begin{array}{cccccc}
1 & 6 & 2 & 3 & 4 & 5 \\
\end{array}
\left[
\begin{array}{cc|cccc}
1 & 0 & 0 & 0 & 0 & 0 \\
0 & 1 & 0 & 0 & 0 & 0 \\
\hline
\frac{1}{4} & 0 & \frac{1}{2} & 0 & \frac{1}{4} & 0 \\
0 & 0 & 0 & 0 & 1 & 0 \\
\frac{1}{16} & \frac{1}{16} & \frac{1}{4} & \frac{1}{8} & \frac{1}{4} & \frac{1}{4} \\
0 & \frac{1}{4} & 0 & 0 & \frac{1}{4} & \frac{1}{2}
\end{array}
\right]
$$

From this rearranged matrix, observe that

$$
Q = \begin{bmatrix}
\frac{1}{2} & 0 & \frac{1}{4} & 0 \\
0 & 0 & 1 & 0 \\
\frac{1}{4} & \frac{1}{8} & \frac{1}{4} & \frac{1}{4} \\
0 & 0 & \frac{1}{4} & \frac{1}{2}
\end{bmatrix}.
$$

49. In Exercise 48, it was shown that the fundamental matrix for this absorbing Markov chain is

$$F = \begin{bmatrix} \frac{8}{3} & \frac{1}{6} & \frac{4}{3} & \frac{2}{3} \\ \frac{4}{3} & \frac{4}{3} & \frac{8}{3} & \frac{4}{3} \\ \frac{4}{3} & \frac{1}{3} & \frac{8}{3} & \frac{4}{3} \\ \frac{2}{3} & \frac{1}{6} & \frac{4}{3} & \frac{8}{3} \end{bmatrix}.$$

If Aa mates with Aa (which corresponds to state 4, which in turn corresponds to row 3 of F), 8/3 pairs of offspring with these genes can be expected before ending up in one of the two absorbing states. This is because 8/3 is the entry in row 3, column 3 of F.

51. **(a)** After the duplication, there are 2n genes and n of them are being selected; this can be done in $\binom{2n}{n}$ different ways. Suppose there are i mutant genes before the duplication and j mutant genes in the next generation. After the duplication, there will be 2i mutant genes, of which j will be selected; this can be done in $\binom{2i}{j}$ different ways. Also, there are 2n - 2i non- mutant genes, of which n - j will be selected; this can be done in $\binom{2n - 2i}{n - j}$ different ways.

Therefore, the probability of a generation with i mutant genes being followed by a generation with j mutant genes, which is the transition probability from state i to state j, is

$$p_{ij} = \frac{\binom{2i}{j}\binom{2n - 2i}{n - j}}{\binom{2n}{n}}.$$

(b) The absorbing states are state 0 and state n. If a generation has no mutant genes, then after duplication there will still be none, and if a generation consists entirely of mutant genes, its successor will also.

(c) Use $p_{ij} = \dfrac{\binom{2i}{j}\binom{2n - 2i}{n - j}}{\binom{2n}{n}}$ with

n = 3 and i = 0, 1, 2, 3 and j = 0, 1, 2, 3 to calculate the entries of the transition matrix. Let $\binom{n}{r} = 0$ when n < r.

$$p_{00} = \frac{\binom{0}{0}\binom{6}{3}}{\binom{6}{3}} = 1, \quad p_{01} = \frac{\binom{0}{1}\binom{6}{2}}{\binom{6}{3}} = 0,$$

$$p_{02} = 0, \quad p_{03} = 0,$$

$$p_{10} = \frac{\binom{2}{0}\binom{4}{3}}{\binom{6}{3}} = \frac{1}{5}, \quad p_{11} = \frac{\binom{2}{1}\binom{4}{2}}{\binom{6}{3}} = \frac{3}{5},$$

$$p_{12} = \frac{\binom{2}{2}\binom{4}{1}}{\binom{6}{3}} = \frac{1}{5}, \quad p_{13} = \frac{\binom{2}{3}\binom{4}{0}}{\binom{6}{3}} = 0,$$

$$p_{20} = \frac{\binom{4}{0}\binom{2}{3}}{\binom{6}{3}} = 0, \quad p_{21} = \frac{\binom{4}{1}\binom{2}{2}}{\binom{6}{3}} = \frac{1}{5},$$

$$p_{22} = \frac{\binom{4}{2}\binom{2}{1}}{\binom{6}{3}} = \frac{3}{5}, \quad p_{23} = \frac{\binom{4}{3}\binom{2}{0}}{\binom{6}{3}} = \frac{1}{5}$$

$$p_{30} = \frac{\binom{6}{0}\binom{0}{3}}{\binom{6}{3}} = 0, \quad p_{31} = 0, \quad p_{32} = 0,$$

$$p_{33} = \frac{\binom{6}{3}\binom{0}{0}}{\binom{6}{3}} = 1$$

The transition matrix is

$$\begin{array}{c} \\ 0 \\ 1 \\ 2 \\ 3 \end{array}\begin{array}{cccc} 0 & 1 & 2 & 3 \end{array} \\ \begin{bmatrix} 1 & 0 & 0 & 0 \\ \frac{1}{5} & \frac{3}{5} & \frac{1}{5} & 0 \\ 0 & \frac{1}{5} & \frac{3}{5} & \frac{1}{5} \\ 0 & 0 & 0 & 1 \end{bmatrix} = P.$$

(d) Rearrange the rows and columns of P.

$$\begin{array}{cccc} 0 & 3 & 1 & 2 \end{array}$$

$$\begin{bmatrix} 1 & 0 & 0 & 0 \\ 0 & 1 & 0 & 0 \\ \frac{1}{5} & 0 & \frac{3}{5} & \frac{1}{5} \\ 0 & \frac{1}{5} & \frac{1}{5} & \frac{3}{5} \end{bmatrix}$$

$$R = \begin{bmatrix} \frac{1}{5} & 0 \\ 0 & \frac{1}{5} \end{bmatrix}, \quad Q = \begin{bmatrix} \frac{3}{5} & \frac{1}{5} \\ \frac{1}{5} & \frac{3}{5} \end{bmatrix}$$

$$F = [I - Q]^{-1}$$

$$= \left(\begin{bmatrix} 1 & 0 \\ 0 & 1 \end{bmatrix} - \begin{bmatrix} \frac{3}{5} & \frac{1}{5} \\ \frac{1}{5} & \frac{3}{5} \end{bmatrix} \right)^{-1}$$

$$= \begin{bmatrix} \frac{2}{5} & -\frac{1}{5} \\ -\frac{1}{5} & \frac{2}{5} \end{bmatrix}^{-1}$$

$$= \begin{bmatrix} \frac{10}{3} & \frac{5}{3} \\ \frac{5}{3} & \frac{10}{3} \end{bmatrix}$$

$$FR = \begin{bmatrix} \frac{10}{3} & \frac{5}{3} \\ \frac{5}{3} & \frac{10}{3} \end{bmatrix}\begin{bmatrix} \frac{1}{5} & 0 \\ 0 & \frac{1}{5} \end{bmatrix}$$

$$= \begin{bmatrix} \frac{2}{3} & \frac{1}{3} \\ \frac{1}{3} & \frac{2}{3} \end{bmatrix}$$

(e) If a set of 3 genes has 1 mutant gene, the probability that the mutant gene will disappear is 2/3, since that is the entry in row 1, column 1 of FR.

(f) If a set of 3 genes has 1 mutant gene, 5 generations would be expected to have 1 mutant gene before either the mutant genes or the nonmutant genes disappear, since that is the sum of the entries in row 1 of F.

Extended Application

1. This verification should be accomplished by computer methods.

2. The entry in row 1, column 1 of the fundamental matrix indicates that there will be about 6.6 visits until there is some tooth decay. Each visit represents 6 months, so 6.6 visits = 39.6 months = 3.3 yr.

3. Use as a computer to show that

$$FR = \begin{bmatrix} .768332 & .231668 \\ .814382 & .185618 \\ .775304 & .224696 \\ .807692 & .192308 \\ .5 & .5 \\ .5 & .5 \\ .5 & .5 \\ .5 & .5 \end{bmatrix}.$$

4. The probability that a healthy tooth
    is eventually lost is $.231668 \approx .23$,
    since that is the entry in row 1,
    column 2 of FR.

5. The four .5 entries in column 2 of
    FR each correspond to a five-digit
    number whose first digit is 1, which
    corresponds to decay on the occlusal
    surface.

CHAPTER 8   TEST

1. Decide which of the following matrices could be transition matrices. Justify your answer.

    (a) $\begin{bmatrix} .1 & .9 \\ 0 & 1 \end{bmatrix}$ **(b)** $\begin{bmatrix} .1 & .2 & .5 & .2 \\ 0 & .8 & .3 & -.1 \\ .1 & .8 & .1 & 0 \\ 0 & 0 & 0 & 1 \end{bmatrix}$ **(c)** $\begin{bmatrix} .21 & 0 & .88 \\ 0 & .45 & .55 \\ .79 & .03 & .18 \end{bmatrix}$

2. Let A be the transition matrix

    $$\begin{bmatrix} \frac{1}{2} & \frac{1}{2} \\ \frac{1}{3} & \frac{2}{3} \end{bmatrix}.$$

    (a) Find the first 3 powers of A.

    (b) Find the probability that state 1 changes to state 2 after 2 repetitions of the experiment.

    (c) Find the probability that state 2 changes to state 1 after 3 repetitions of the experiment.

3. A survey conducted for General Motors revealed that 40% of GM buyers would buy GM again while 75% of Toyota buyers would buy Toyotas again. Suppose the initial distribution vector is [100   50] with GM coming first.

    (a) Write the transition matrix for this problem.

    (b) Find the distribution vector after 1 time period.

    (c) Find the distribution vector after 2 time periods.

4. Find the equilibrium vector for the regular Markov matrix $\begin{bmatrix} .2 & .8 \\ .4 & .6 \end{bmatrix}$.

5. In the small Illinois town of Red Bud, there are three political parties:
    the Corn Party, the Wheat Party, and the Tea Party. Each year 50% of the
    Corn Party members switch to the Wheat Party while the rest remain loyal.
    Among Wheat Party members, 25% switch to Corn, and another 25% switch to
    Tea, and the rest remain loyal. 50% of the Tea Party members switch to
    Wheat and the rest remain loyal.

    (a) Draw the transition diagram for this problem.

    (b) Write the transition matrix.

    (c) Show that the matrix is regular.

    (d) Find the long-range prediction for the proportion
    of voters in each party.

6. Let $P = \begin{bmatrix} 1 & 0 & 0 \\ 0 & 1 & 0 \\ \frac{1}{4} & \frac{1}{2} & \frac{1}{4} \end{bmatrix}$ be the transition matrix for an absorbing Markov chain.

    (a) Find all absorbing states.

    (b) Find the fundamental matrix F.

    (c) Find the matrix FR.

    (d) What is the long-range probability of the chain terminating in state 1?

CHAPTER 8   TEST ANSWERS

1. **(a)** This could be a transition matrix since it is a square matrix with all entries between 0 and 1, inclusive, and the sum of the entries in each row is 1.

(b) This could not be a transition matrix because it has a negative entry.

(c) This could not be a transition matrix because the sum of the entries in the first row is not 1.

2. **(a)** $A = \begin{bmatrix} \frac{1}{2} & \frac{1}{2} \\ \frac{1}{3} & \frac{2}{3} \end{bmatrix}$, $A^2 = \begin{bmatrix} \frac{5}{12} & \frac{7}{12} \\ \frac{7}{18} & \frac{11}{18} \end{bmatrix}$ **(b)** 7/12 **(c)** 43/108

$A^3 = \begin{bmatrix} \frac{29}{72} & \frac{43}{72} \\ \frac{43}{108} & \frac{33}{54} \end{bmatrix}$

3. **(a)** $A = \begin{bmatrix} .4 & .6 \\ .25 & .75 \end{bmatrix}$ **(b)** [52.5   97.5] **(c)** [45.375   104.625]

4. [1/3   2/3]

5. **(a)**

(b)

	Corn	Wheat	Tea
Corn	.5	.5	0
Wheat	.25	.5	.25
Tea	0	.5	.5

(c) $A^2 = \begin{bmatrix} .375 & .5 & .125 \\ .25 & .5 & .25 \\ .125 & .25 & .375 \end{bmatrix}$, which has all positive entries.

(d) [.25   .5   .25]

6. **(a)** States 1 and 2 **(b)** F = [4/3] **(c)** FR = [1/3   2/3] **(d)** 1/3

CHAPTER 9   GAME THEORY

Section 9.1

1. **(a)** Buy speculative; she thinks the market will go up and $30,000 is the highest possible net amount.

    (b) Buy blue-chip; she thinks the market will go down, and $18,000 is better than $11,000.

    (c) If there is a .7 probability the market will go up, then there is a .3 probability it will go down. Find her expected profit for each strategy.

    Blue-chip:
    $(.7)(25,000) + (.3)(18,000) = \$22,900$
    Speculative:
    $(.7)(30,000) + (.3)(11,000) = \$24,300$

    Therefore she should buy speculative and her expected profit is $24,300.

    (d) Find her expected profit for each strategy.

    Blue-chip:
    $(.2)(25,000) + (.8)(18,000) = \$19,400$

    Speculative:
    $(.2)(30,000) + (.8)(11,000) = \$14,800$

    Therefore she should buy blue-chip and her expected profit is $19,400.

3. **(a)** Set up in the stadium; she doesn't think it will rain and $1500 is the highest possible net profit.

    (b) Set up in the gym; the worst that can happen is a profit of $1000.

(c) If there is a .6 probability of rain, then there is a .4 probability of no rain. Find her expected profit for each strategy.

Stadium: $.6(-1550) + .4(1500) = -\$330$
  Gym: $.6(1000) + .4(1000) = \$1000$
  Both: $.6(750) + .4(1400) = \$1010$

She should set up both for a maximum expected profit of $1010.

5. **(a)** The payoff matrix is as follows.

	Better	Not Better
Market	$50,000	-$25,000
Don't Market	-$40,000	-$10,000

    (b) Find the expected profit under the 2 strategies.

    Market product:

    $(.4)(50,000) + (.6)(-25,000)$
    $= \$5000$

    Don't market:

    $(.4)(-40,000) + (.6)(-10,000)$
    $= -\$22,000$

    They should market the product and make a profit of $5000 since that is better than losing $22,000.

7. **(a)** The payoff matrix is as follows.

	Strike	No Strike
Bid $30,000	-$5500	$4500
Bid $40,000	$4500	$0

(b) Find his expected earnings under each stategy.

Bid $30,000:

$(.6)(-5500) + (.4)(4500) = -\$1500$

Bid $40,000:

$(.6)(4500) + (.4)(0) = \$2700$

He should bid $40,000 and make a profit of $2700, since that is better than losing $1500.

9. Find the expected utility under each strategy.

Jobs:

$(.35)(25) + (.65)(-10) = 2.25$

Environment:

$(.35)(-15) + (.65)(30) = 14.25$

She should emphasize the environment. The expected utility of this strategy is 14.25.

Section 9.2

$$\begin{array}{c} & & \text{B} \\ & & 1 \quad\ 2 \quad\ 3 \\ 1. \quad \text{A} & \begin{array}{c} 1 \\ 2 \\ 3 \end{array} & \begin{bmatrix} 6 & -4 & 0 \\ 3 & -2 & 6 \\ -1 & 5 & 11 \end{bmatrix} \end{array}$$

Consider the strategy (1, 1).
The first row, first column entry is 6, indicating a payoff of $6 from B to A.

3. Consider the strategy (2, 2).
The second row, second column entry is -2, indicating a payoff of $2 from A to B.

5. Consider the strategy (3, 1).
The third row, first column entry is -1, indicating a payoff of $1 from A to B.

7. Yes, each entry in column 2 is smaller than the corresponding entry in column 3, so column 2 dominates column 3.

9. $\begin{bmatrix} 0 & -2 & 8 \\ 3 & -1 & -9 \end{bmatrix}$

Column 2 dominates column 1, so remove column 1 to obtain

$$\begin{bmatrix} -2 & 8 \\ -1 & -9 \end{bmatrix}.$$

11. $\begin{bmatrix} 1 & 4 \\ 4 & -1 \\ 3 & 5 \\ -4 & 0 \end{bmatrix}$

Row 3 dominates rows 1 and 4, so remove rows 1 and 4 to obtain

$$\begin{bmatrix} 4 & -1 \\ 3 & 5 \end{bmatrix}.$$

13. $\begin{bmatrix} 8 & 12 & -7 \\ -2 & 1 & 4 \end{bmatrix}$

Column 1 dominates column 2, so remove column 2 to obtain

$$\begin{bmatrix} 8 & -7 \\ -2 & 4 \end{bmatrix}.$$

15.
$$\begin{bmatrix} \underset{\underline{\textcircled{3}}}{} & \textcircled{5} \\ 2 & \underline{-5} \end{bmatrix}$$

Underline the smallest number in each row and circle the largest value in each column.
The 3 at (1, 1) is the smallest number in its row and the largest number in its column, so the saddle point is 3 at (1, 1). This game is strictly determined and its value is 3.

17.
$$\begin{bmatrix} 3 & -4 & \textcircled{1} \\ 5 & \textcircled{3} & \underline{-2} \end{bmatrix}$$

Underline the smallest number in each row and circle the largest value in each column; in this matrix, the two categorizations do not overlap.
There is no saddle point. This game is not strictly determined.

19.
$$\begin{bmatrix} \underline{-6} & 2 \\ -1 & \underline{-10} \\ \textcircled{3} & \textcircled{5} \end{bmatrix}$$

The 3 at (3, 1) is the smallest number in its row and the largest number in its column, so the saddle point is 3 at (3, 1). This game is strictly determined and its value is 3.

21.
$$\begin{bmatrix} 2 & 3 & \textcircled{\underline{1}} \\ -1 & \textcircled{4} & \underline{-7} \\ 5 & 2 & 0 \\ \textcircled{8} & \underline{-4} & -1 \end{bmatrix}$$

The 1 at (1, 3) is the smallest number in its row and the largest number in its column, so the saddle point is 1 at (1, 3). This game is strictly determined and its value is 1.

23.
$$\begin{bmatrix} -6 & 1 & \textcircled{4} & \textcircled{2} \\ \textcircled{9} & \textcircled{3} & \underline{-8} & -7 \end{bmatrix}$$

There is no saddle point. This game is not strictly determined.

25. Focus on any single column of the payoff matrix. In that column, suppose the row 1 entry is x_1, the row 2 entry is x_2, and the row 3 entry is x_3.
Since row 1 dominates row 2, $x_1 > x_2$, and since row 2 dominates row 3, $x_2 > x_3$. By transitivity, it follows that $x_1 > x_3$. The same phenomenon involving these three rows occurs in every column, so row 1 dominates row 3.
We have shown that, whenever row 1 dominates row 2 and row 2 dominates row 3, it must also be true that row 1 dominates row 3.

27.

		B	
	City 1	City 2	City 3
A City 1	5	−2	6
City 2	7	⑤	9
City 3	3	−3	5

To get the entries in the above matrix, look, for example, at the entry in row 2, column 1. If merchant A locates in city 2 and merchant B in city 1, then merchant A will get 80% of the business in city 2, 20% in city 3, and 60% in city 1. Taking into account the fraction of the population living in each city, we get

$$.80(.45) + .20(.30) + .60(.25) = .57.$$

Thus, merchant A gets 57% of the total business. Now 57% is 7 percentage points above 50%, so the entry in row 2, column 1 is +7. Likewise for row 3, column 1, we get

$$.80(.25) + .20(.30) + .60(.45) = .53 = 53\%,$$

which is 3 percentage points above 50%. The other entries are found in a similar manner. (Note that all diagonal entries are 5 since 55% is 5 percentage points above 50%.) The 5 at (2, 2) is the smallest entry in its row and the largest in its column, so the saddle point is the 5 at (2, 2) and the value of the game is 5.

29. $\begin{bmatrix} ③ & -8 & \boxed{⑨} \\ 0 & ⑥ & \underline{-12} \\ -8 & 4 & \underline{-10} \end{bmatrix}$

(where ⑨ = −9)

Underline the smallest number in each row and circle the largest value in each column.

−9 is the smallest entry in its row and the largest in its column. The saddle point is −9 at (1, 3), and the value of the game is −9.

31. The payoff matrix is as follows.

	Stone	Scissors	Paper
Stone	0	①	−1
Scissors	−1	0	①
Paper	①	−1	0

Underline the smallest number in each row and circle the largest value in each column; in this matrix, the two categorizations do not overlap. The game is not strictly determined since it does not have a saddle point.

Section 9.3

1. (a) $[.5 \quad .5]\begin{bmatrix} 3 & -4 \\ -5 & 2 \end{bmatrix}\begin{bmatrix} .3 \\ .7 \end{bmatrix}$

$= [.5 \quad .5]\begin{bmatrix} -1.9 \\ -.1 \end{bmatrix}$

$= [-.95 - .05]$

$= [-1]$

The expected value is −1.

(b) $[.1 \quad .9]\begin{bmatrix} 3 & -4 \\ -5 & 2 \end{bmatrix}\begin{bmatrix} .3 \\ .7 \end{bmatrix}$

$$= [.1 \quad .9]\begin{bmatrix} -1.9 \\ -.1 \end{bmatrix} = [-.28]$$

The expected value is $-.28$.

(c) $[.8 \quad .2]\begin{bmatrix} 3 & -4 \\ -5 & 2 \end{bmatrix}\begin{bmatrix} .3 \\ .7 \end{bmatrix}$

$$= [.8 \quad .2]\begin{bmatrix} -1.9 \\ -.1 \end{bmatrix} = [-1.54]$$

The expected value is -1.54.

(d) $[.2 \quad .8]\begin{bmatrix} 3 & -4 \\ -5 & 2 \end{bmatrix}\begin{bmatrix} .3 \\ .7 \end{bmatrix}$

$$= [.2 \quad .8]\begin{bmatrix} -1.9 \\ -.1 \end{bmatrix} = [-.46]$$

The expected value is $-.46$.

3. $\begin{bmatrix} 5 & 1 \\ 3 & 4 \end{bmatrix}$

There are no saddle points. For A, the optimum strategy is

$$p_1 = \frac{4-3}{5-3-1+4} = \frac{1}{5},$$

$$p_2 = 1 - \frac{1}{5} = \frac{4}{5}.$$

For player B, the optimum strategy is

$$q_1 = \frac{4-1}{5-3-1+4} = \frac{3}{5},$$

$$q_2 = 1 - \frac{3}{5} = \frac{2}{5}.$$

The value of the game is

$$\frac{(5)(4)-(3)(1)}{5-3-1+4}$$

$$= \frac{20-3}{5} = \frac{17}{5}.$$

5. $\begin{bmatrix} -2 & 0 \\ 3 & -4 \end{bmatrix}$

There are no saddle points. For player A, the optimum strategy is

$$p_1 = \frac{-4-3}{-2-3-0-4} = \frac{-7}{-9} = \frac{7}{9},$$

$$p_2 = 1 - \frac{7}{9} = \frac{2}{9}.$$

For player B, the optimum strategy is

$$q_1 = \frac{-4-0}{-2-3-0-4} = \frac{-4}{-9} = \frac{4}{9},$$

$$q_2 = 1 - \frac{4}{9} = \frac{5}{9}.$$

The value of the game is

$$\frac{-2(-4)-(0)(3)}{-2-3-0-4} = -\frac{8}{9}.$$

7. $\begin{bmatrix} 4 & -3 \\ -1 & 7 \end{bmatrix}$

There are no saddle points. For player A, the optimum strategy is

$$p_1 = \frac{7-(-1)}{4-(-1)-(-3)+7} = \frac{8}{15},$$

$$p_2 = 1 - \frac{8}{15} = \frac{7}{15}.$$

For player B, the optimum strategy is

$$q_1 = \frac{7-(-3)}{15} = \frac{10}{15} = \frac{2}{3},$$

$$q_2 = 1 - \frac{2}{3} = \frac{1}{3}.$$

The value of the game is

$$\frac{(4)(7)-(-3)(-1)}{15} = \frac{25}{15} = \frac{5}{3}.$$

9.
$$\begin{bmatrix} -2 & \frac{1}{2} \\ 0 & -3 \end{bmatrix}$$

There are no saddle points. For player A, the optimum strategy is

$$p_1 = \frac{-3 - 0}{-2 - 0 - \frac{1}{2} - 3} = \frac{-3}{\frac{-11}{2}}$$

$$= \frac{6}{11},$$

$$p_2 = 1 - \frac{6}{11} = \frac{5}{11}.$$

For player B, the optimum strategy is

$$q_1 = \frac{-3 - \frac{1}{2}}{-2 - 0 - \frac{1}{2} - 3} = \frac{-\frac{7}{2}}{\frac{-11}{2}} = \frac{7}{11},$$

$$q_2 = 1 - \frac{7}{11} = \frac{4}{11}.$$

The value of the game is

$$\frac{-2(-3) - \left(\frac{1}{2}\right)(0)}{-2 - 0 - \frac{1}{2} - 3} = \frac{6}{\frac{-11}{2}} = -\frac{12}{11}.$$

11.
$$\begin{bmatrix} \frac{8}{3} & -\frac{1}{2} \\ \frac{3}{4} & -\frac{5}{12} \end{bmatrix}$$

The game is strictly determined since it has a saddle point at (2, 2). The value of the game is $-\frac{5}{12}$.

13.
$$\begin{bmatrix} -1 & 2 \\ 3 & 1 \end{bmatrix}$$

There are no saddle points. For player A, the optimum strategy is

$$p_1 = \frac{1 - 3}{-1 - 3 - 2 + 1} = \frac{-2}{-5} = \frac{2}{5},$$

$$p_2 = 1 - \frac{2}{5} = \frac{3}{5}.$$

For player B, the optimum strategy is

$$q_1 = \frac{1 - 2}{1 - 3 - 2 + 1} = \frac{-1}{-5} = \frac{1}{5},$$

$$q_2 = 1 - \frac{1}{5} = \frac{4}{5}.$$

The value of the game is

$$\frac{-1(1) - (2)(3)}{-1 - 3 - 2 + 1} = \frac{-7}{-5} = \frac{7}{5}.$$

15.
$$\begin{bmatrix} -4 & 9 \\ 3 & -5 \\ 8 & 7 \end{bmatrix}$$

Row 3 dominates row 2, so remove row 2. This gives the matrix

$$\begin{bmatrix} -4 & 9 \\ 8 & 7 \end{bmatrix}.$$

For player A, the optimum strategy is

$$p_1 = \frac{7 - 8}{-4 - 8 - 9 + 7} = \frac{-1}{-14} = \frac{1}{14},$$

$$p_2 = 0 \text{ (row 2 was removed)},$$

$$p_3 = 1 - \frac{1}{14} = \frac{13}{14}.$$

For player B, the optimum strategy is

$$q_1 = \frac{7 - 9}{-4 - 8 - 9 + 7} = \frac{-2}{-14} = \frac{1}{7},$$

$$q_2 = 1 - \frac{1}{7} = \frac{6}{7}.$$

The value of the game is

$$\frac{-4(7) - (9)(8)}{-4 - 8 - 9 + 7} = \frac{-100}{-14} = \frac{50}{7}.$$

17.
$$\begin{bmatrix} 8 & 6 & 3 \\ -1 & -2 & 4 \end{bmatrix}$$

Column 2 dominates column 1, so remove column 1. This gives the matrix

$$\begin{bmatrix} 6 & 3 \\ -2 & 4 \end{bmatrix}.$$

For player A, the optimum strategy is

$$p_1 = \frac{4 - (-2)}{6 + 2 - 3 + 4} = \frac{6}{9} = \frac{2}{3},$$

$$p_2 = 1 - \frac{2}{3} = \frac{1}{3}.$$

For player B, the optimum strategy is

$$q_1 = 0 \text{ (column 1 was removed)},$$

$$q_2 = \frac{4 - 3}{6 + 2 - 3 + 4} = \frac{1}{9},$$

$$q_3 = 1 - \frac{1}{9} = \frac{8}{9}.$$

The value of the game is

$$\frac{6(4) - (3)(-2)}{6 + 2 - 3 + 4} = \frac{30}{9} = \frac{10}{3}.$$

19. $\begin{bmatrix} 9 & -1 & 6 \\ 13 & 11 & 8 \\ 6 & 0 & 9 \end{bmatrix}$

Row 2 dominates row 1, so remove row 1. This gives the matrix

$$\begin{bmatrix} 13 & 11 & 8 \\ 6 & 0 & 9 \end{bmatrix}.$$

Now, column 2 dominates column 1, so remove column 1. This gives the matrix

$$\begin{bmatrix} 11 & 8 \\ 0 & 9 \end{bmatrix}.$$

For player A, the optimum strategy is

$$p_1 = 0 \text{ (row 1 was removed)},$$

$$p_2 = \frac{9 - 0}{11 - 0 - 8 + 9} = \frac{9}{12} = \frac{3}{4},$$

$$p_3 = 1 - \frac{3}{4} = \frac{1}{4}.$$

For player B, the optimum strategy is

$$q_1 = 0, \text{ (column 1 was removed)},$$

$$q_2 = \frac{9 - 8}{11 - 0 - 8 + 9} = \frac{1}{12},$$

$$q_3 = 1 - \frac{1}{12} = \frac{11}{12}.$$

The value of the game is

$$\frac{11(9) - (8)(0)}{11 - 0 - 8 + 9} = \frac{99}{12} = \frac{33}{4}.$$

21. In a non–strictly–determined game, there is no saddle point. Let $M = \begin{bmatrix} a_{11} & a_{12} \\ a_{21} & a_{22} \end{bmatrix}$ be the payoff matrix of the game. Assume that player B chooses column 1 with probability p_1. The expected value for B, assuming A plays row 1, is E_1, where

$$E_1 = a_{11} \cdot p_1 + a_{12} \cdot (1 - p_1).$$

The expected value for B if A plays row 2 is E_2, where

$$E_2 = a_{21} \cdot p_1 + a_{22} \cdot (1 - p_1).$$

The optimum strategy for player B is found by letting $E_1 = E_2$.

$a_{11} \cdot p_1 + a_{12} \cdot (1 - p_1)$
$\quad = a_{21} \cdot p_1 + a_{22}(1 - p_1)$

$a_{11} \cdot p_1 + a_{12} - a_{12} \cdot p_1$
$\quad = a_{21} \cdot p_1 + a_{22} - a_{22} \cdot p_1$

$a_{11} \cdot p_1 - a_{21} \cdot p_1 - a_{12} \cdot p_1 + a_{22} \cdot p_1$
$\quad = a_{22} - a_{12}$

$p_1(a_{11} - a_{21} - a_{12} + a_{22})$
$\quad = a_{22} - a_{12}$

$$p_1 = \frac{a_{22} - a_{12}}{a_{11} - a_{21} - a_{12} + a_{22}}$$

Since $p_2 = 1 - p_1$,

$$p_2 = 1 - \frac{a_{22} - a_{12}}{a_{11} - a_{21} - a_{12} + a_{22}}$$

$$= \frac{a_{11} - a_{21} - a_{12} + a_{22} - (a_{22} - a_{12})}{a_{11} - a_{21} - a_{12} + a_{22}}$$

$$= \frac{a_{11} - a_{21}}{a_{11} - a_{21} - a_{12} + a_{22}}.$$

These are the formulas given in the text for p_1 and p_2.

25.

Bates

		T.V.	Radio
Allied	T.V.	1.0	$-.7$
	Radio	$-.5$	$.5$

The optimum strategy for Allied is

$$p_1 = \frac{.5 - (-.5)}{1 - (-.5) - (-.7) + (.5)}$$

$$= \frac{1}{2.7} = \frac{10}{27},$$

$$p_2 = 1 - \frac{10}{27} = \frac{17}{27}.$$

Allied should use T.V. with probability $10/27$ and use radio with probability $17/27$.

The value of the game is

$$\frac{(1)(.5) - (-.7)(-.5)}{2.7} = \frac{.15}{2.7} = \frac{1}{18},$$

which represents increased sales of

$$1,000,000\left(\frac{1}{18}\right) \approx 1,000,000(.055556)$$

$$= \$55,556.$$

27.

Selling during:

		Rain	Shine
Buying for:	Rain	250	-150
	Shine	-150	350

The best mixed strategy for Merrill is

$$p_1 = \frac{350 - (-150)}{250 - (-150) - (-150) + 350}$$

$$= \frac{500}{900} = \frac{5}{9},$$

$$p_2 = 1 - \frac{5}{9} = \frac{4}{9}.$$

He should invest in rainy day goods $5/9$ of the time and in sunny day goods $4/9$ of the time.

The value of the game is

$$\frac{(250)(350) - (-150)^2}{900} = \frac{65,000}{900}$$

$$= \frac{650}{9}$$

$$\approx \$72.22$$

Thus, his profit is $\$72.22$.

29. The payoff matrix is as follows.

Jamie

		Pounce	Freeze
Euclid	Pounce	3	1
	Freeze	-2	2

The optimum strategy for Euclid is

$$p_1 = \frac{2 - (-2)}{3 - (-2) - 1 + 2} = \frac{4}{6} = \frac{2}{3},$$

$$p_2 = 1 - \frac{2}{3} = \frac{1}{3}.$$

Euclid should pounce $2/3$ of the time and freeze $1/3$ of the time.

The optimum strategy for Jamie is

$$q_1 = \frac{2 - 1}{3 - (-2) - 1 + 2} = \frac{1}{6},$$

$$q_2 = 1 - \frac{1}{6} = \frac{5}{6}.$$

Jamie should pounce $1/6$ of the time and freeze $5/6$ of the time.

The value of the game is

$$\frac{3(2) - (-2)(1)}{3 - (-2) - 1 + 2} = \frac{8}{6} = \frac{4}{3}.$$

31. (a) The payoff matrix is as follows.

Number of Fingers
B Shows

$$\begin{array}{c} \text{Number of} \\ \text{Fingers} \\ \text{A Shows} \end{array} \begin{array}{c} 1 \\ 2 \end{array} \begin{array}{cc} 1 & 2 \\ \left[\begin{array}{cc} 2 & -3 \\ -3 & 4 \end{array} \right] \end{array}$$

(b) For player A, the optimum strategy is

$$p_1 = \frac{4 - (-3)}{2 - (-3) - (-3) + 4} = \frac{7}{12},$$

$$p_2 = 1 - \frac{7}{12} = \frac{5}{12}.$$

For player B, the optimum strategy is

$$q_1 = \frac{4 - (-3)}{2 - (-3) - (-3) + 4} = \frac{7}{12},$$

$$q_2 = 1 - \frac{7}{12} = \frac{5}{12}.$$

This means that each player should show 1 finger 7/12 of the time and 2 fingers 5/12 of the time. The value of the game is

$$\frac{2(4) - (-3)(-3)}{2 - (-3) - (-3) + 4} = -\frac{1}{12}.$$

Section 9.4

1. $\begin{bmatrix} 1 & 2 \\ 3 & 1 \end{bmatrix}$

To find the optimum strategy for player A,

minimize $w = x + y$

subject to: $x + 3y \geq 1$

$\qquad\qquad\quad 2x + y \geq 1$

with $\qquad\qquad x \geq 0, \; y \geq 0.$

Solve this linear programming problem by the graphical method. Sketch the feasible region.

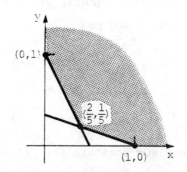

The region is unbounded, with corner points $(0, 1)$, $\left(\frac{2}{5}, \frac{1}{5}\right)$, and $(1, 0)$.

Corner Point	Value of $w = x + y$
$(0, 1)$	$0 + 1 = 1$
$\left(\frac{2}{5}, \frac{1}{5}\right)$	$\frac{2}{5} + \frac{1}{5} = \frac{3}{5}$
$(1, 0)$	$1 + 0 = 1$

The minimum value is $w = \frac{3}{5}$ at the point where $x = \frac{2}{5}$, $y = \frac{1}{5}$. Thus, the value of the game is $g = \frac{1}{w} = \frac{5}{3}$, and the optimum strategy for A is

$$p_1 = gx = \left(\frac{5}{3}\right)\left(\frac{2}{5}\right) = \frac{2}{3},$$

$$p_2 = gy = \left(\frac{5}{3}\right)\left(\frac{1}{5}\right) = \frac{1}{3}.$$

To find the optimum strategy for player B,

maximize $z = x + y$

subject to: $x + 2y \leq 1$

$\qquad 3x + y \leq 1$

with $x \geq 0, y \geq 0.$

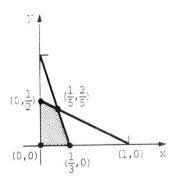

Corner Point	Value of $z = x + y$
$\left(0, \frac{1}{2}\right)$	$0 + \frac{1}{2} = \frac{1}{2}$
$\left(\frac{1}{5}, \frac{2}{5}\right)$	$\frac{1}{5} + \frac{2}{5} = \frac{3}{5}$
$\left(\frac{1}{3}, 0\right)$	$\frac{1}{3} + 0 = \frac{1}{3}$
$(0, 0)$	$0 + 0 = 0$

The maximum value is $z = \frac{3}{5}$ at the

point where $x = \frac{1}{5}, y = \frac{2}{5}.$ The value

of the game is $g = \frac{1}{z} = \frac{5}{3}$ (agreeing

with our earlier findings), and the

optimum strategy for B is

$$q_1 = gx = \left(\frac{5}{3}\right)\left(\frac{1}{5}\right) = \frac{1}{3} \text{ and}$$
$$q_2 = gy = \left(\frac{5}{3}\right)\left(\frac{2}{5}\right) = \frac{2}{3}.$$

3. $\begin{bmatrix} 4 & -2 \\ -1 & 6 \end{bmatrix}$

Get rid of negative numbers by adding

2 to each entry, to obtain

$\begin{bmatrix} 6 & 0 \\ 1 & 8 \end{bmatrix}.$

(The 2 will have to be subtracted

away later, after the calculations

have been performed.)

To find the optimum strategy for

player A,

minimize $w = x + y$

subject to: $6x + y \geq 1$

$\qquad 8y \geq 1$

with $x \geq 0, y \geq 0.$

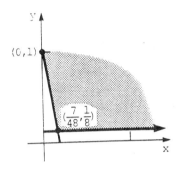

Corner Point	Value of $w = x + y$
$(0, 1)$	$0 + 1 = 1$
$\left(\frac{7}{48}, \frac{1}{8}\right)$	$\frac{7}{48} + \frac{1}{8} = \frac{13}{48}$

The minimum value is $w = \frac{13}{48}$ at

$\left(\frac{7}{48}, \frac{1}{8}\right).$ Thus, $g = \frac{1}{w} = \frac{48}{13}$, and the

optimum strategy for A is

$$p_1 = gx = \left(\frac{48}{13}\right)\left(\frac{7}{48}\right) = \frac{7}{13},$$
$$p_2 = gy = \left(\frac{48}{13}\right)\left(\frac{1}{8}\right) = \frac{6}{13}.$$

The value of the game is $\frac{48}{13} - 2 = \frac{22}{13}.$

To find the optimum strategy for

player B,

maximize $z = x + y$

subject to: $6x \qquad \le 1$

$\qquad\qquad x + 8y \le 1$

with $x \ge 0,\ y \ge 0.$

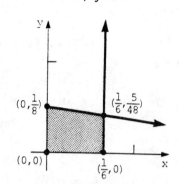

Corner Point	Value of $z = x + y$
$\left(0, \dfrac{1}{8}\right)$	$0 + \dfrac{1}{8} = \dfrac{1}{8}$
$\left(\dfrac{1}{6}, \dfrac{25}{48}\right)$	$\dfrac{1}{6} + \dfrac{5}{48} = \dfrac{13}{48}$
$\left(\dfrac{1}{6}, 0\right)$	$\dfrac{1}{6} + 0 = \dfrac{1}{6}$
$(0, 0)$	$0 + 0 = 0$

The maximum value is $z = \dfrac{13}{48}$ at $\left(\dfrac{1}{6}, \dfrac{5}{48}\right)$. The optimum strategy for B is

$$q_1 = gx = \left(\frac{48}{13}\right)\left(\frac{1}{6}\right) = \frac{8}{13},$$

$$q_2 = gy = \left(\frac{48}{13}\right)\left(\frac{5}{48}\right) = \frac{5}{13}.$$

5. $\begin{bmatrix} 8 & -7 \\ -2 & 4 \end{bmatrix}$

Add 7 to each entry, to obtain

$\begin{bmatrix} 15 & 0 \\ 5 & 11 \end{bmatrix}.$

To find the optimum strategy for player A,

minimize $w = x + y$

subject to: $15x + 5y \ge 1$

$\qquad\qquad\qquad 11y \ge 1$

with $x \ge 0,\ y \ge 0.$

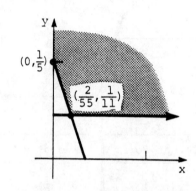

Corner Point	Value of $w = x + y$
$\left(0, \dfrac{1}{5}\right)$	$0 + \dfrac{1}{5} = \dfrac{1}{5}$
$\left(\dfrac{2}{55}, \dfrac{1}{11}\right)$	$\dfrac{2}{55} + \dfrac{1}{11} = \dfrac{7}{55}$

The minimum value is $w = \dfrac{7}{55}$ at $\left(\dfrac{2}{55}, \dfrac{1}{11}\right)$. Thus, $g = \dfrac{1}{w} = \dfrac{55}{7}$, and the optimum strategy for A is

$$p_1 = gx = \left(\frac{55}{7}\right)\left(\frac{2}{55}\right) = \frac{2}{7},$$

$$p_2 = gy = \left(\frac{55}{7}\right)\left(\frac{1}{11}\right) = \frac{5}{7}.$$

The value of the game is $\dfrac{55}{7} - 7 = \dfrac{6}{7}$.

To find the optimum strategy for player B,

maximize $z = x + y$

subject to: $15x \qquad\quad \le 1$

$\qquad\qquad 5x + 11y \le 1$

with $x \ge 0,\ y \ge 0.$

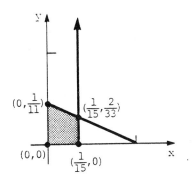

Corner Point	Value of $z = x + y$
$(0, \frac{1}{11})$	$0 + \frac{1}{11} = \frac{1}{11}$
$(\frac{1}{15}, \frac{2}{33})$	$\frac{1}{15} + \frac{2}{33} = \frac{21}{165} = \frac{7}{55}$
$(\frac{1}{15}, 0)$	$\frac{1}{15} + 0 = \frac{1}{15}$
$(0, 0)$	$0 + 0 = 0$

The maximum value is $z = \frac{7}{55}$ at

$(\frac{1}{15}, \frac{2}{33})$.

The optimum strategy for B is

$$q_1 = gx = \left(\frac{55}{7}\right)\left(\frac{1}{15}\right) = \frac{11}{21},$$

$$q_2 = gy = \left(\frac{55}{7}\right)\left(\frac{2}{33}\right) = \frac{10}{21}.$$

7. $\begin{bmatrix} 3 & -4 & 1 \\ 5 & 3 & -2 \end{bmatrix}$

Column 1 is dominated by the other
two columns, so delete it.

$\begin{bmatrix} -4 & 1 \\ 3 & -2 \end{bmatrix}$

Add 4 to each entry, to obtain

$\begin{bmatrix} 0 & 5 \\ 7 & 2 \end{bmatrix}$.

The linear programming problem to be
solved is as follows.

Maximize $\quad z = x_2 + x_3$

subject to: $\qquad 5x_3 \le 1$

$\qquad\qquad 7x_2 + 2x_3 \le 1$

with $\qquad x_2 \ge 0,\ x_3 \ge 0.$

Use the simplex method to solve the
problem.

The initial tableau is

$$\begin{array}{ccccc} x_2 & x_3 & x_4 & x_5 & z \\ \end{array}$$
$$\left[\begin{array}{ccccc|c} 0 & 5 & 1 & 0 & 0 & 1 \\ \textcircled{7} & 2 & 0 & 1 & 0 & 1 \\ \hline -1 & -1 & 0 & 0 & 1 & 0 \end{array}\right].$$

Pivot on the circled entry.

$$\begin{array}{ccccc} x_2 & x_3 & x_4 & x_5 & z \\ \end{array}$$
$$\left[\begin{array}{ccccc|c} 0 & \textcircled{5} & 1 & 0 & 0 & 1 \\ 7 & 2 & 0 & 1 & 0 & 1 \\ \hline 0 & -5 & 0 & 1 & 7 & 1 \end{array}\right]$$
$7R_3 + R_2 \to R_3$

Pivot again.

$$\begin{array}{ccccc} x_2 & x_3 & x_4 & x_5 & z \\ \end{array}$$
$$\begin{array}{l} \\ -2R_1 + 5R_2 \to R_2 \\ R_1 + R_3 \to R_3 \end{array} \left[\begin{array}{ccccc|c} 0 & 5 & 1 & 0 & 0 & 1 \\ 35 & 0 & -2 & 5 & 0 & 3 \\ \hline 0 & 0 & 1 & 1 & 7 & 2 \end{array}\right]$$

Create a 1 in the columns corresponding to x_2, x_3, and z.

$$\begin{array}{ccccc} x_2 & x_3 & x_4 & x_5 & z \\ \end{array}$$
$$\begin{array}{l} \frac{1}{5}R_1 \to R_1 \\ \\ \frac{1}{35}R_2 \to R_2 \\ \\ \frac{1}{7}R_3 \to R_3 \end{array} \left[\begin{array}{ccccc|c} 0 & 1 & \frac{1}{5} & 0 & 0 & \frac{1}{5} \\ 1 & 0 & -\frac{2}{35} & \frac{1}{7} & 0 & \frac{3}{35} \\ \hline 0 & 0 & \frac{1}{7} & \frac{1}{7} & 1 & \frac{2}{7} \end{array}\right]$$

From this final tableau, we have

$x_2 = \frac{3}{35}$, $x_3 = \frac{1}{5}$, $y_1 = \frac{1}{7}$, $y_2 = \frac{1}{7}$,

$z = \frac{2}{7}$. Note that $g = \frac{1}{z} = \frac{7}{2}$.

The optimum strategy for player A is

$$p_1 = gy_1 = \left(\frac{7}{2}\right)\left(\frac{1}{7}\right) = \frac{1}{2},$$

$$p_2 = gy_2 = \left(\frac{7}{2}\right)\left(\frac{1}{7}\right) = \frac{1}{2}.$$

The optimum strategy for player B is

$q_1 = 0$ (column 1 was removed),

$$q_2 = gx_2 = \left(\frac{7}{2}\right)\left(\frac{3}{35}\right) = \frac{3}{10},$$

$$q_3 = gx_3 = \left(\frac{7}{2}\right)\left(\frac{1}{5}\right) = \frac{7}{10}.$$

The value of the game is $\frac{7}{2} - 4 = -\frac{1}{2}$.

9. $\begin{bmatrix} -1 & 2 & 4 \\ 3 & -2 & 0 \end{bmatrix}$

Delete column 3, since it is dominated by column 2.

$\begin{bmatrix} -1 & 2 \\ 3 & -2 \end{bmatrix}$

Add 2 to each entry, to obtain

$\begin{bmatrix} 1 & 4 \\ 5 & 0 \end{bmatrix}.$

The linear programming problem to be solved is as follows.

Maximize $z = x_1 + x_2$

subject to: $x_1 + 4x_2 \le 1$

$\qquad\qquad 5x_1 \qquad\quad \le 1$

with $x_1 \ge 0,\ x_2 \ge 0.$

The initial tableau is

$$\begin{array}{ccccc} x_1 & x_2 & x_4 & x_5 & z \\ \end{array}$$
$$\begin{bmatrix} 1 & 4 & 1 & 0 & 0 & 1 \\ \boxed{5} & 0 & 0 & 1 & 0 & 1 \\ -1 & -1 & 0 & 0 & 1 & 0 \end{bmatrix}.$$

Pivot on each circled entry.

$$\begin{array}{cccccc} & x_1 & x_2 & x_4 & x_5 & z \\ \end{array}$$

$5R_1 - R_2 \to R_1 \quad \begin{bmatrix} 0 & 20 & 5 & -1 & 0 & 4 \\ 5 & 0 & 0 & 1 & 0 & 1 \\ 0 & -5 & 0 & 1 & 5 & 1 \end{bmatrix}$
$5R_3 + R_2 \to R_3$

$$\begin{array}{ccccc} x_1 & x_2 & x_4 & x_5 & z \\ \end{array}$$

$\qquad\qquad\qquad \begin{bmatrix} 0 & 20 & 5 & -1 & 0 & 4 \\ 5 & 0 & 0 & 1 & 0 & 1 \\ 0 & 0 & 5 & 3 & 20 & 8 \end{bmatrix}$
$4R_3 + R_1 \to R_3$

Create a 1 in the columns corresponding to x_1, x_2, and z.

$$\begin{array}{ccccc} x_1 & x_2 & x_4 & x_5 & z \\ \end{array}$$

$\frac{1}{20}R_1 \to R_1 \begin{bmatrix} 0 & 1 & \frac{1}{4} & -\frac{1}{20} & 0 & \frac{1}{5} \\ 1 & 0 & 0 & \frac{1}{5} & 0 & \frac{1}{5} \\ 0 & 0 & \frac{1}{4} & \frac{3}{20} & 1 & \frac{2}{5} \end{bmatrix}$
$\frac{1}{5}R_2 \to R_2$
$\frac{1}{20}R_3 \to R_3$

From this final tableau, we have

$x_1 = \frac{1}{5}$, $x_2 = \frac{1}{5}$, $y_1 = \frac{1}{4}$. $y_2 = \frac{3}{20}$, and

$z = \frac{2}{5}$.

Note that $g = \frac{1}{z} = \frac{5}{2}$.

The optimum strategy for player A is

$$p_1 = gy_1 = \left(\frac{5}{2}\right)\left(\frac{1}{4}\right) = \frac{5}{8},$$

$$p_2 = gy_2 = \left(\frac{5}{2}\right)\left(\frac{3}{20}\right) = \frac{3}{8}.$$

The optimum strategy for player B is

$$q_1 = gx_1 = \left(\frac{5}{2}\right)\left(\frac{1}{5}\right) = \frac{1}{2},$$

$$q_2 = gx_2 = \left(\frac{5}{2}\right)\left(\frac{1}{5}\right) = \frac{1}{2},$$

$q_3 = 0$ (column 3 was removed).

The value of the game is $\frac{5}{2} - 2 = \frac{1}{2}$.

11. $\begin{bmatrix} 1 & 0 & -1 \\ -1 & 0 & 1 \\ 2 & -1 & 2 \end{bmatrix}$

Add 1 to each entry, to obtain

$\begin{bmatrix} 2 & 1 & 0 \\ 0 & 1 & 2 \\ 3 & 0 & 3 \end{bmatrix}$.

The linear programming problem to be solved is as follows.

Maximize $z = x_1 + x_2 + x_3$

subject to: $2x_1 + x_2 \qquad\quad \le 1$

$\qquad\qquad\quad x_2 + 2x_3 \le 1$

$\qquad\quad 3x_1 + \qquad 3x_3 \le 1$

with $x_1 \ge 0,\ x_2 \ge 0,\ x_3 \ge 0$.

The initial tableau is

$\begin{array}{cccccc} x_1 & x_2 & x_3 & x_4 & x_5 & x_6 & z \end{array}$

$\begin{bmatrix} 2 & 1 & 0 & 1 & 0 & 0 & 0 & 1 \\ 0 & 1 & 2 & 0 & 1 & 0 & 0 & 1 \\ ③ & 0 & 3 & 0 & 0 & 1 & 0 & 1 \\ \hline -1 & -1 & -1 & 0 & 0 & 0 & 1 & 0 \end{bmatrix}.$

Pivot on each circled entry.

$\begin{array}{cccccccc} & x_1 & x_2 & x_3 & x_4 & x_5 & x_6 & z \end{array}$

$\begin{array}{l} 3R_1 - 2R_3 \to R_1 \\ \\ \\ 3R_4 + R_3 \to R_4 \end{array} \begin{bmatrix} 0 & ③ & -6 & 3 & 0 & -2 & 0 & 1 \\ 0 & 1 & 2 & 0 & 1 & 0 & 0 & 1 \\ 3 & 0 & 3 & 0 & 0 & 1 & 0 & 1 \\ \hline 0 & -3 & 0 & 0 & 0 & 1 & 3 & 1 \end{bmatrix}$

$\begin{array}{cccccccc} & x_1 & x_2 & x_3 & x_4 & x_5 & x_6 & z \end{array}$

$\begin{array}{l} 3R_2 - R_1 \to R_2 \\ \\ \\ R_1 + R_4 \to R_4 \end{array} \begin{bmatrix} 0 & 3 & -6 & 3 & 0 & -2 & 0 & 1 \\ 0 & 0 & ⑫ & -3 & 3 & 2 & 0 & 2 \\ 3 & 0 & 3 & 0 & 0 & 1 & 0 & 1 \\ \hline 0 & 0 & -6 & 3 & 0 & -1 & 3 & 2 \end{bmatrix}$

$\begin{array}{cccccccc} & x_1 & x_2 & x_3 & x_4 & x_5 & x_6 & z \end{array}$

$\begin{array}{l} 2R_1 + R_2 \to R_1 \\ \\ \\ 4R_3 - R_2 \to R_3 \\ \\ 2R_4 + R_2 \to R_4 \end{array} \begin{bmatrix} 0 & 6 & 0 & 3 & 3 & -2 & 0 & 4 \\ 0 & 0 & 12 & -3 & 3 & 2 & 0 & 2 \\ 12 & 0 & 0 & 3 & -3 & 2 & 0 & 2 \\ \hline 0 & 0 & 0 & 3 & 3 & 0 & 6 & 6 \end{bmatrix}$

Create a 1 in the columns corresponding to x_1, x_2, x_3, and z.

$\begin{array}{ccccccc} x_1 & x_2 & x_3 & x_4 & x_5 & x_6 & z \end{array}$

$\begin{array}{l} \frac{1}{6}R_1 \to R_1 \\ \\ \frac{1}{12}R_2 \to R_2 \\ \\ \frac{1}{12}R_3 \to R_3 \\ \\ \frac{1}{6}R_4 \to R_4 \end{array} \begin{bmatrix} 0 & 1 & 0 & \frac{1}{2} & \frac{1}{2} & -\frac{1}{3} & 0 & \frac{2}{3} \\ 0 & 0 & 1 & -\frac{1}{4} & \frac{1}{4} & \frac{1}{6} & 0 & \frac{1}{6} \\ 1 & 0 & 0 & \frac{1}{4} & -\frac{1}{4} & \frac{1}{6} & 0 & \frac{1}{6} \\ \hline 0 & 0 & 0 & \frac{1}{2} & \frac{1}{2} & 0 & 1 & 1 \end{bmatrix}$

From this final tableau, we have

$x_1 = \frac{1}{6},\ x_2 = \frac{2}{3},\ x_3 = \frac{1}{6},\ y_1 = \frac{1}{2},$

$y_2 = \frac{1}{2},\ y_3 = 0,$ and $z = 1.$

Note that $g = \frac{1}{z} = 1.$

The optimum strategy for player A is

$p_1 = gy_1 = (1)\left(\frac{1}{2}\right) = \frac{1}{2},$

$p_2 = gy_2 = (1)\left(\frac{1}{2}\right) = \frac{1}{2},$

$p_3 = gy_3 = (1)(0) = 0.$

The optimum strategy for player B is

$q_1 = gx_1 = (1)\left(\frac{1}{6}\right) = \frac{1}{6},$

$q_2 = gx_2 = (1)\left(\frac{2}{3}\right) = \frac{2}{3},$

$q_3 = gx_3 = (1)\left(\frac{1}{6}\right) = \frac{1}{6}.$

The value of the game is $1 - 1 = 0.$

13. The payoff matrix is as follows.

	Strike	No Strike
Bid $30,000	-$5500	$4500
Bid $40,000	$4500	$0

Add $5500 to each entry, to obtain

$$\begin{bmatrix} 0 & 10,000 \\ 10,000 & 5500 \end{bmatrix}.$$

To find the optimum strategy for the contractor,

minimize $w = x + y$

subject to: $10,000y \geq 1$

$10,000x + 5500y \geq 1$

with $x \geq 0, \ y \geq 0.$

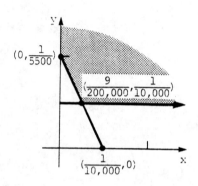

Corner Point	Value of $w = x + y$
$(0, \dfrac{1}{5500})$	$0 + \dfrac{1}{5500} = \dfrac{1}{5500}$
$(\dfrac{9}{200,000}, \dfrac{1}{10,000})$	$\dfrac{9}{200,000} + \dfrac{1}{10,000} = \dfrac{29}{200,000}$

The minimum value is $w = \dfrac{29}{200,000}$ at $(\dfrac{9}{200,000}, \dfrac{1}{10,000})$. Thus, $g = \dfrac{1}{w} = \dfrac{200,000}{29}$, and the optimum strategy for the contractor is

$$p_1 = gx = \left(\frac{200,000}{29}\right)\left(\frac{9}{200,000}\right) = \frac{9}{29},$$

$$p_2 = gy = \left(\frac{200,000}{29}\right)\left(\frac{1}{10,000}\right) = \frac{20}{29}.$$

That is, the contractor should bid $30,000 with probability $\dfrac{9}{29}$ and bid $40,000 with probability $\dfrac{20}{29}$. The value of the game is

$$\frac{200,000}{29} - 5500 \approx \$1396.55.$$

15. (a) The payoff matrix is as follows.

Original Imitators

		Atlanta	Boston	Cleveland
	Atlanta	$5000	$10,000	$10,000
General Items	Boston	$8000	$4000	$8000
	Cleveland	$6000	$6000	$3000

Note that $5000 = \frac{1}{2}(10,000)$, $4000 = \frac{1}{2}(8000)$, and $3000 = \frac{1}{2}(6000)$ are the reduced profits for General Items when the two companies run ads in the same city.

(b) The linear programming problem to be solved is as follows.

Maximize $\quad z = x_1 + x_2 + x_3$

subject to: $\quad 5000x_1 + 10,000x_2 + 10,000x_3 \leq 1$

$\qquad\qquad 8000x_1 + \quad 4000x_2 + \quad 8000x_3 \leq 1$

$\qquad\qquad 6000x_1 + \quad 6000x_2 + \quad 3000x_3 \leq 1$

with $\qquad x_1 \geq 0,\ x_2 \geq 0,\ x_3 \geq 0.$

The initial tableau is

	x_1	x_2	x_3	x_4	x_5	x_6	z	
	5000	10,000	10,000	1	0	0	0	1
	(8000)	4000	8000	0	1	0	0	1
	6000	6000	3000	0	0	1	0	1
	−1	−1	−1	0	0	0	1	0

Pivot on each circled entry.

	x_1	x_2	x_3	x_4	x_5	x_6	z	
$-5R_2 + 8R_1 \to R_1$	0	(60,000)	40,000	8	−5	0	0	3
	8000	4000	8000	0	1	0	0	1
$-3R_2 + 4R_3 \to R_3$	0	12,000	−12,000	0	−3	4	0	1
$R_2 + 8000R_4 \to R_4$	0	−4000	0	0	1	1	8000	1

	x_1	x_2	x_3	x_4	x_5	x_6	z	
	0	60,000	40,000	8	−5	0	0	3
$-R_1 + 15R_2 \to R_2$	120,000	0	80,000	−8	20	0	0	12
$-R_1 + 5R_3 \to R_3$	0	0	−100,000	−8	−10	20	0	2
$R_1 + 15R_4 \to R_4$	0	0	40,000	8	10	0	120,000	18

Create a 1 in the columns corresponding to x_1, x_2, x_6, and z.

$$
\begin{array}{c}
\frac{1}{60,000}R_1 \to R_1 \\[4pt]
\frac{1}{120,000}R_2 \to R_2 \\[4pt]
\frac{1}{20}R_3 \to R_3 \\[4pt]
\frac{1}{120,000}R_4 \to R_4
\end{array}
\quad
\begin{array}{ccccccc|c}
x_1 & x_2 & x_3 & x_4 & x_5 & x_6 & z & \\
0 & 1 & \frac{2}{3} & \frac{1}{7500} & -\frac{1}{12,000} & 0 & 0 & \frac{1}{20,000} \\[4pt]
0 & 0 & \frac{2}{3} & -\frac{1}{15,000} & \frac{1}{6000} & 0 & 0 & \frac{1}{10,000} \\[4pt]
0 & 0 & -5000 & -\frac{2}{5} & -\frac{1}{2} & 1 & 0 & \frac{1}{10} \\[4pt]
0 & 0 & \frac{1}{3} & \frac{1}{15,000} & \frac{1}{12,000} & 0 & 1 & \frac{3}{20,000}
\end{array}
$$

From this final tableau, we have

$$x_1 = \frac{1}{10,000}, \ x_2 = \frac{1}{20,000}, \ x_3 = 0, \ y_1 = \frac{1}{15,000}, \ y_2 = \frac{1}{12,000}, \ y_3 = 0, \ \text{and}$$

$$z = \frac{3}{20,000}.$$

Note that $g = \frac{1}{z} = \frac{20,000}{3} \approx 6666.67$, so the value of the game is \$6666.67.

The optimum strategy for General Items is

$$p_1 = gy_1 = \left(\frac{20,000}{3}\right)\left(\frac{1}{15,000}\right) = \frac{4}{9},$$

$$p_2 = gy_2 = \left(\frac{20,000}{3}\right)\left(\frac{1}{12,000}\right) = \frac{5}{9},$$

$$p_3 = gy_3 = \left(\frac{20,000}{3}\right)(0) = 0.$$

That is, General Items should advertise in Atlanta with probability $\frac{4}{9}$, in Boston with probability $\frac{5}{9}$, and never in Cleveland.

The optimum strategy for Original Imitators is

$$q_1 = gx_1 = \left(\frac{20,000}{3}\right)\left(\frac{1}{10,000}\right) = \frac{2}{3},$$

$$q_2 = gx_2 = \left(\frac{20,000}{3}\right)\left(\frac{1}{20,000}\right) = \frac{1}{3},$$

$$q_3 = gx_3 = \left(\frac{20,000}{3}\right)(0) = 0.$$

That is, Original Imitators should advertise in Atlanta with probability $\frac{2}{3}$, in Boston with probability $\frac{1}{3}$, and never in Cleveland.

17. $\begin{bmatrix} 50 & 0 \\ 10 & 40 \end{bmatrix}$

To find the student's optimum

strategy,

minimize $w = x + y$

subject to: $50x + 10y \geq 1$

$40y \geq 1$

with $x \geq 0,\ y \geq 0$.

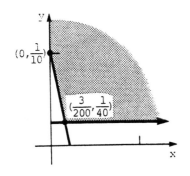

Corner Point	Value of $w = x + y$
$(0, \frac{1}{10})$	$0 + \frac{1}{10} = \frac{1}{10}$
$(\frac{3}{200}, \frac{1}{40})$	$\frac{3}{200} + \frac{1}{40} = \frac{1}{25}$

The minimum value is $w = \frac{1}{25}$ at

$(\frac{3}{200}, \frac{1}{40})$.

Thus, the value of the game is

$g = \frac{1}{w} = 25$, and the optimum strategy

for the student is

$p_1 = gx = (25)\left(\frac{3}{200}\right) = \frac{3}{8}$,

$p_2 = gy = (25)\left(\frac{1}{40}\right) = \frac{5}{8}$.

That is, the student should choose

the calculator with probability $\frac{3}{8}$

and the book with probability $\frac{5}{8}$.

19. (a) The payoff matrix is as follows.

	Rock	Scissors	Paper
Rock	0	1	-1
Scissors	-1	0	1
Paper	1	-1	0

Add 1 to each entry, to obtain

$$\begin{bmatrix} 1 & 2 & 0 \\ 0 & 1 & 2 \\ 2 & 0 & 1 \end{bmatrix}.$$

The linear programming problem to be

solved is

maximize $w = x_1 + x_2 + x_3$

subject to: $x_1 + 2x_2 \leq 1$

$x_2 + 2x_3 \leq 1$

$2x_1 + x_3 \leq 1$

with $x_1 \geq 0,\ x_2 \geq 0,\ x_3 \geq 0$.

The initial tableau is

x_1	x_2	x_3	x_4	x_5	x_6	z	
1	②	0	1	0	0	0	1
0	1	2	0	1	0	0	1
2	0	1	0	0	1	0	1
-1	-1	-1	0	0	0	1	0

Pivot on each circled entry.

	x_1	x_2	x_3	x_4	x_5	x_6	z	
	1	2	0	1	0	0	0	1
$2R_2 - R_1 \rightarrow R_2$	-1	0	④	-1	2	0	0	1
	2	0	1	0	0	1	0	1
$2R_4 + R_1 \rightarrow R_4$	-1	0	-2	1	0	0	2	1

	x_1	x_2	x_3	x_4	x_5	x_6	z	
	1	2	0	1	0	0	0	1
	-1	0	4	-1	2	0	0	1
$4R_3 - R_2 \rightarrow R_3$	⑨	0	0	1	-2	4	0	3
$2R_4 + R_2 \rightarrow R_4$	-3	0	0	1	2	0	4	3

$$
\begin{array}{c}
\\
9R_1 - R_3 \to R_1 \\
9R_2 + R_3 \to R_2 \\
\\
3R_4 + R_3 \to R_4
\end{array}
\quad
\begin{array}{ccccccc|c}
x_1 & x_2 & x_3 & x_4 & x_5 & x_6 & z & \\
0 & 18 & 0 & 8 & 2 & -4 & 0 & 6 \\
0 & 0 & 36 & -8 & 16 & 4 & 0 & 12 \\
9 & 0 & 0 & 1 & -2 & 4 & 0 & 3 \\
0 & 0 & 0 & 4 & 4 & 4 & 12 & 12
\end{array}
$$

Create a 1 in the columns corresponding to x_1, x_2, x_3, and z.

$$
\begin{array}{c}
\\
\frac{1}{18}R_1 \to R_1 \\[4pt]
\frac{1}{36}R_2 \to R_2 \\[4pt]
\frac{1}{9}R_3 \to R_3 \\[4pt]
\frac{1}{12}R_4 \to R_4
\end{array}
\quad
\begin{array}{ccccccc|c}
x_1 & x_2 & x_3 & x_4 & x_5 & x_6 & z & \\
0 & 1 & 0 & \frac{4}{9} & \frac{1}{9} & -\frac{2}{9} & 0 & \frac{1}{3} \\[4pt]
0 & 0 & 1 & -\frac{2}{9} & \frac{4}{9} & \frac{1}{9} & 0 & \frac{1}{3} \\[4pt]
1 & 0 & 0 & \frac{1}{9} & -\frac{2}{9} & \frac{4}{9} & 0 & \frac{1}{3} \\[4pt]
0 & 0 & 0 & \frac{1}{3} & \frac{1}{3} & \frac{1}{3} & 1 & 1
\end{array}
$$

From this final tableau, we have

$$x_1 = \frac{1}{3}, \ x_2 = \frac{1}{3}, \ x_3 = \frac{1}{3},$$

$$y_1 = \frac{1}{3}, \ y_2 = \frac{1}{3}, \ y_3 = \frac{1}{3}, \text{ and } z = 1.$$

Note that $g = \frac{1}{z} = 1$.

The optimum strategy for player A is

$$p_1 = gy_1 = (1)\left(\frac{1}{3}\right) = \frac{1}{3},$$

$$p_2 = gy_2 = (1)\left(\frac{1}{3}\right) = \frac{1}{3},$$

$$p_3 = gy_3 = (1)\left(\frac{1}{3}\right) = \frac{1}{3}.$$

The optimum strategy for player B is

$$q_1 = gx_1 = (1)\left(\frac{1}{3}\right) = \frac{1}{3},$$

$$q_2 = gx_2 = (1)\left(\frac{1}{3}\right) = \frac{1}{3},$$

$$q_3 = gx_3 = (1)\left(\frac{1}{3}\right) = \frac{1}{3}.$$

The value of the game is $1 - 1 = 0$.

(b) The game is symmetric in that neither player has an advantage, and each choice is as strong as every other choice.

21. **(a)** The payoff matrix is as follows.

<u>Kije</u>

		(3, 0)	(0, 3)	(2, 1)	(1, 2)
	(4, 0)	4	0	2	1
	(0, 4)	0	4	1	2
Blotto	(3, 1)	1	−1	3	0
	(1, 3)	−1	1	0	3
	(2, 2)	−2	−2	2	2

Note that the strategy (3, 1) for Colonel Blotto means that he sends 3 regiments to the first post and 1 regiment to the second post; the strategy (0, 3) for Captain Kije means that he sends 0 regiments to the first post and 3 regiments to the second post.

Row 1, column 1 of the payoff matrix means (4, 0) for Blotto and (3, 0) for Kije. Blotto sends more regiments to the first post, so he wins 1 point, plus 3 points for capturing Kije's 3 regiments. Neither leader sends any regiments to the second post, so no points are won there. The payoff here is 1 + 3 = 4 points to Blotto.

Row 4, column 2 of the matrix means (1, 3) for Blotto and (0, 3) for Kije. Blotto earns a point for capturing the first post, but Kije sends no regiments there so there are no points for capturing regiments. Both leaders send 3 regiments to the second post, so there is a standoff and no additional points. The payoff here is 1 point to Blotto.

Row 5, column 1 of the matrix means (2, 2) for Blotto and (3, 0) for Kije. Kije earns 1 post-capturing point and 2 regiment-capturing points for the first post, while Blotto earns 1 post-capturing point for the second post. The net result is a payoff of 2 points to Kije, represented in the matrix as −2.

Continue in this manner to obtain all of the entries of the payoff matrix.

(b) Add 2 to each entry of the payoff matrix, to obtain

$$\begin{bmatrix} 6 & 2 & 4 & 3 \\ 2 & 6 & 3 & 4 \\ 3 & 1 & 5 & 2 \\ 1 & 3 & 2 & 5 \\ 0 & 0 & 4 & 4 \end{bmatrix}.$$

The linear programming problem to be solved is

maximize $\quad z = x_1 + x_2 + x_3 + x_4$

subject to: $6x_1 + 2x_2 + 4x_3 + 3x_4 \leq 1$

$\qquad\qquad 2x_1 + 6x_2 + 3x_3 + 4x_4 \leq 1$

$\qquad\qquad 3x_1 + x_2 + 5x_3 + 2x_4 \leq 1$

$\qquad\qquad x_1 + 3x_2 + 2x_3 + 5x_4 \leq 1$

$\qquad\qquad\qquad\qquad 4x_3 + 4x_4 \leq 1$

with $\qquad x_1 \geq 0, \ x_2 \geq 0, \ x_3 \geq 0, \ x_4 \geq 0.$

The initial tableau is

x_1	x_2	x_3	x_4	x_5	x_6	x_7	x_8	x_9	z	
⑥	2	4	3	1	0	0	0	0	0	1
2	6	3	4	0	1	0	0	0	0	1
3	1	5	2	0	0	1	0	0	0	1
1	3	2	5	0	0	0	1	0	0	1
0	0	4	4	0	0	0	0	1	0	1
−1	−1	−1	−1	0	0	0	0	0	1	0

.

Pivot on each circled entry.

	x_1	x_2	x_3	x_4	x_5	x_6	x_7	x_8	x_9	z	
	6	2	4	3	1	0	0	0	0	0	1
$3R_2 - R_1 \rightarrow R_2$	0	⑯	5	9	−1	3	0	0	0	0	2
$2R_3 - R_1 \rightarrow R_3$	0	0	6	1	−1	0	2	0	0	0	1
$6R_4 - R_1 \rightarrow R_4$	0	16	8	27	−1	0	0	6	0	0	5
	0	0	4	4	0	0	0	0	1	0	1
$6R_6 + R_1 \rightarrow R_6$	0	−4	−2	−3	1	0	0	0	0	6	1

	x_1	x_2	x_3	x_4	x_5	x_6	x_7	x_8	x_9	z	
$8R_1 - R_2 \rightarrow R_1$	48	0	27	15	9	−3	0	0	0	0	6
	0	16	5	9	−1	3	0	0	0	0	2
	0	0	⑥	1	−1	0	2	0	0	0	1
$R_4 - R_2 \rightarrow R_4$	0	0	3	18	0	−3	0	6	0	0	3
	0	0	4	4	0	0	0	0	1	0	1
$4R_6 + R_2 \rightarrow R_6$	0	0	−3	−3	3	3	0	0	0	24	6

$$\begin{array}{c} \\ 2R_1 - 9R_3 \rightarrow R_1 \\ 6R_2 - 5R_3 \rightarrow R_2 \\ \\ 2R_4 - R_3 \rightarrow R_4 \\ 3R_5 - 2R_3 \rightarrow R_5 \\ 2R_6 + R_3 \rightarrow R_6 \end{array} \quad \begin{array}{cccccccccc} x_1 & x_2 & x_3 & x_4 & x_5 & x_6 & x_7 & x_8 & x_9 & z \end{array}$$

$$\left[\begin{array}{cccccccccc|c} 96 & 0 & 0 & 21 & 27 & -6 & -18 & 0 & 0 & 0 & 3 \\ 0 & 96 & 0 & 49 & -1 & 18 & -10 & 0 & 0 & 0 & 7 \\ 0 & 0 & 6 & 1 & -1 & 0 & 2 & 0 & 0 & 0 & 1 \\ 0 & 0 & 0 & 35 & 1 & -6 & -2 & 12 & 0 & 0 & 5 \\ 0 & 0 & 0 & \boxed{10} & 2 & 0 & -4 & 0 & 3 & 0 & 1 \\ 0 & 0 & 0 & -5 & 5 & 6 & 2 & 0 & 0 & 48 & 13 \end{array}\right]$$

$$\begin{array}{c} \\ 10R_1 - 21R_5 \rightarrow R_1 \\ 10R_2 - 49R_5 \rightarrow R_2 \\ 10R_3 - R_5 \rightarrow R_3 \\ 2R_4 - 7R_5 \rightarrow R_4 \\ \\ 2R_6 + R_5 \rightarrow R_6 \end{array} \quad \begin{array}{cccccccccc} x_1 & x_2 & x_3 & x_4 & x_5 & x_6 & x_7 & x_8 & x_9 & z \end{array}$$

$$\left[\begin{array}{cccccccccc|c} 960 & 0 & 0 & 0 & 228 & -60 & -96 & 0 & -63 & 0 & 9 \\ 0 & 960 & 0 & 0 & -108 & 180 & 96 & 0 & -147 & 0 & 21 \\ 0 & 0 & 60 & 0 & -12 & 0 & 24 & 0 & -3 & 0 & 9 \\ 0 & 0 & 0 & 0 & -12 & -12 & 24 & 24 & -21 & 0 & 3 \\ 0 & 0 & 0 & 10 & 2 & 0 & -4 & 0 & 3 & 0 & 1 \\ 0 & 0 & 0 & 0 & 12 & 12 & 0 & 0 & 3 & 96 & 27 \end{array}\right]$$

Create a 1 in the columns corresponding to x_1, x_2, x_3, x_4, x_8, and z.

$$\begin{array}{c} \\ \frac{1}{960}R_1 \rightarrow R_1 \\[4pt] \frac{1}{960}R_2 \rightarrow R_2 \\[4pt] \frac{1}{60}R_3 \rightarrow R_3 \\[4pt] \frac{1}{24}R_4 \rightarrow R_4 \\[4pt] \frac{1}{10}R_5 \rightarrow R_5 \\[4pt] \frac{1}{96}R_6 \rightarrow R_6 \end{array} \quad \begin{array}{cccccccccc} x_1 & x_2 & x_3 & x_4 & x_5 & x_6 & x_7 & x_8 & x_9 & z \end{array}$$

$$\left[\begin{array}{cccccccccc|c} 1 & 0 & 0 & 0 & \frac{19}{80} & -\frac{1}{16} & -\frac{1}{10} & 0 & -\frac{21}{320} & 0 & \frac{3}{320} \\[4pt] 0 & 1 & 0 & 0 & -\frac{9}{80} & \frac{3}{16} & \frac{1}{10} & 0 & -\frac{49}{320} & 0 & \frac{7}{320} \\[4pt] 0 & 0 & 1 & 0 & -\frac{1}{5} & 0 & \frac{2}{5} & 0 & -\frac{1}{20} & 0 & \frac{3}{20} \\[4pt] 0 & 0 & 0 & 0 & -\frac{1}{2} & -\frac{1}{2} & 1 & 1 & -\frac{7}{8} & 0 & \frac{1}{8} \\[4pt] 0 & 0 & 0 & 1 & \frac{1}{5} & 0 & -\frac{2}{5} & 0 & \frac{3}{10} & 0 & \frac{1}{10} \\[4pt] 0 & 0 & 0 & 0 & \frac{1}{8} & \frac{1}{8} & 0 & 0 & \frac{1}{32} & 1 & \frac{9}{32} \end{array}\right]$$

From this final tableau, we have

$$x_1 = \frac{3}{320}, \quad x_2 = \frac{7}{320}, \quad x_3 = \frac{3}{20}, \quad x_4 = \frac{1}{10},$$

$$y_1 = \frac{1}{8}, \quad y_2 = \frac{1}{8}, \quad y_3 = 0, \quad y_4 = 0, \quad y_5 = \frac{1}{32}, \quad \text{and} \quad z = \frac{9}{32}.$$

Note that $g = \frac{1}{z} = \frac{32}{9}$.

The optimum strategy for Colonel Blotto is

$$p_1 = gy_1 = \left(\frac{32}{9}\right)\left(\frac{1}{8}\right) = \frac{4}{9},$$

$$p_2 = gy_2 = \left(\frac{32}{9}\right)\left(\frac{1}{8}\right) = \frac{4}{9},$$

$$p_3 = gy_3 = \left(\frac{32}{9}\right)(0) = 0,$$

$$p_4 = gy_4 = \left(\frac{32}{9}\right)(0) = 0,$$

$$p_5 = gy_5 = \left(\frac{32}{9}\right)\left(\frac{1}{32}\right) = \frac{1}{9}.$$

That is, Blotto uses strategies (4, 0) and (0, 4) with probability 4/9 each, strategy (2, 2) with probability 1/9, and never sends 3 regiments to one post and 1 to the other.

The optimum strategy for Captain Kije is

$$q_1 = gx_1 = \left(\frac{32}{9}\right)\left(\frac{3}{320}\right) = \frac{1}{30},$$

$$q_2 = gx_2 = \left(\frac{32}{9}\right)\left(\frac{7}{320}\right) = \frac{7}{90},$$

$$q_3 = gx_3 = \left(\frac{32}{9}\right)\left(\frac{3}{20}\right) = \frac{8}{15},$$

$$q_4 = gx_4 = \left(\frac{32}{9}\right)\left(\frac{1}{10}\right) = \frac{16}{45}.$$

That is, Kije uses strategy (3, 0) with probability 1/30, strategy (0, 3) with probability 7/90, strategy (2, 1) with probability 8/15, and strategy (1, 2) with probability 16/45.

The value of the game is

$$\frac{32}{9} - 2 = \frac{14}{9}.$$

(c) Let $B = \begin{bmatrix} q_1 \\ q_2 \\ q_3 \\ q_4 \end{bmatrix}$ be any stragegy that Captain Kije could use, which means that it is a probability vector and $q_1 + q_2 + q_3 + q_4 = 1$. If Colonel Blotto uses the strategy $A = \begin{bmatrix} \frac{4}{9} & \frac{4}{9} & 0 & 0 & \frac{1}{9} \end{bmatrix}$ that was found in part (b), then

$$AMB = \begin{bmatrix} \frac{4}{9} & \frac{4}{9} & 0 & 0 & \frac{1}{9} \end{bmatrix} \begin{bmatrix} 4 & 0 & 2 & 1 \\ 0 & 4 & 1 & 2 \\ 1 & -1 & 3 & 0 \\ -1 & 1 & 0 & 3 \\ -2 & -2 & 2 & 2 \end{bmatrix} \begin{bmatrix} q_1 \\ q_2 \\ q_3 \\ q_4 \end{bmatrix}$$

$$= \begin{bmatrix} \frac{14}{9} & \frac{14}{9} & \frac{14}{9} & \frac{14}{9} \end{bmatrix} \begin{bmatrix} q_1 \\ q_2 \\ q_3 \\ q_4 \end{bmatrix}$$

$$= \begin{bmatrix} \frac{14}{9}q_1 + \frac{14}{9}q_2 + \frac{14}{9}q_3 + \frac{14}{9}q_4 \end{bmatrix}$$

$$= \begin{bmatrix} \frac{14}{9}(q_1 + q_2 + q_3 + q_4) \end{bmatrix}$$

$$= \begin{bmatrix} \frac{14}{9}(1) \end{bmatrix}$$

$$= \begin{bmatrix} \frac{14}{9} \end{bmatrix}.$$

Therefore, the value of the game is $\frac{14}{9}$ regardless of what strategy Captain Kije uses.

Chapter 9 Review Exercises

3. $\begin{bmatrix} -2 & 5 & -6 & 3 \\ 0 & -1 & 7 & 5 \\ 2 & 6 & -4 & 4 \end{bmatrix}$

The entry at $(1, 1)$ is -2, indicating that the payoff is $2 from A to B.

5. The entry at $(2, 3)$ is 7, indicating that the payoff is $7 from B to A.

7. Row 3 dominates row 1 and column 1 dominates column 4.

9. $\begin{bmatrix} -11 & 6 & 8 & 9 \\ -10 & -12 & 3 & 2 \end{bmatrix}$

Column 1 dominates both column 3 and column 4. Remove the dominated columns to obtain

$$\begin{bmatrix} -11 & 6 \\ -10 & -12 \end{bmatrix}.$$

11. $\begin{bmatrix} -2 & 4 & 1 \\ 3 & 2 & 7 \\ -8 & 1 & 6 \\ 0 & 3 & 9 \end{bmatrix}$

Row 2 dominates row 3. Remove row 3 to obtain

$$\begin{bmatrix} -2 & 4 & 1 \\ 3 & 2 & 7 \\ 0 & 3 & 9 \end{bmatrix}.$$

Column 1 dominates column 3. Remove column 3 to obtain

$$\begin{bmatrix} -2 & 4 \\ 3 & 2 \\ 0 & 3 \end{bmatrix}.$$

13. $\begin{bmatrix} \underline{-2} & 3 \\ \underline{-4} & ⑤ \end{bmatrix}$

Underline the smallest number in each row and circle the largest value in each column. The -2 at $(1, 1)$ is the smallest number in its row and the largest number in its column, so the saddle point is -2 at $(1, 1)$. The value of the game is -2.

15. $\begin{bmatrix} \underline{-4} & -1 \\ 6 & \underline{⓪} \\ ⑧ & \underline{3} \end{bmatrix}$

The 0 at $(2, 2)$ is the smallest number in its row and the largest number in its column, so the saddle point is 0 at $(2, 2)$. The value of the game is 0, and so it is a fair game.

17. $\begin{bmatrix} ⑧ & 1 & \underline{-7} & 2 \\ -1 & ④ & \underline{⊖③} & ③ \end{bmatrix}$

The -3 at $(2, 3)$ is the smallest number in its row and the largest number in its column, so the saddle point is -3 at $(2, 3)$. The value of the game is -3.

19. $\begin{bmatrix} 1 & 0 \\ -2 & 3 \end{bmatrix}$

The optimum strategy for player A is

$$p_1 = \frac{3 - (-2)}{1 - (-2) - 0 + 3} = \frac{5}{6},$$

$$p_2 = 1 - \frac{5}{6} = \frac{1}{6}.$$

The optimum strategy for player B is

$$q_1 = \frac{3 - (0)}{1 - (-2) - 0 + 3} = \frac{3}{6} = \frac{1}{2},$$

$$q_2 = 1 - \frac{1}{2} = \frac{1}{2}.$$

The value of the game is

$$\frac{(1)(3) - (0)(-2)}{6} = \frac{1}{2}.$$

21. $\begin{bmatrix} -3 & 5 \\ 1 & 0 \end{bmatrix}$

The optimum strategy for player A is

$$p_1 = \frac{0 - 1}{-3 - 1 - 5 + 0} = \frac{-1}{-9} = \frac{1}{9},$$

$$p_2 = 1 - \frac{1}{9} = \frac{8}{9}.$$

The optimum strategy for player B is

$$q_1 = \frac{0 - 5}{-3 - 1 - 5 + 0} = \frac{-5}{-9} = \frac{5}{9},$$

$$q_2 = 1 - \frac{5}{9} = \frac{4}{9}.$$

The value of the game is

$$\frac{-3(0) - (5)(1)}{-3 - 1 - 5 + 0} = \frac{-5}{-9} = \frac{5}{9}.$$

23. $\begin{bmatrix} -4 & 8 & 0 \\ -2 & 9 & -3 \end{bmatrix}$

Column 1 dominates column 2. Remove column 2, to obtain

$$\begin{bmatrix} -4 & 0 \\ -2 & -3 \end{bmatrix}.$$

The optimum strategy for player A is

$$p_1 = \frac{-3 + 2}{-4 + 2 - 0 - 3} = \frac{-1}{-5} = \frac{1}{5},$$

$$p_2 = 1 - \frac{1}{5} = \frac{4}{5}.$$

The optimum strategy for player B is

$$q_1 = \frac{-3 - 0}{-4 + 2 - 0 - 3} = \frac{-3}{-5} = \frac{3}{5},$$

$$q_2 = 0 \text{ (column 2 was removed)},$$

$$q_3 = 1 - \frac{3}{5} = \frac{2}{5}.$$

The value of the game is

$$\frac{-4(-3) - (0)(-2)}{-4 + 2 - 0 - 3} = \frac{12}{-5} = -\frac{12}{5}.$$

25. $\begin{bmatrix} 2 & -1 \\ -4 & 5 \\ -1 & -2 \end{bmatrix}$

Row 1 dominates row 3. Remove row 3 to obtain

$$\begin{bmatrix} 2 & -1 \\ -4 & 5 \end{bmatrix}.$$

The optimum strategy for player A is

$$p_1 = \frac{5 + 4}{2 + 4 + 1 + 5} = \frac{9}{12} = \frac{3}{4},$$

$$p_2 = 1 - \frac{3}{4} = \frac{1}{4},$$

$$p_3 = 0 \text{ (row 3 was removed)}.$$

The optimum strategy for player B is

$$q_1 = \frac{5 + 1}{2 + 4 + 1 + 5} = \frac{6}{12} = \frac{1}{2},$$

$$q_2 = 1 - \frac{1}{2} = \frac{1}{2}.$$

The value of the game is

$$\frac{2(5) - (-1)(-4)}{2 + 4 + 1 + 5} = \frac{6}{12} = \frac{1}{2}.$$

27.
$$\begin{bmatrix} -4 & 2 \\ 3 & -5 \end{bmatrix}$$

Get rid of negative numbers by adding 5 to each entry, to obtain

$$\begin{bmatrix} 1 & 7 \\ 8 & 0 \end{bmatrix}.$$

To find the optimum strategy for
player A,

minimize $w = x + y$
subject to: $x + 8y \geq 1$
            $7x \quad\quad \geq 1$
with $x \geq 0,\ y \geq 0.$

Solve this linear programming problem by the graphical method.   Sketch
the feasible region.

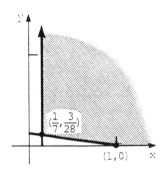

The region is unbounded, with corner
points $(\frac{1}{7}, \frac{3}{28})$ and $(1, 0)$.

Corner Point	Value of $w = x + y$
$(\frac{1}{7}, \frac{3}{28})$	$\frac{1}{7} + \frac{3}{28} = \frac{1}{4}$
$(1, 0)$	$1 + 0 = 1$

The minimum value is $w = \frac{1}{4}$ at
$(\frac{1}{7}, \frac{3}{28})$.

Thus, $g = \frac{1}{w} = 4$, and the optimum
strategy for A is

$$p_1 = gx = (4)\left(\frac{1}{7}\right) = \frac{4}{7},$$

$$p_2 = gy = (4)\left(\frac{3}{28}\right) = \frac{3}{7}.$$

To find the optimum strategy for
player B,

maximize $z = x + y$
Subject to: $x + 7y \leq 1$
            $8x \quad\quad \leq 1$
with $x \geq 0,\ y \geq 0.$

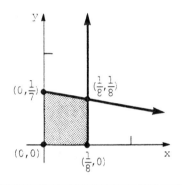

Corner Point	Value of $z = x + y$
$(0, \frac{1}{7})$	$0 + \frac{1}{7} = \frac{1}{7}$
$(\frac{1}{8}, \frac{1}{8})$	$\frac{1}{8} + \frac{1}{8} = \frac{1}{4}$
$(\frac{1}{8}, 0)$	$\frac{1}{8} + 0 = \frac{1}{8}$
$(0, 0)$	$0 + 0 = 0$

The maximum value is $z = \frac{1}{4}$ at
$(\frac{1}{8}, \frac{1}{8})$.

The optimum strategy for B is

$$q_1 = gx = (4)\left(\frac{1}{8}\right) = \frac{1}{2},$$

$$q_2 = gy = (4)\left(\frac{1}{8}\right) = \frac{1}{2}.$$

The value of the game is $4 - 5 = -1$.

29. $\begin{bmatrix} 1 & 0 \\ -3 & 4 \end{bmatrix}$

Add 3 to each entry, to obtain

$$\begin{bmatrix} 4 & 3 \\ 0 & 7 \end{bmatrix}.$$

To find the optimum strategy for player A,

minimize $w = x + y$

subject to: $4x \quad\quad \geq 1$

$\quad\quad\quad\quad 3x + 7y \geq 1$

with $x \geq 0, \; y \geq 0$.

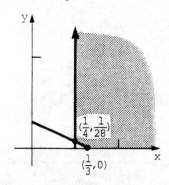

Corner Point	Value of $w = x + y$
$\left(\frac{1}{4}, \frac{1}{28}\right)$	$\frac{1}{4} + \frac{1}{28} = \frac{2}{7}$
$\left(\frac{1}{3}, 0\right)$	$\frac{1}{3} + 0 = \frac{1}{3}$

The minimum value is $w = \frac{2}{7}$ at $\left(\frac{1}{4}, \frac{1}{28}\right)$. Thus, $g = \frac{1}{w} = \frac{7}{2}$, and the optimum strategy for A is

$$p_1 = gx = \left(\frac{7}{2}\right)\left(\frac{1}{4}\right) = \frac{7}{8},$$

$$p_2 = gy = \left(\frac{7}{2}\right)\left(\frac{1}{28}\right) = \frac{1}{8}.$$

To find the optimum strategy for player B,

maximize $z = x + y$

subject to: $4x + 3y \leq 1$

$\quad\quad\quad\quad\quad 7y \leq 1$

with $x \geq 0, \; y \geq 0$.

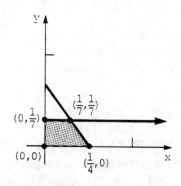

Corner Point	Value of $z = x + y$
$\left(0, \frac{1}{7}\right)$	$0 + \frac{1}{7} = \frac{1}{7}$
$\left(\frac{1}{7}, \frac{1}{7}\right)$	$\frac{1}{7} + \frac{1}{7} = \frac{2}{7}$
$\left(\frac{1}{4}, 0\right)$	$\frac{1}{4} + 0 = \frac{1}{4}$
$(0, 0)$	$0 + 0 = 0$

The maximum value is $z = \frac{2}{7}$ at $\left(\frac{1}{7}, \frac{1}{7}\right)$. The optimum strategy for B is

$$q_1 = gx = \left(\frac{7}{2}\right)\left(\frac{1}{7}\right) = \frac{1}{2},$$

$$q_2 = gy = \left(\frac{7}{2}\right)\left(\frac{1}{7}\right) = \frac{1}{2}.$$

The value of the game is $\frac{7}{2} - 3 = \frac{1}{2}$.

31. $\begin{bmatrix} 4 & 3 & 1 \\ -1 & 0 & 2 \end{bmatrix}$

Add 1 to each entry, to obtain

$$\begin{bmatrix} 5 & 4 & 2 \\ 0 & 1 & 3 \end{bmatrix}.$$

The linear programming problem to be solved is as follows.

Maximize $z = x_1 + x_2 + x_3$

subject to: $5x_1 + 4x_2 + 2x_3 \leq 1$

$x_2 + 3x_3 \leq 1$

with $x_1 \geq 0, \ x_2 \geq 0, \ x_3 \geq 0$.

Use the simplex method to solve the problem. The initial tableau is

	x_1	x_2	x_3	x_4	x_5	z	
	5	④	2	1	0	0	1
	0	1	3	0	1	0	1
	-1	-1	-1	0	0	1	0

Pivot on each circled entry.

	x_1	x_2	x_3	x_4	x_5	z	
	5	4	2	1	0	0	1
$4R_2 - R_1 \to R_2$	-5	0	⑩	-1	4	0	3
$4R_3 + R_1 \to R_3$	1	0	-2	1	0	4	1

	x_1	x_2	x_3	x_4	x_5	z	
$5R_1 - R_2 \to R_1$	30	20	0	6	-4	0	2
	-5	0	10	-1	4	0	3
$5R_3 + R_2 \to R_3$	0	0	0	4	4	20	8

Create a 1 in the columns corresponding to x_2, x_3, and z.

	x_1	x_2	x_3	x_4	x_5	z	
$\frac{1}{20}R_1 \to R_1$	$\frac{3}{2}$	1	0	$\frac{3}{10}$	$-\frac{1}{5}$	0	$\frac{1}{10}$
$\frac{1}{10}R_2 \to R_2$	$-\frac{1}{2}$	0	1	$-\frac{1}{10}$	$\frac{2}{5}$	0	$\frac{3}{10}$
$\frac{1}{20}R_3 \to R_3$	0	0	0	$\frac{1}{5}$	$\frac{1}{5}$	1	$\frac{2}{5}$

From this final tableau, we have

$$x_1 = 0, \ x_2 = \frac{1}{10}, \ x_3 = \frac{3}{10},$$

$$y_1 = \frac{1}{5}, \ y_2 = \frac{1}{5}, \text{ and } z = \frac{2}{5}.$$

Note that $g = \dfrac{1}{z} = \dfrac{5}{2}$.

The optimum strategy for player A is

$$p_1 = gy_1 = \left(\frac{5}{2}\right)\left(\frac{1}{5}\right) = \frac{1}{2},$$

$$p_2 = gy_2 = \left(\frac{5}{2}\right)\left(\frac{1}{5}\right) = \frac{1}{2}.$$

The optimum strategy for player B is

$$q_1 = gx_1 = \left(\frac{5}{2}\right)(0) = 0,$$

$$q_2 = gx_2 = \left(\frac{5}{2}\right)\left(\frac{1}{10}\right) = \frac{1}{4},$$

$$q_3 = gx_3 = \left(\frac{5}{2}\right)\left(\frac{3}{10}\right) = \frac{3}{4}.$$

The value of the game is $\dfrac{5}{2} - 1 = \dfrac{3}{2}$.

33. $\begin{bmatrix} -2 & 1 & 0 \\ 2 & 0 & -2 \\ 0 & -1 & 3 \end{bmatrix}$

Add 2 to each entry, to obtain

$$\begin{bmatrix} 0 & 3 & 2 \\ 4 & 2 & 0 \\ 2 & 1 & 5 \end{bmatrix}.$$

The problem to be solved is as follows.

Maximize $z = x_1 + x_2 + x_3$

subject to: $3x_2 + 2x_3 \leq 1$

$4x_1 + 2x_2 \leq 1$

$2x_1 + x_2 + 5x_3 \leq 1$

with $x_1 \geq 0, \ x_2 \geq 0, \ x_3 \geq 0$.

The initial simplex tableau is

$$
\begin{array}{ccccccc|c}
x_1 & x_2 & x_3 & x_4 & x_5 & x_6 & z & \\
0 & 3 & 2 & 1 & 0 & 0 & 0 & 1 \\
\textcircled{4} & 2 & 0 & 0 & 1 & 0 & 0 & 1 \\
2 & 1 & 5 & 0 & 0 & 1 & 0 & 1 \\
-1 & -1 & -1 & 0 & 0 & 0 & 1 & 0
\end{array}
$$

Pivot on each circled entry.

$$
\begin{array}{rcccccccc}
& x_1 & x_2 & x_3 & x_4 & x_5 & x_6 & z & \\
& 0 & 3 & 2 & 1 & 0 & 0 & 0 & 1 \\
& 4 & 2 & 0 & 0 & 1 & 0 & 0 & 1 \\
2R_3 - R_2 \to R_3 & 0 & 0 & \textcircled{10} & 0 & -1 & 2 & 0 & 1 \\
4R_4 + R_2 \to R_4 & 0 & -2 & -4 & 0 & 1 & 0 & 4 & 1
\end{array}
$$

$$
\begin{array}{rcccccccc}
& x_1 & x_2 & x_3 & x_4 & x_5 & x_6 & z & \\
5R_1 - R_3 \to R_1 & 0 & \textcircled{15} & 0 & 5 & 1 & -2 & 0 & 4 \\
& 4 & 2 & 0 & 0 & 1 & 0 & 0 & 1 \\
& 0 & 0 & 10 & 0 & -1 & 2 & 0 & 1 \\
2R_3 + 5R_4 \to R_4 & 0 & -10 & 0 & 0 & 3 & 4 & 20 & 7
\end{array}
$$

$$
\begin{array}{rcccccccc}
& x_1 & x_2 & x_3 & x_4 & x_5 & x_6 & z & \\
& 0 & 15 & 0 & 5 & 1 & -2 & 0 & 4 \\
15R_2 - 2R_1 \to R_2 & 60 & 0 & 0 & -10 & 13 & 4 & 0 & 7 \\
& 0 & 0 & 10 & 0 & -1 & 2 & 0 & 1 \\
2R_1 + 3R_4 \to R_4 & 0 & 0 & 0 & 10 & 11 & 8 & 60 & 29
\end{array}
$$

Create a 1 in the columns corresponding to x_1, x_2, x_3, and z.

$$
\begin{array}{rcccccccc}
& x_1 & x_2 & x_3 & x_4 & x_5 & x_6 & z & \\
\frac{1}{15}R_1 \to R_1 & 0 & 1 & 0 & \frac{1}{3} & \frac{1}{15} & -\frac{2}{15} & 0 & \frac{4}{15} \\
\frac{1}{60}R_2 \to R_2 & 1 & 0 & 0 & -\frac{1}{6} & \frac{13}{60} & \frac{1}{15} & 0 & \frac{7}{60} \\
\frac{1}{10}R_3 \to R_3 & 0 & 0 & 1 & 0 & -\frac{1}{10} & \frac{1}{5} & 0 & \frac{1}{10} \\
\frac{1}{60}R_4 \to R_4 & 0 & 0 & 0 & \frac{1}{6} & \frac{11}{60} & \frac{2}{15} & 1 & \frac{29}{60}
\end{array}
$$

From this final tableau, we have

$$
x_1 = \frac{7}{60}, \quad x_2 = \frac{4}{15}, \quad x_3 = \frac{1}{10},
$$

$$
y_1 = \frac{1}{6}, \quad y_2 = \frac{11}{60}, \quad y_3 = \frac{2}{15}, \quad \text{and } z = \frac{29}{60}.
$$

Note that $g = \frac{1}{z} = \frac{60}{29}$.

The optimum strategy for player A is

$$p_1 = gy_1 = \left(\frac{60}{29}\right)\left(\frac{1}{6}\right) = \frac{10}{29},$$

$$p_2 = gy_2 = \left(\frac{60}{29}\right)\left(\frac{11}{60}\right) = \frac{11}{29},$$

$$p_3 = gy_3 = \left(\frac{60}{29}\right)\left(\frac{2}{15}\right) = \frac{8}{29}.$$

The optimum strategy for player B is

$$q_1 = gx_1 = \left(\frac{60}{29}\right)\left(\frac{7}{60}\right) = \frac{7}{29},$$

$$q_2 = gx_2 = \left(\frac{60}{29}\right)\left(\frac{4}{15}\right) = \frac{16}{29},$$

$$q_3 = gx_3 = \left(\frac{60}{29}\right)\left(\frac{1}{10}\right) = \frac{6}{29}.$$

The value of the game is $\frac{60}{29} - 2 = \frac{2}{29}$.

37.

Management

	Friendly	Hostile
Labor Friendly	$600	$800
Hostile	$400	$950

Be hostile; then he has a chance at the $950 wage increase, which is the largest possible increase.

39. If there is a .7 chance that the company will be hostile, then there is a .3 chance that it will be friendly. Find the expected payoff for each strategy.

Friendly: $(.3)(600) + (.7)(800) = \$740$
Hostile: $(.3)(400) + (.7)(950) = \$785$

Therefore, he should be hostile. The expected payoff is $785.

41.
$$\begin{bmatrix} \boxed{600} & 800 \\ \hline 400 & \boxed{950} \end{bmatrix}$$

The 600 at (1, 1) is the smallest number in its row and the largest

number in its column, so it is a saddle point and the game is strictly determined. Labor and management should both always be friendly, and the value of the game is 600.

Opponent

43.

	Favors	Waffles	Opposes
Candidate Favors	0	−1000	−4000
Waffles	1000	0	−500
Opposes	5000	2000	0

As a pessimist, the candidate would look at the worst possible outcome from each strategy, and choose the best of them. If she favors the factory, she could lose 4000 votes; if she waffles, she could lose 500 votes; and if she opposes the issue, she could lose 0 votes. The best of these options is to lose 0 votes, so she should oppose the factory.

45. Find the expected payoffs (in vote changes) for each strategy.

Favors:
$$0(0) + .7(-1000) + .3(-4000) = -1900$$

Waffles:
$$0(1000) + .7(0) + .3(-500) = -1500$$

Opposes:
$$0(5000) + .7(2000) + .3(0) = 1400$$

The candidate should oppose the factory for an expected gain of 1400 votes.

Economy

47.

	Inflationary	Stable
Stocks Blue-Chip	$2800	$3200
Growth	$5000	−$2000

The optimum strategy for Hector is

$$p_1 = \frac{-2000 - 5000}{2800 - 5000 - 3200 + (-2000)} = \frac{35}{37},$$

$$p_2 = 1 - \frac{35}{37} = \frac{2}{37}.$$

That is, Hector should invest in blue-chip stocks with probability $\frac{35}{37}$ and growth stocks with probability $\frac{2}{37}$.

The value of the game is

$$\frac{2800(-2000) - 5000(3200)}{2800 - 5000 - 3200 + (-2000)} = \frac{-21,600,000}{-7400} \approx \$2918.92.$$

49. $\begin{bmatrix} 2800 & 3200 \\ 5000 & -2000 \end{bmatrix}$

Add 2000 to each entry, to obtain

$$\begin{bmatrix} 4800 & 5200 \\ 700\mathord{'} & 0 \end{bmatrix}.$$

The linear programming problem to be solved is

maximize $z = x_1 + x_2$

subject to: $4800x_1 + 5200x_2 \le 1$

$\quad\quad\quad\quad 7000x \quad\quad\quad \le 1$

with $x_1 \ge 0,\ x_2 \ge 0.$

The initial tableau is as follows.

x_1	x_2	x_3	x_4	z	
4800	(5200)	1	0	0	1
7000	0	0	1	0	1
−1	−1	0	0	1	0

Pivot on each circled entry.

	x_1	x_2	x_3	x_4	z	
	4800	5200	1	0	0	1
	(7000)	0	0	1	0	1
$5200R_3 + R_1 \to R_3$	−400	0	1	0	5200	1

	x_1	x_2	x_3	x_4	z	
$35R_1 - 24R_2 \to R_1$	0	182,000	35	−24	0	11
	7000	0	0	1	0	1
$2R_2 + 35R_3 \to R_3$	0	0	35	2	182,000	37

Create a 1 in the columns corresponding to x_1, x_2, and z.

$$
\begin{array}{c}
\frac{1}{182{,}000}R_1 \to R_1 \\[4pt]
\frac{1}{7000}R_2 \to R_2 \\[4pt]
\frac{1}{182{,}000}R_3 \to R_3
\end{array}
\quad
\begin{array}{ccccc}
x_1 & x_2 & x_3 & x_4 & z \\
\end{array}
\left[
\begin{array}{ccccc|c}
0 & 1 & \frac{1}{5200} & -\frac{3}{22{,}750} & 0 & \frac{11}{182{,}000} \\[6pt]
1 & 0 & 0 & \frac{1}{7000} & 0 & \frac{1}{7000} \\[6pt]
0 & 0 & \frac{1}{5200} & \frac{1}{91{,}000} & 1 & \frac{37}{182{,}000}
\end{array}
\right]
$$

From this final tableau, we have

$$x_1 = \frac{1}{7000}, \quad x_2 = \frac{11}{182{,}000},$$

$$y_1 = \frac{1}{5200}, \quad y_2 = \frac{1}{91{,}000}, \text{ and } z = \frac{37}{182{,}000}.$$

Note that $g = \frac{1}{z} = \frac{182{,}000}{37}$.

The optimum strategy for Hector is

$$p_1 = gy_1 = \left(\frac{182{,}000}{37}\right)\left(\frac{1}{5200}\right) = \frac{35}{37},$$

$$p_2 = gy_2 = \left(\frac{182{,}000}{37}\right)\left(\frac{1}{91{,}000}\right) = \frac{2}{37}.$$

That is, he should invest in blue-chip stocks with probability $\frac{35}{37}$ and growth stocks with probabilitly $\frac{2}{37}$.

The value of the game is

$$\frac{182{,}000}{37} - 2000 \approx \$2918.92.$$

Rontovia

51.

		Attack 1	Attack 2
Ravogna	Defend 1	4	1
	Defend 2	3	4

To find the optimum strategy for Ravogna,

$$\begin{aligned}
\text{minimize} \quad & w = x + y \\
\text{subject to:} \quad & 4x + 3y \geq 1 \\
& x + 4y \geq 1 \\
\text{with} \quad & x \geq 0, \; y \geq 0.
\end{aligned}$$

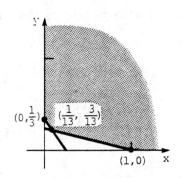

Corner Point	Value of $w = x + y$
$(0, \frac{1}{3})$	$0 + \frac{1}{3} = \frac{1}{3}$
$(\frac{1}{13}, \frac{3}{13})$	$\frac{1}{13} + \frac{3}{13} = \frac{4}{13}$
$(1, 0)$	$1 + 0 = 1$

The minimum value is $w = \frac{4}{13}$ at $(\frac{1}{13}, \frac{3}{13})$. Thus, $g = \frac{1}{w} = \frac{13}{4}$ is the value of the game. The optimum strategy for Ravogna is

$$p_1 = gx = \left(\frac{13}{4}\right)\left(\frac{1}{13}\right) = \frac{1}{4},$$

$$p_2 = gy = \left(\frac{13}{4}\right)\left(\frac{3}{13}\right) = \frac{3}{4}.$$

That is, Ravogna should defend installation #1 with probability $\frac{1}{4}$ and installatio #2 with probability $\frac{3}{4}$.

To find the optimum strategy for Rontovia,

$$\text{maximize} \quad z = x + y$$
$$\text{subject to:} \quad 4x + y \leq 1$$
$$3x + 4y \leq 1$$
$$\text{with} \quad x \geq 0, y \geq 0.$$

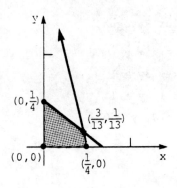

Corner Point	Value of z = x + y
$(0, \frac{1}{4})$	$0 + \frac{1}{4} = \frac{1}{4}$
$(\frac{3}{13}, \frac{1}{13})$	$\frac{3}{13} + \frac{1}{13} = \frac{4}{13}$
$(\frac{1}{4}, 0)$	$\frac{1}{4} + 0 = \frac{1}{4}$
$(0, 0)$	$0 + 0 = 0$

The maximum value is $z = \frac{4}{13}$ at $(\frac{3}{13}, \frac{1}{13})$. The optimum strategy for Rontovia is

$$q_1 = gx = \left(\frac{13}{4}\right)\left(\frac{3}{13}\right) = \frac{3}{4},$$

$$q_2 = gy = \left(\frac{13}{4}\right)\left(\frac{1}{13}\right) = \frac{1}{4}.$$

That is, Rontovia should attack installation #1 with probability $\frac{3}{4}$ and installation #2 with probability $\frac{1}{4}$.

Extended Application

1. For action a_1 (offer a standard policy):

$$E_1 = [P(s_1|T_2)](20M) + [P(s_2|T_2)](13M) + [P(s_3|T_2)](3M)$$
$$= .8411(20M) + .1309(13M) + .0280(3M)$$
$$= 16.822M + 1.7017M + .084M$$
$$= 18.61M$$

For action a_2 (offer a substandard policy):

$$E_2 = [P(s_1|T_2)](-50) + [P(s_2|T_2)](20m) + [P(s_3|T_2)](10M)$$
$$= .8411(-50) + .1309(20M) + .0280(10M)$$
$$= -42.055 + 2.618M + .28M$$
$$= 2.898M - 42.055$$

For action a_3 (offer a substandard policy):

$$E_3 = .8411(-50) + .1309(-50) + .0280(20M)$$
$$= -42.055 - 6.545 + .56M$$
$$= .56M - 48.6$$

2. For action a_1 (offer a standard policy):

$$E_1 = [P(s_1|T_1)](20M) + P[(s_2|T_1)](13M) + P[(s_3|T_1)](3M)$$
$$= (.9695)(20M) + (.0251)(13M) + (.0054)(3M)$$
$$= 19.7325M$$

For action a_2 (offer a substandard policy):

$$E_2 = [P(s_1|T_1)](-50) + P[(s_2|T_1)](20M) + P[(s_3|T_1)](10M)$$
$$= (.9695)(-50) + (.0251)(20M) + (.0054)(10M)$$
$$= .556M - 48.475$$

For action a_3 (offer a substandard policy):

$$E_3 = (.9695)(-50) + (.0251)(-50) + (.0054)(20M)$$
$$= .108M - 49.73$$

3. For action a_1 (offer a standard policy):

$$E_1 = .90(20M) + .07(13M) + .03(3M)$$
$$= 18M + .91M + .09M$$
$$= 19M$$

For action a_2 (offer a standard policy):

$$E_2 = .90(-50) + .07(20M) + .03(10M)$$
$$= -45 + 1.4M + .3M$$
$$= 1.7M - 45$$

For action a_3 (offer a substandard policy):

$$E_3 = .90(-50) + .07(-50) + .03(20M)$$
$$= -45 - 3.5 + .6M$$
$$= .6M - 48.5$$

CHAPTER 9   TEST

1. The Dean of Students at the local college must decide on the discipline rules for the upcoming year. She can choose to be Strict or Wimpy. Her success depends on whether the dorms are full of Rowdies or Studious folk.

$$\begin{array}{c} \\ \text{Strict} \\ \text{Wimpy} \end{array} \begin{array}{cc} \text{Rowdies} & \text{Studious} \\ \begin{bmatrix} 100 & -30 \\ -100 & 200 \end{bmatrix} \end{array}$$

   (a) What course of action would an optimist take?

   (b) What course of action would a pessimist take?

   (c) What course of action would yield the largest expected payoff if the probability of Rowdies in the dorm is .6?

2. A politician is faced with three possible levels of opposition to his policies: Light, Medium, and Heavy. He must decide on three levels of advertising to combat this opposition: Low, Medium, or High. His payoff matrix is given below.

$$\begin{array}{c} \\ \text{Low} \\ \text{Medium} \\ \text{High} \end{array} \begin{array}{ccc} \text{Light} & \text{Medium} & \text{Heavy} \\ \begin{bmatrix} 100 & -30 & -60 \\ -10 & 75 & -70 \\ -50 & -20 & 50 \end{bmatrix} \end{array}$$

   (a) What would an optimist do?

   (b) What would a pessimist do?

   (c) The probability of heavy opposition is .3, and of medium opposition is .4. What is the best course of action?

3. Remove any dominated strategies from the payoff matrix.

$$\begin{bmatrix} 2 & 3 & 0 & -1 \\ 1 & 2 & -2 & -5 \\ 0 & 3 & 1 & 0 \\ -1 & 0 & 0 & 1 \end{bmatrix}.$$

4. Find the saddle point, if it exists, in the matrix

$$\begin{bmatrix} 1 & 20 & 3 \\ -1 & 1 & 0 \\ -3 & 0 & 1 \end{bmatrix}.$$

5. For each game below, find the optimum strategy for each player and the value
   of the game.

   (a) $\begin{bmatrix} 1 & -2 \\ 2 & 3 \end{bmatrix}$ **(b)** $\begin{bmatrix} 6 & -3 \\ 5 & 9 \end{bmatrix}$

6. Consider the game

   $$\begin{bmatrix} -3 & 2 \\ 4 & -5 \end{bmatrix}.$$

   (a) Find the optimum strategy for player A using the graphical method.

   (b) Find the optimum strategy for player B using the simplex method.

   (c) Find the value of the game.

CHAPTER 9 TEST ANSWERS

1. **(a)** Wimpy **(b)** Strict **(c)** Strict

2. **(a)** Low advertising **(b)** High advertising **(c)** Medium advertising

3. $\begin{bmatrix} 2 & 0 & -1 \\ 0 & 1 & 0 \\ -1 & 0 & 1 \end{bmatrix}$ 4. Saddle point 1 at (1, 1)

5. **(a)** Player A: $p_1 = 0$, $p_2 = 1$

   Player B: $q_1 = 1$, $q_2 = 0$

   Value of game = 2

   (b) Player A: $p_1 = \dfrac{4}{13}$, $p_2 = \dfrac{9}{13}$

   Player B: $q_1 = \dfrac{12}{13}$, $q_2 = \dfrac{1}{13}$

   Value of game = $\dfrac{69}{13}$

6. **(a)** Player A: $p_1 = \dfrac{9}{14}$, $p_2 = \dfrac{5}{14}$

   (b) Player B: $q_1 = \dfrac{1}{2}$, $q_2 = \dfrac{1}{2}$

   (c) Value of game = $-\dfrac{1}{2}$

CHAPTER 10   MATHEMATICS OF FINANCE

Section 10.1

1. The interest rate and time period determine the amount of interest on a fixed principal. In the formula I = Prt, if P remains constant then the value of I will be affected by the values of r and t.

3. $I = Prt = 3850(.09)\left(\frac{8}{12}\right) = \$231.00$

5. $I = Prt = 3724(.084)\left(\frac{11}{12}\right) = \$286.75$

7. $I = Prt = 2930.42(.119)\left(\frac{123}{360}\right)$
   $= \$119.15$

9. There are $14 + 30 + 30 + 30 + 30 = 134$ days, since a 30–day month is assumed.

   $I = Prt = 5408(.12)\left(\frac{134}{360}\right) = \$241.56$

11. There are $9 + 31 + 30 + 31 = 101$ days.

    $I = Prt = 11,000(.10)\left(\frac{101}{365}\right) = \$304.38$

13. There are $18 + 31 + 30 + 31 + 31 + 28 + 31 + 30 + 31 + 30 + 30 = 321$ days.

    $I = Prt = 37,098(.112)\left(\frac{321}{365}\right)$
    $= \$3654.10$

17. $P = \dfrac{A}{1 + rt}$

    $= \dfrac{48,000}{1 + .05\left(\frac{9}{12}\right)}$

    $= \$46,265.06$

19. $P = \dfrac{A}{1 + rt}$

    $= \dfrac{29,764}{1 + .072\left(\frac{310}{360}\right)}$

    $= \$28,026.37$

21. $P = A(1 - rt)$

    $= 9450\left[1 - .10\left(\frac{7}{12}\right)\right]$

    $\approx 9450(.9417)$

    $\approx \$8898.75$

23. $P = A(1 - rt)$

    $= 50,900\left[1 - .082\left(\frac{238}{360}\right)\right]$

    $\approx 50,900(.9458)$

    $\approx \$48,140.65$

25. The discount is

    $6200(.10)\left(\frac{8}{12}\right) \approx 413.33.$

    The proceeds are found by subtracting the discount from the original amount.

    $6200 - 413.33 = 5786.67$

    Use I = Prt with I = 413.33, P = 5786.67, and $t = \frac{8}{12}$, and solve for r.

    $413.33 = 5786.67(r)\left(\frac{8}{12}\right)$

    $.107 \approx r$

The interest rate paid by the borrower is about 10.7%.

27. The discount is

$$58,000(.108)\left(\frac{9}{12}\right) = 4698$$

and the proceeds are

$$58,000 - 4698 = 53,302.$$

Use $I = Prt$ with $I = 4698$, $P = 53,302$, and $t = \frac{9}{12}$, and solve for r.

$$4698 = 53,302(r)\left(\frac{9}{12}\right)$$

$$.118 \approx r$$

The interest rate paid by the borrower is about 11.8%.

29. Use the formula for future value.

$$A = P(1 + rt)$$

$$= 25,900\left[1 + (.084)\left(\frac{11}{12}\right)\right]$$

$$= 25,900(1.077)$$

$$= \$27,894.30$$

She repaid $27,894.30.

31. The interest is

$$101,133.33 - 100,000 = 1133.33.$$

Use the formula for simple interest.

$$I = Prt$$

$$1133.33 = 100,000(r)\left(\frac{60}{360}\right)$$

$$.068 \approx r$$

The interest rate was about 6.8%.

33. The future value is $7(5104) = 35,728$. Use the formula for present value.

$$P = \frac{A}{1 + rt}$$

$$= \frac{35,728}{1 + (.0642)\left(\frac{7}{12}\right)}$$

$$\approx 34,438.29.$$

They should deposit about $34,438.29.

35. The proceeds are $6100. Use the formula for proceeds and solve for A.

$$P = A(1 - rt)$$

$$A = \frac{P}{1 - rt}$$

$$= \frac{6100}{1 - .188\left(\frac{7}{12}\right)}$$

$$\approx 6550.92$$

The amount of the loan is about $6550.92.

37. The interest per share is

$$(24 - 22) + .50 = 2.50.$$

Use the simple interest formula with $I = 2.50$, $P = 22$, and $t = 1$, and solve for r.

$$I = Prt$$

$$2.50 = 22(r)(1)$$

$$.114 \approx r$$

The interest rate is about 11.4%.

39. Find the maturity value of the loan, the amount the contractor must pay to the plumber.

$$A = P(1 + rt)$$
$$= 13,500\left[1 + .09\left(\frac{9}{12}\right)\right]$$
$$= 13,500(1.0675)$$
$$= 14,411.25$$

Three months after the note is signed, there are six more months before the loan is payable. The bank will apply its discount rate to the 14,411.25, to obtain

Amount of discount
$$= 14,411.25(.101)\left(\frac{6}{12}\right)$$
$$\approx 727.77.$$

The plumber will receive from the bank

$$\$14,411.25 - \$727.77 = \$13,683.48,$$

which will be enough to pay the $13,582 bill.

Section 10.2

3. The formula for the compound amount is

$$A = P(1 + i)^n.$$

To find A, substitute P = 1000, i = .06, and n = 8.

$$A = 1000(1 + .06)^8$$
$$= 1000(1.06)^8$$
$$\approx 1000(1.59385)$$
$$= 1593.85$$

The compound amount is $1593.85.

5. To find the compound amount A, substitute P = 470, $i = \frac{.10}{2} = .05$, and n = 12 · 2 = 24 in the formula for the compound amount.

$$A = P(1 + i)^n$$
$$= 470(1.05)^{24}$$
$$\approx 470(3.22510)$$
$$= \$1515.80$$

7. P = 6500, $i = \frac{.12}{4}$, n = 6 · 4

$$A = P(1 + i)^n$$
$$= 6500\left(1 + \frac{.12}{4}\right)^{6(4)}$$
$$= 6500(1.03)^{24}$$
$$\approx 6500(2.03279)$$
$$= \$13,213.14$$

9. First find the compound amount.

$$A = P(1 + i)^n$$
$$= 6000(1 + .08)^8$$
$$= 6000(1.08)^8$$
$$\approx 6000(1.85093)$$
$$= \$11,105.58$$

To find the amount of interest earned, subtract the initial deposit from the compound amount.

Amount of interest
$$= A - P$$
$$= \$11,105.58 - \$6000$$
$$= \$5105.58$$

11. $A = P(1 + i)^n$
$$= 43,000\left(1 + \frac{.10}{2}\right)^{2(9)}$$
$$= 43,000(1.05)^{18}$$
$$\approx 43,000(2.40662)$$
$$= \$103,484.66$$

Amount of interest

$$= \$103{,}484.66 - \$43{,}000$$

$$= \$60{,}484.66$$

13. $A = P(1 + i)^n$

$$A = 2196.58\left(1 + \frac{.108}{4}\right)^{4(4)}$$

$$= 2196.58(1.027)^{16}$$

$$\approx 2196.58(1.53153)$$

$$\approx 3364.14$$

Amount of interest

$$= \$3364.14 - \$2196.58$$

$$= \$1167.56$$

15. $A = 4500$, $i = .08$, $n = 9$

Substitute these values in the formula for present value of an amount at compound interest.

$$P = \frac{A}{(1 + i)^n}$$

$$= \frac{4500}{(1.08)^9}$$

$$\approx \frac{4500}{1.99900}$$

$$\approx \$2251.12$$

17. $P = \dfrac{A}{(1 + i)^n}$

$$= \frac{15{,}902.74}{(1 + .098)^7}$$

$$\approx \frac{15{,}902.74}{1.92405}$$

$$\approx \$8265.24$$

19. $P = \dfrac{A}{(1 + i)^n}$

$$= \frac{2000}{\left(1 + \frac{.09}{2}\right)^{8(2)}}$$

$$= \frac{2000}{(1.045)^{16}}$$

$$\approx \frac{2000}{2.02237}$$

$$\approx \$988.94$$

21. $P = \dfrac{A}{(1 + i)^n}$

$$= \frac{8000}{\left(1 + \frac{.10}{4}\right)^{5(4)}}$$

$$= \frac{8800}{(1.025)^{20}}$$

$$\approx \$5370.38$$

25. Substitute $r = .04$ and $m = 2$ in the formula for effective rate.

$$r_e = \left(1 + \frac{r}{m}\right)^m - 1$$

$$= \left(1 + \frac{.04}{2}\right)^2 - 1$$

$$= (1.02)^2 - 1$$

$$= 1.0404 - 1$$

$$= .0404$$

The effective rate is 4.04%.

27. $r = .08$, $m = 2$

$$r_e = \left(1 + \frac{r}{m}\right)^m - 1$$

$$= \left(1 + \frac{.08}{2}\right)^2 - 1$$

$$= (1.04)^2 - 1$$

$$= .0816$$

The effective rate is 8.16%.

29. $r = .12$, $m = 2$

$$r_e = \left(1 + \frac{r}{m}\right)^m - 1$$

$$= \left(1 + \frac{.12}{2}\right)^2 - 1$$

$$= (1.06)^2 - 1$$

$$= .1236$$

The effective rate is 12.36%.

31. Substitute $P = 3000$, $i = \frac{.06}{4}$, and $n = 18(4)$ in the formula for the compound amount.

$$A = P(1 + i)^n$$

$$= 3000\left(1 + \frac{.06}{4}\right)^{18(4)}$$

$$= 3000(1.015)^{72}$$

$$\approx \$8763.47$$

33. Substitute $P = 78,000$, $i = .03$, and $n = 12$ in the formula for the compound amount.

$$A = P(1 + i)^n$$

$$= 78,000(1.03)^{12}$$

$$\approx 111,000$$

The average price will be about $111,000.

35. $P = 50,000$, $i = \frac{.12}{12}$, $n = 12(4)$

First find the compound amount.

$$A = P(1 + i)^n$$

$$= 50,000\left(1 + \frac{.12}{12}\right)^{12(4)}$$

$$= 50,000(1.01)^{48}$$

$$\approx \$80,611.30$$

Amount of interest

$$= A - P$$

$$= \$80,611.30 - \$50,000$$

$$= \$30,611.30$$

37. Substitute $A = 2.9$ million, $i = \frac{.08}{12}$, and $n = 12(5)$ in the formula for present value with compound interest.

$$P = \frac{A}{(1 + i)^n}$$

$$= \frac{2.9}{\left(1 + \frac{.08}{12}\right)^{12(5)}}$$

$$\approx \frac{2.9}{(1.00667)^{60}}$$

$$\approx 1.946$$

They should invest about $1.946 million now.

39. Substitute $P = 10,000$, $i = \frac{.06}{2}$, and $n = 2(3)$ in the formula for the compound amount.

$$A = P(1 + i)^n$$

$$= 10,000\left(1 + \frac{.06}{2}\right)^{2(3)}$$

$$= 10,000(1.03)^6$$

$$\approx 11,940.52$$

She should contribute about $11,940.52 in 3 yr.

41. To find the number of years it will take prices to double at 4% annual inflation, find n in the equation

$$2 = (1 + .04)^n,$$

which simplifies to

$$2 = (1.04)^n.$$

By trying various values of n, find that n = 18 is approximately correct, because

$$1.04^{18} \approx 2.0258 \approx 2.$$

Prices will double in about 18 yr.

43. To find the number of years it will be until the generating capacity will need to be doubled, find n in the equation

$$2 = (1 + .06)^n,$$

which simplifies to

$$2 = (1.06)^n.$$

By trying various values of n, find that n = 12 is approximately correct, because $1.06^{12} \approx 2.0122 \approx 2$. The generating capacity will need to be doubled in about 12 yr.

45. P = 150,000, i = −.024, n = 2

$$\begin{aligned} A &= P(1 + i)^n \\ &= 150,000(1 - .024)^2 \\ &= 150,000(.976)^2 \\ &= \$142,886.40 \end{aligned}$$

47. P = 150,000, i = −.024, n = 8

$$\begin{aligned} A &= P(1 + i)^n \\ &= 150,000(1 - .024)^8 \\ &= 150,000(.976)^8 \\ &= \$123,506.50 \end{aligned}$$

49. First consider the case of earning interest at a rate of k per annum compounded quarterly for all eight years and earning \$2203.76 interest on the \$1000 investment.

$$2203.76 = 1000\left(1 + \frac{k}{4}\right)^{8(4)}$$

$$2.20376 = \left(1 + \frac{k}{4}\right)^{32}$$

Use a calculator to raise both sides to the power $\frac{1}{32}$.

$$1.025 = 1 + \frac{k}{4}$$

$$.025 = \frac{k}{4}$$

$$.1 = k$$

Next consider the actual investments. The \$1000 was invested for the first five years at a rate of j per annum compounded semiannually.

$$A = 1000\left(1 + \frac{j}{2}\right)^{5(2)}$$

$$A = 1000\left(1 + \frac{j}{2}\right)^{10}$$

This amount was then invested for the remaining three years at k = .1 per annum compounded quarterly for a final compound amount of \$1990.76.

$$1990.76 = A\left(1 + \frac{.1}{4}\right)^{3(4)}$$

$$1990.76 = A(1.025)^{12}$$

$$1480.24 \approx A$$

Recall that $A = 1000\left(1 + \frac{j}{2}\right)^{10}$ and substitute this value into the above equation.

$$1480.24 = 1000\left(1 + \frac{j}{2}\right)^{10}$$

$$1.48024 = \left(1 + \frac{j}{2}\right)^{10}$$

Use a calculator to raise both sides to the power $\frac{1}{10}$.

$$1.04 \approx 1 + \frac{j}{2}$$

$$.04 = \frac{j}{2}$$

$$.08 = j$$

The ratio of k to j is

$$\frac{k}{j} = \frac{.1}{.08} = \frac{10}{8} = \frac{5}{4}.$$

Section 10.3

1. a = 3, r = 2

The first five terms are

$$3, \; 3(2), \; 3(2)^2, \; 3(2)^3, \; 3(2)^4$$

or 3, 6, 12, 24, 48.

The fifth term is 48.

3. a = -8, r = 3

The first five terms are

$$-8, \; -8(3), \; -8(3)^2, \; -8(3)^3, \; -8(3)^4$$

or -8, -24, -72, -216, -648.

The fifth term is -648.

5. a = 1, r = -3

The first five terms are

$$1, \; 1(-3), \; 1(-3)^2, \; 1(-3)^3, \; 1(-3)^4$$

or 1, -3, 9, -27, 81.

The fifth term is 81.

7. a = 1024, r = $\frac{1}{2}$

The first five terms are

$$1024, \; 1024\left(\frac{1}{2}\right), \; 1024\left(\frac{1}{2}\right)^2, \; 1024\left(\frac{1}{2}\right)^3,$$

$$1024\left(\frac{1}{2}\right)^4$$

or 1024, 512, 256, 128, 64.

The fifth term is 64.

9. a = 1, r = 2, n = 4

To find the sum of the first 4 terms, S_4, use the formula for the sum of the first n terms of geometric sequence.

$$S_n = \frac{a(r^n - 1)}{r - 1}$$

$$S_4 = \frac{1(2^4 - 1)}{2 - 1}$$

$$= \frac{16 - 1}{1} = 15$$

11. a = 5, r = $\frac{1}{5}$, n = 4

$$S_n = \frac{a(r^n - 1)}{r - 1}$$

$$S_4 = \frac{5\left[\left(\frac{1}{5}\right)^4 - 1\right]}{\frac{1}{5} - 1}$$

$$= \frac{5\left(-\frac{624}{625}\right)}{-\frac{4}{5}}$$

$$= \frac{-\frac{624}{125}}{-\frac{4}{5}}$$

$$= \left(-\frac{624}{125}\right)\left(-\frac{5}{4}\right)$$

$$= \frac{156}{25}$$

13. $a = 128$, $r = -\frac{3}{2}$, $n = 4$

$$S_n = \frac{a(r^n - 1)}{r - 1}$$

$$S_4 = \frac{128\left[\left(-\frac{3}{2}\right)^4 - 1\right]}{-\frac{3}{2} - 1}$$

$$= \frac{128\left(\frac{65}{16}\right)}{-\frac{5}{2}}$$

$$= -208$$

15.

$$s_{\overline{n}|i} = \frac{(1 + i)^n - 1}{i}$$

$$s_{\overline{12}|.05} = \frac{(1 + .05)^{12} - 1}{.05}$$

$$\approx 15.91713$$

17.

$$s_{\overline{n}|i} = \frac{(1 + i)^n - 1}{i}$$

$$s_{\overline{16}|.04} = \frac{(1.04)^{16} - 1}{.04}$$

$$\approx 21.82453$$

19.

$$s_{\overline{n}|i} = \frac{(1 + i)^n - 1}{i}$$

$$s_{\overline{20}|.01} = \frac{(1.01)^{20} - 1}{.01}$$

$$\approx 22.01900$$

23. $R = 1000$, $i = .06$, $n = 5$

Use the formula for the future value of an ordinary annuity.

$$S = R \cdot s_{\overline{n}|i}$$

$$= 1000 \cdot s_{\overline{5}|.06}$$

$$= 1000 \cdot \frac{(1.06)^5 - 1}{.06}$$

$$\approx 1000(5.63709)$$

$$= \$5637.09$$

25. $R = 29{,}500$, $i = .05$, $n = 15$

$$S = R \cdot s_{\overline{n}|i}$$

$$= 29{,}500 \cdot s_{\overline{15}|.05}$$

$$\approx \$636{,}567.63$$

27. $R = 3700$, $i = \frac{.08}{2} = .04$,

$n = 2 \cdot 11 = 22$

$$S = R \cdot s_{\overline{n}|i}$$

$$= 3700 \cdot s_{\overline{22}|.04}$$

$$\approx \$126{,}717.49$$

29. $R = 4600$, $i = \frac{.08}{4} = .02$,

$n = 9 \cdot 4 = 36$

$$S = R \cdot s_{\overline{n}|i}$$

$$= 4600 \cdot s_{\overline{36}|.02}$$

$$\approx \$239{,}174.10$$

31. $R = 42{,}000$, $i = \frac{.10}{2} = .05$,

$n = 12 \cdot 2 = 24$

$$S = R \cdot s_{\overline{n}|i}$$

$$= 42{,}000 \cdot s_{\overline{24}|.05}$$

$$\approx \$1{,}869{,}084.00$$

33. $R = 1400$, $i = .08$, $n = 10$

To find the value of an annuity due, use the formula for the future value of an ordinary annuity, but include one additional time period and subtract the amount of one payment.

$S = R \cdot s_{\overline{n+1}|i} - R$

$\quad = 1400 \cdot s_{\overline{11}|.08} - 1400$

$\quad \approx \$21{,}903.68$

35. $R = 4000$, $i = .06$, $n = 11$

$S = R \cdot s_{\overline{n+1}|i} - R$

$\quad = 4000 \cdot s_{\overline{12}|.06} - 4000$

$\quad \approx \$63{,}479.76$

37. $R = 750$, $i = \dfrac{.06}{12} = .005$,

$n = 15 \cdot 12 = 180$

$S = R \cdot s_{\overline{n+1}|i} - R$

$\quad = 750 \cdot s_{\overline{181}|.005} - 750$

$\quad \approx \$219{,}204.60$

39. $R = 1500$, $i = \dfrac{.06}{2} = .03$,

$n = 11 \cdot 2 = 22$

$S = R \cdot s_{\overline{n+1}|i} - R$

$\quad = 1500 \cdot s_{\overline{23}|.03} - 1500$

$\quad \approx \$47{,}179.32$

41. $S = 100{,}000$, $i = \dfrac{.08}{2} = .04$,

$n = 9 \cdot 2 = 18$

Let R represent the amount to be deposited into the sinking fund each year. Solve the formula $S = R \cdot s_{\overline{n}|i}$ for R and proceed.

$R = \dfrac{S}{s_{\overline{n}|i}}$

$\quad = \dfrac{100{,}000}{s_{\overline{18}|.04}}$

$\quad \approx \$3899.32$

43. $S = 8500$, $i = .08$, $n = 7$

$R = \dfrac{S}{s_{\overline{n}|i}}$

$\quad = \dfrac{8500}{s_{\overline{7}|.08}}$

$\quad \approx \$952.62$

45. $S = 75{,}000$, $i = \dfrac{.06}{2} = .03$,

$n\left(4\tfrac{1}{2}\right)(2) = 9$

$R = \dfrac{S}{s_{\overline{n}|i}}$

$\quad = \dfrac{75{,}000}{s_{\overline{9}|.03}}$

$\quad \approx \$7382.54$

47. $S = 25{,}000$, $i = \dfrac{.05}{4} = .0125$,

$n = \left(3\tfrac{1}{2}\right)(4) = 14$

$R = \dfrac{S}{s_{\overline{n}|i}}$

$\quad = \dfrac{25{,}000}{s_{\overline{14}|.0125}}$

$\quad \approx \$1645.13$

49. $S = 9000$, $i = \frac{.12}{12} = .01$,

$n = \left(2\frac{1}{2}\right)(12) = 30$

$R = \dfrac{S}{s_{\overline{n}|i}}$

$= \dfrac{9000}{s_{\overline{30}|}.01}$

$\approx \$258.73$

51. $R = 12,000$, $i = .08$, $n = 9$

(a) $S = R \cdot s_{\overline{n}|i}$

$= 12,000 \cdot s_{\overline{9}|}.08$

$\approx \$149,850.69$

(b) $R = 12,000$, $i = .06$, $n = 9$

$S = R \cdot s_{\overline{n}|i}$

$= 12,000 \cdot s_{\overline{9}|}.06$

$\approx \$137,895.79$

(c) The amount that would be lost is the difference of the above two amounts, which is

$\$149,850.69 - \$137,895.79$

$= \$11,954.90.$

53. $R = 80$, $i = \frac{.075}{12} = .00625$,

$n = 3 \cdot 12 + 9 = 45$

Because the deposits are made at the beginning of each month, this is an annuity due.

$S = R \cdot s_{\overline{n+1}|i} - R$

$= 80 \cdot s_{\overline{46}|}.00625 - R$

≈ 4168.30

There will be about $4168.30 in the account.

55. For the first 15 yr, we have an ordinary annuity with $R = 1000$, $i = \frac{.08}{4} = .02$, and $n = 15 \cdot 4 = 60$. The amount on deposit after 15 yr is

$S = R \cdot s_{\overline{n}|i}$

$= 1000 \cdot s_{\overline{60}|}.02$

$\approx 114,051.54.$

For the remaining 5 yr, this amount earns compound interest at 8% compounded quarterly.
To find the final amount on deposit, use the formula for the compound amount, with $P = 114,051.54$, $i = \frac{.08}{4} = .02$, and $n = 5 \cdot 4 = 20$.

$A = P(1 + i)^n$

$= 114,051.54(1.02)^{20}$

$\approx 169,474.59$

The man will have about $169,474.59 in the account when he retires.

57. For the first 8 yr, we have an annuity due, with $R = 2435$, $i = \frac{.06}{2} = .03$, and $n = 8 \cdot 2 = 16$. The amount on deposit after 8 yr is

$S = R \cdot s_{\overline{n+1}|i} - R$

$= 2435 \cdot s_{\overline{17}|}.03 - 2435$

$\approx 2435(21.76159) - 2435$

$= 52.989.47 - 2435$

$= \$50,554.47.$

For the remaining 5 yr, this amount earns compound interest at 6% compounded semiannually. To find the final amount on deposit, use the formula for the compound amount, with

$P = 50,554.47$, $i = \dfrac{.06}{2} = .03$, and $n = 5 \cdot 2 = 10$.

$$A = P(1 + i)^n$$
$$= 50,554.47(1.03)^{10}$$
$$\approx 67,940.98$$

The final amount on deposit will be about \$67,940.98.

59. (a) $S = 10,000$, $i = \dfrac{.08}{4} = .02$, $n = 8 \cdot 4 = 32$

Let R represent the amount of each payment.

$$S = R \cdot s_{\overline{n}|i}$$
$$10,000 = R \cdot s_{\overline{32}|.02}$$
$$R \approx \frac{10,000}{44.22703}$$
$$\approx 226.11$$

If the money is deposited at 8% compounded quarterly, Berkowitz's quarterly deposit will need to be about \$226.11.

(b) $S = 10,000$, $i = \dfrac{.06}{4} = .015$, $n = 8 \cdot 4 = 32$

Let R represent the amount of each payment.

$$S = R \cdot s_{\overline{n}|i}$$
$$10,000 = R \cdot s_{\overline{32}|.015}$$
$$R \approx \frac{10,000}{40.68829}$$
$$\approx 245.77$$

If the money is deposited at 6% compounded quarterly, Berkowitz's quarterly deposit will need to be about \$245.77.

61. $S = 18,000$, $i = \dfrac{.05}{4} = .0125$, $n = 6 \cdot 4 = 24$

Let R represent the amount of each payment.

$$S = R \cdot s_{\overline{n}|i}$$
$$18,000 = R \cdot s_{\overline{24}|.0125}$$
$$R = \frac{18,000}{s_{\overline{24}|.0125}}$$
$$\approx \frac{18,000}{27.78808}$$
$$\approx 647.76$$

She must deposit about \$647.76 at the end of each quarter.

63. $R = \dfrac{2000}{2} = 1000$, $i = \dfrac{.08}{2} = .04$, $n = 25 \cdot 2 = 50$

$$S = R \cdot s_{\overline{n}|i}$$
$$= 1000 \cdot s_{\overline{50}|.04}$$
$$\approx 152,667.08$$

There will be about \$152,667.08 in the IRA.

65. $R = \dfrac{2000}{2} = 1000$, $i = \dfrac{.10}{2} = .05$,

$n = 25 \cdot 2 = 50$

$$S = R \cdot s_{\overline{n}|i}$$

$$= 1000 \cdot s_{\overline{50}|.05}$$

$$\approx 209{,}348.00$$

There will be about \$209,348 in the IRA.

67. This exercise should be solved by computer methods. The solution will vary according to the computer program that is used. The answers are as follows.

(a) The buyer's quarterly interest payment will be \$1200.

(b) The buyer's semiannual payments into the sinking fund will be \$3511.58 for each of the first 13 payments and \$3511.59 for the last payment.

(c) A table showing the amount in the sinking fund after each deposit is as follows.

Payment Number	Amount of Deposit	Interest Earned	Total
1	\$3511.58	\$0	\$3511.58
2	\$3511.58	\$105.35	\$7128.51
3	\$3511.58	\$213.86	\$10,853.95
4	\$3511.58	\$325.62	\$14,691.15
5	\$3511.58	\$440.73	\$18,643.46
6	\$3511.58	\$559.30	\$22,714.34
7	\$3511.58	\$681.43	\$26,907.35
8	\$3511.58	\$807.22	\$31,226.15
9	\$3511.58	\$936.78	\$35,674.51
10	\$3511.58	\$1070.24	\$40,256.33
11	\$3511.58	\$1207.69	\$44,975.60
12	\$3511.58	\$1349.27	\$49,836.45
13	\$3511.58	\$1495.09	\$54,843.12
14	\$3411.59	\$1645.29	\$60,000.00

Section 10.4

1. $\dfrac{1 - (1 + i)^{-n}}{i}$

is represented by $a_{\overline{n}|i}$, and it is choice (c).

3. $a_{\overline{n}|i} = \dfrac{1 - (1 + i)^{-n}}{i}$

$$a_{\overline{15}|.06} = \dfrac{1 - (1 + .06)^{-15}}{.06}$$

$$= \dfrac{1 - (1.06)^{-15}}{.06}$$

$$\approx 9.71255$$

5. $a_{\overline{n}|i} = \dfrac{1 - (1 + i)^{-n}}{i}$

$$a_{\overline{18}|.04} = \dfrac{1 - (1.04)^{-18}}{.04}$$

$$\approx 12.65930$$

7. $a_{\overline{n}|i} = \dfrac{1 - (1 + i)^{-n}}{i}$

$$a_{\overline{16}|.01} = \dfrac{1 - (1.01)^{-16}}{.01}$$

$$\approx 14.71787$$

11. $R = 1400$, $i = .08$, $n = 8$

Use the formula for the present value of an annuity.

$$P = R \cdot a_{\overline{n}|i}$$

$$P = 1400 \cdot a_{\overline{8}|.08}$$

$$\approx 1400(5.746639)$$

$$\approx \$8045.30$$

13. $R = 50{,}000$, $i = \dfrac{.08}{4} = .02$,

$n = 10 \cdot 4 = 40$

$P = R \cdot a_{\overline{n}|i}$

$P = 50{,}000 \cdot a_{\overline{40}|.02}$

$\approx 50{,}000(27.35548)$

$\approx \$1{,}367{,}774.00$

15. $R = 18{,}579$, $i = \dfrac{.094}{2} = .047$,

$n = 8 \cdot 2 = 16$

$P = R \cdot a_{\overline{n}|i}$

$P = 18{,}579 \cdot a_{\overline{16}|.047}$

$\approx 18{,}579(11.072953)$

$\approx \$205{,}724.40$

17. The lump sum is the same as the present value of the annuity.

$R = 10{,}000$, $i = .05$, $n = 15$

$P = R \cdot a_{\overline{n}|i}$

$P = 10{,}000 \cdot a_{\overline{15}|.05}$

$\approx 10{,}000(10.37966)$

$\approx \$103{,}796.60$

21. $P = 41{,}000$, $i = \dfrac{.10}{2} = .05$, $n = 10$

Let R be the amount of each payment.

$P = R \cdot a_{\overline{n}|i}$

$R = \dfrac{P}{a_{\overline{n}|i}}$

$R = \dfrac{41{,}000}{a_{\overline{10}|.05}}$

$\approx \dfrac{41{,}000}{7.72173}$

$\approx \$5309.69$

23. $P = 140{,}000$, $i = \dfrac{.12}{4} = .03$, $n = 15$

Let R be the amount of each payment.

$R = \dfrac{P}{a_{\overline{n}|i}}$

$R = \dfrac{140{,}000}{a_{\overline{15}|.03}}$

$\approx \dfrac{140{,}000}{11.93794}$

$\approx \$11{,}727.32$

25. $P = 5500$, $i = \dfrac{.12}{12} = .01$, $n = 24$

Let R be the amount of each payment.

$R = \dfrac{P}{a_{\overline{n}|i}}$

$R = \dfrac{5500}{a_{\overline{24}|.01}}$

$\approx \dfrac{5500}{21.24339}$

$\approx \$258.90$

27. The Portion to Principal column of the table indicates that $87.10 of the 11th payment of $88.85 is used to reduce the debt.

29. The amount of interest paid in the last 4 months of the loan is

$3.47 + 2.61 + 1.75 + .88 = \$8.71.$

31. $4000 deposited every 6 months for 10 yr at 6% compounded semiannually will be worth $4000 \cdot s_{\overline{20}|.03} \approx \$107{,}481.48.$

(Note that $i = \frac{.06}{2}$ and

$n = 10 \cdot 2 = 20$.) For the lump sum

investment of x dollars, use

$i = \frac{.08}{4} = .02$ and $n = 10 \cdot 4 = 40$ in

the formula $A = P(1 + i)^n$.

Our unknown amount x will be worth

$x(1.02)^{40}$, so

$$x(1.02)^{40} = 107{,}481.48$$
$$x \approx 48{,}677.34.$$

About \$48,677.34 should be invested

today.

33. $P = 170{,}892$, $i = \frac{.0811}{12} = .006758$,

$n = 30 \cdot 12 = 360$

$$R = \frac{P}{a_{\overline{n}| i}}$$

$$R = \frac{170{,}892}{a_{\overline{360}| .006758}}$$

$$\approx \frac{170{,}892}{134.87644}$$

$$\approx \$1267.07$$

35. $P = 96{,}511$, $i = \frac{.0957}{12} = .007975$,

$n = 25 \cdot 12 = 300$

$$R = \frac{P}{a_{\overline{n}| i}}$$

$$R = \frac{96{,}511}{a_{\overline{300}| .007975}}$$

$$\approx \frac{96{,}511}{113.82165}$$

$$\approx \$847.91$$

37. **(a)** $P = 6000$, $i = \frac{.12}{12} = .01$,

$n = 4 \cdot 12 = 48$

Let R be the amount of each payment.

$$R = \frac{P}{a_{\overline{n}| i}}$$

$$R = \frac{6000}{a_{\overline{48}| .01}}$$

$$\approx \frac{6000}{37.97396}$$

$$\approx \$158.00$$

(b) 48 payments of \$158.00 are made,

and 48(\$158.00) = \$7584. The total

amount of interest Le will pay is

\$7584 − \$6000 = \$1584.00.

39. This is just the present value of the

annuity with $R = 50{,}000$, $i = .06$, and

$n = 20$.

$$P = R \cdot a_{\overline{n}| i}$$

$$P = 50{,}000 \cdot a_{\overline{20}| .06}$$

$$\approx 50{,}000(11.46992)$$

$$\approx 573{,}496.00$$

The lump sum that the management must

invest is about \$573,496.00.

41. $P = 72{,}000$, $i = \frac{.10}{2} = .05$, $n = 9$

$R = \frac{72{,}000}{a_{\overline{9}| .05}} \approx \$10{,}129.69$ is the amount

of each payment.

Payment Number	Amount of Payment	Interest for Period	Portion to Principal	Principal at End of Period
0	————	————	————	$72,000.00
1	$10,129.69	$3600.00	$6529.69	$65,470.31
2	$10,129.69	$3273.52	$6856.17	$58,614.14
3	$10,129.69	$2930.71	$7198.98	$51,415.16
4	$10,129.69	$2570.76	$7558.93	$43,856.23

43. The loan is for 14,000 + 7200 − 1200 = $20,000.

$P = 20,000$, $i = \dfrac{.12}{2} = .06$, $n = 5 \cdot 2 = 10$

$R = \dfrac{20,000}{a_{\overline{10}|}.06} \approx \$2717.36$ is the amount of each payment.

Payment Number	Amount of Payment	Interest for Period	Portion to Principal	Principal at End of Period
0	————	————	————	$20,000.00
1	$2717.36	$1200.00	$1517.36	$18,482.64
2	$2717.36	$1108.96	$1608.40	$16,874.24
3	$2717.36	$1012.45	$1704.91	$15,169.33
4	$2717.36	$910.16	$1807.20	$13,362.13

45. This is an amortization problem with $P = 25,000$. R represents the amount of each annual withdrawal.

(a) $i = .06$, $n = 8$

$$R = \frac{P}{a_{\overline{n}|}i}$$

$$R = \frac{25,000}{a_{\overline{8}|}.06} \approx \frac{25,000}{6.20979} \approx 4025.90$$

He will be able to withdraw about $4025.90 per month for the 8 yr.

(b) i = .06, n = 12

$$R = \frac{P}{a_{\overline{n}|i}}$$

$$R = \frac{25,000}{a_{\overline{12}|.06}}$$

$$\approx \frac{25,000}{8.38384}$$

$$\approx 2981.93$$

He will be able to withdraw about $2981.93 per month for the 12 yr.

47. This exercise should be solved by computer methods. The solution will vary according to the computer program that is used. The amortization schedule is as follows.

Payment Number	Amount of Payment	Interest for Period	Portion to Principal	Principal at End of Period
1	$5783.49	$3225.54	$2557.95	$35,389.55
2	$5783.49	$3008.11	$2775.38	$32,614.17
3	$5783.49	$2772.20	$3011.29	$29,602.88
4	$5783.49	$2516.24	$3267.25	$26,335.63
5	$5783.49	$2238.53	$3544.96	$22,790.67
6	$5783.49	$1937.21	$3846.28	$18,944.39
7	$5783.49	$1610.27	$4173.22	$14,771.17
8	$5783.49	$1255.55	$4527.94	$10,243.22
9	$5783.49	$870.67	$4912.82	$5,330.41
10	$5783.49	$453.08	$5530.41	$0

Chapter 10 Review Exercises

1. $I = Prt$

$\qquad = 15,903(.08)\left(\dfrac{8}{12}\right)$

$\qquad = \$848.16$

3. $I = Prt$

$\qquad = 42,368(.0522)\left(\dfrac{5}{12}\right)$

$\qquad \approx \$921.50$

7. $P = \dfrac{A}{1 + rt}$

$\qquad = \dfrac{459.57}{1 + .045\left(\dfrac{7}{12}\right)}$

$\qquad \approx \$447.81$

9. $P = A(1 - rt)$

$\qquad = 802.34\left[1 - (.086)\left(\dfrac{11}{12}\right)\right]$

$\qquad \approx \$739.09$

11. For a given amount of money at a given interest rate for a given time period greater than 1, compound interest produces more interest than simple interest.

13. $A = P(1 + i)^n$

$\qquad = 19,456.11\left(1 + \dfrac{.12}{2}\right)^{2(7)}$

$\qquad = 19,456.11(1.06)^{14}$

$\qquad \approx 19,456.11(2.26090)$

$\qquad \approx \$43,988.40$

15. $A = P(1 + i)^n$

$\qquad = 57,809.34\left(1 + \dfrac{.12}{4}\right)^{4(5)}$

$\qquad = 57,809.34(1.03)^{20}$

$\qquad \approx 57,809.34(1.80611)$

$\qquad \approx \$104,410.10$

17. $A = P(1 + i)^n$

$\qquad = 12,699.36\left(1 + \dfrac{.10}{2}\right)^{2(7)}$

$\qquad = 12,699.36(1.05)^{14}$

$\qquad \approx 12,699.36(1.97993)$

$\qquad \approx \$25,143.86$

The interest earned is

$\qquad \$25,143.86 - \$12,699.36$

$\qquad = \$12,444.50.$

19. $A = P(1 + i)^n$

$\qquad = 34,677.23\left(1 + \dfrac{.0972}{12}\right)^{32}$

$\qquad \approx 34,677.23(1.29454)$

$\qquad \approx \$44,891.08$

The interest earned is

$\qquad \$44,891.08 - \$34,677.23$

$\qquad = \$10,213.85.$

21. $P = \dfrac{A}{(1 + i)^n}$

$\qquad = \dfrac{17,650}{\left(1 + \dfrac{.08}{4}\right)^{4(4)}}$

$\qquad = \dfrac{17,650}{(1.02)^{16}}$

$\qquad \approx \dfrac{17,650}{1.37279}$

$\qquad \approx \$12,857.07$

23. $P = \dfrac{A}{(1 + i)^n}$

$= \dfrac{2388.90}{\left(1 + \frac{.0593}{12}\right)^{44}}$

$\approx \dfrac{2388.90}{(1.00494)^{44}}$

$\approx \dfrac{2388.90}{1.24222}$

$\approx \$1923.09$

25. $a = 4,\ r = \dfrac{1}{2}$

The first five terms are

$4,\ 4\left(\dfrac{1}{2}\right),\ 4\left(\dfrac{1}{2}\right)^2,\ 4\left(\dfrac{1}{2}\right)^3,\ 4\left(\dfrac{1}{2}\right)^4$

or $4,\ 2,\ 1,\ \dfrac{1}{2},\ \dfrac{1}{4}.$

27. $a = -2,\ r = -2$

The first five terms are

$-2,\ -2(-2),\ -2(-2)^2,\ -2(-2)^3,\ -2(-2)^4$

or $-2,\ 4,\ -8,\ 16,\ -32.$

The fifth term is -32.

29. $a = 8000,\ r = -\dfrac{1}{2},\ n = 5$

$S_n = \dfrac{a(r^n - 1)}{r - 1}$

$S_5 = \dfrac{8000\left[\left(-\frac{1}{2}\right)^5 - 1\right]}{-\frac{1}{2} - 1}$

$= \dfrac{8000\left(-\frac{33}{32}\right)}{-\frac{3}{2}}$

$= \dfrac{-8250}{-\frac{3}{2}}$

$= (-8250)\left(-\dfrac{2}{3}\right)$

$= 5500$

31. $s_{\overline{n}|i} = \dfrac{(1 + i)^n - 1}{i}$

$s_{\overline{20}|}.05 = \dfrac{(1.05)^{20} - 1}{.05}$

$\approx \dfrac{1.6532977}{.05}$

$= 33.06595$

33. $R = 500,\ i = \dfrac{.06}{2} = .03,$

$n = 8 \cdot 2 = 16$

This is an ordinary annuity.

$S = R \cdot s_{\overline{n}|i}$

$S = R \cdot s_{\overline{16}|}.03$

$\approx 500(20.15688)$

$\approx \$10,078.44$

35. $R = 4000,\ i = \dfrac{.06}{4} = .015,$

$n = 7 \cdot 4 = 28$

This is an ordinary annuity.

$S = R \cdot s_{\overline{n}|i}$

$S = R \cdot s_{\overline{28}|}.015$

$\approx 4000(34.48148)$

$\approx \$137,925.91$

37. $R = 672,\ i = \dfrac{.08}{4} = .02,$

$n = 7 \cdot 4 = 28$

This is an annuity due, so we calculate $s_{\overline{n}|i}$ for one additional payment.

$R \cdot s_{\overline{29}|}.02 \approx 672(38.79223)$

$\approx \$26,068.38$

Now subtract the amount of one payment to find the future value.

$$S = \$26,068.38 - \$672 = \$25,396.38$$

41. $S = 57,000$, $i = \dfrac{.06}{2} = .03$,

$n = \left(8\frac{1}{2}\right)(2) = 17$

Let R represent the amount of each paymemt.

$$S = R \cdot s_{\overline{n}|i}$$

$$57,000 = R \cdot s_{\overline{17}|}.03$$

$$R = \frac{57,000}{s_{\overline{17}|}.03}$$

$$\approx \frac{57,000}{21.76159}$$

$$\approx \$2619.29$$

43. $S = 1,056,788$, $i = \dfrac{.0812}{12} \approx .0067667$,

$n = \left(4\frac{1}{2}\right)(12) = 54$

Let R represent the amount of each payment.

$$S = R \cdot s_{\overline{n}|i}$$

$$1,056,788 = R \cdot s_{\overline{54}|}.0067667$$

$$R = \frac{1,056,788}{s_{\overline{54}|}.0067667}$$

$$\approx \frac{1,056,788}{64.92885}$$

$$\approx \$16,277.35$$

45. $R = 1500$, $i = \dfrac{.08}{4} = .02$,

$n = 7(4) = 28$

$$P = R \cdot a_{\overline{n}|i}$$

$$P = 1500 \cdot a_{\overline{28}|}.02$$

$$\approx 1500(21.28127)$$

$$\approx \$31,921.91$$

47. $R = 877.34$, $i = \dfrac{.094}{12} \approx .007833$,

$n = 17$

$$P = R \cdot a_{\overline{n}|i}$$

$$P = 877.34 \cdot a_{\overline{17}|}.007833$$

$$\approx 877.34(15.85875)$$

$$\approx \$13,913.48$$

49. $P = 80,000$, $i = .08$, $n = 9$

Let R represent the amount of each payment.

$$P = R \cdot a_{\overline{n}|i}$$

$$80,000 = R \cdot a_{\overline{9}|}.08$$

$$R = \frac{80,000}{a_{\overline{9}|}.08}$$

$$\approx \frac{80,000}{6.24689}$$

$$= \$12,806.38$$

51. $P = 32,000$, $i = \dfrac{.094}{4} = .0235$, $n = 17$

Let R represent the amount of each payment.

$$P = R \cdot a_{\overline{n}|i}$$

$$32,000 = R \cdot a_{\overline{17}|}.0235$$

$$R = \frac{32,000}{a_{\overline{17}|}.0235}$$

$$\approx \frac{32,000}{13.88246}$$

$$\approx \$2305.07$$

53. $P = 56,890$, $i = \frac{.1074}{12} = .00895$,

$n = 25 \cdot 12 = 300$

Let R represent the amount of each payment.

$$P = R \cdot a_{\overline{n}|i}$$

$$56,890 = R \cdot a_{\overline{300}|.00895}$$

$$R = \frac{56,890}{a_{\overline{300}|.00895}}$$

$$\approx \frac{56,890}{104.0178}$$

$$\approx \$546.93$$

55. $896.06 of the fifth payment of $1022.64 is interest.

57. In the first 3 months of the loan, the total amount of interest paid is

$$899.58 + 898.71 + 897.83$$
$$= \$2696.12.$$

59. Find the maturity value of the loan, the amount Tom Wilson must pay his mother.

$$A = P(1 + rt)$$

$$= 5800\left[1 + .10\left(\tfrac{10}{12}\right)\right]$$

$$\approx 5800(1.08333)$$

$$\approx 6283.33$$

The bank will apply its discount rate to the 6283.33, to obtain

Amount of discount

$$= 6283.33(.1345)\left(\tfrac{3}{12}\right)$$

$$\approx 211.28.$$

Tom's mother will receive from the bank

$$\$6283.33 - \$211.28 = \$6072.05,$$

which is enough to buy the furniture.

61. $I = Prt$

$$t = \frac{I}{Pr}$$

$$= \frac{3255}{28,000(.115)}$$

$$\approx 1.011$$

The loan is for about 1.011 yr; convert this to months.

$$(1.011 \text{ yr})\left(\frac{12 \text{ mo}}{1 \text{ yr}}\right) \approx 12.13 \text{ mo}$$

The loan is for about 12.13 months.

63. $A = 7500$, $i = \frac{.10}{2} = .05$, $n = 3(2) = 6$

Let P represent the lump sum.

$$A = P(1 + i)^n$$

$$P = \frac{A}{(1 + i)^n}$$

$$= \frac{7500}{(1.05)^6}$$

$$\approx \frac{7500}{1.34}$$

$$\approx 5596.62$$

She should deposit about $5596.62 today.

65. $R = 5000$, $i = \frac{.10}{2} = .05$,

$$n = \left(7\tfrac{1}{2}\right)(2) = 15$$

This is an ordinary annuity.

$$S = R \cdot s_{\overline{n}|i}$$

$$S = 5000 \cdot s_{\overline{15}|.05}$$

$$\approx 5000(21.57856)$$

$$\approx \$107,892.82$$

The amount of interest earned is

$$\$107,892.82 - 15(\$5000)$$

$$= \$32,892.82.$$

67. $P = 48,000$, $i = .10$, $n = 7$

Let R represent the amount of each payment.

$$P = R \cdot a_{\overline{n}|i}$$

$$R = \frac{P}{a_{\overline{n}|i}}$$

$$= \frac{48,000}{a_{\overline{7}|.10}}$$

$$\approx \frac{48,000}{4.86842}$$

$$\approx 9859.46$$

The owner should deposit about $9859.46 at the end of each year.

69. $P = 115,700$, $i = \frac{.105}{12} = .00875$,

$n = 300$

Let R represent the amount of each payment.

$$R = \frac{P}{a_{\overline{n}|i}}$$

$$R = \frac{115,700}{a_{\overline{300}|.00875}}$$

$$\approx \frac{115,700}{105.91182}$$

$$\approx 1092.42$$

Each monthly payment will be about $1092.42. The total amount of interest will be

$$300(\$1092.42) - \$115,700$$

$$= \$212,026.$$

71. The death benefit grows to

$$10,000(1.05)^7 \approx 14,071.00.$$

This 14,071 is the present value of an annuity due with $P = 14,071$,

$i = \frac{.03}{12} = .0025$, and $n = 120$. Let x

represent the amount of each monthly payment.

$$P = R \cdot a_{\overline{n+1}|i} - R$$

$$14,071 = x \cdot a_{\overline{121}|.0025} - x$$

$$14,071 = (a_{\overline{121}|.0025} - 1)x$$

$$14,071 \approx (104.301 - 1)x$$

$$14,071 \approx 103.301x$$

$$135 \approx x$$

Each payment is about $135, which corresponds to choice (d).

Extended Application

Note: Since the data for Exercises 1-4 is given to the nearest dollar, answers should be rounded to the nearest dollar.

1. Cash flow

$$= -.52(6228 + 2976) + .52(26,251)$$

$$+ .48(10,778)$$

$$= -4786.08 + 13,650.52 + 5173.44$$

$$= \$14,038$$

2. Cash flow
   = −.52(6228 + 2976) + .52(26,251)
     + .48(1347)
   = −4786.08 + 13,650.52 + 646.56
   = $9511

3. Cash flow
   = −.52(6386 + 2870) + .52(26,251)
     + .48(0)
   = −4813.12 + 13,650.52 + 0
   = $8837

4. Cash flow
   = −.52(6228 + 2976) + .52(10,618)
     + .48(6736)
   = −4786.08 + 5521.36 + 3233.28
   = $3968

CHAPTER 10   TEST

1. Find the simple interest on $1252 at 5% for 11 months.

2. Using a 360 day year, find the simple interest on $12,000 at 6.25% for 170 days.

3. Find the present value for the following future amount:

   $14,500 in 9 months at 7.2% simple interest.

4. Find the compound amount if $7000 is deposited for 8 yr in an account paying 6% per year compounded quarterly.

5. Find the present value of $8000 at 8.1% compounded monthly for 2 yr.

6. Find the fifth term of the geometric sequence with a = 4.7, r = 2.

7. Ralph deposits $100 at the end of each month for 3 yr in an account paying 6% compounded monthly. He then uses the money from this account to buy a certificate of deposit which pays 7.5% compounded annually. If he redeems the certicicate 4 yr later, how much money will he receive?

8. What amount must be deposited at the end of each quarter at 8% compounded quarterly to have $10,000 in 10 yr?

9. Find the present value of an ordinary annuity with payments of $800 per month at 6% compounded monthly for 4 yr.

10. Ms Morroco borrows $12,000 at 8% compounded quarterly, to be paid off with equal quarterly payments over 2 yr.

    (a) What quarterly payment is needed to amortize this loan?

    (b) Prepare an amortization table for the first 4 payments.

CHAPTER 10   TEST ANSWERS

1. $57.38   2. $354.17   3. $13,757.12   4. $11.272.27

5. $6807.23   6. 75.2   7. $5253.21   8. $165.56

9. $34,064.25

10. **(a)** $1638.12

(b)

Payment Number	Amount of Payment	Interest for Period	Portion to Principal	Principal at End of Period
0	——	——	——	$12,000.00
1	$1638.12	$240.00	$1398.12	$10,601.88
2	$1638.12	$212.04	$1426.08	$9175.80
3	$1638.12	$183.52	$1454.60	$7721.19
4	$1638.12	$154.42	$1483.70	$6237.49

NOTES

NOTES